Polymers from Renewable Resources

ACS SYMPOSIUM SERIES **764**

Polymers from Renewable Resources

Biopolyesters and Biocatalysis

Carmen Scholz, EDITOR
University of Alabama

Richard A. Gross, EDITOR
Polytechnic University

American Chemical Society, Washington, DC

Library of Congress Cataloging-in-Publication Data

Scholz, Carmen, 1963–
    Polymers from renewable resources : biopolyesters and biocatalysts / Carmen Scholz,
Richard A. Gross.
       p.   cm.—(ACS symposium series ; 764)

    Includes bibliographical references and index.

    ISBN 0–8412–3646–1

    1. Polyesters—Congresses. 2. Polymers—Biodegradation—Congresses. 3. Polymers—
Biotechnology—Congresses.
    I. Gross, Richard A., 1957– II. American Chemical Society. Meeting (216th : 1998 :
Boston, Mass.) III. Title. IV. Series.

TP1180.P6 S36   2000
668.9—dc21   00–28897

The paper used in this publication meets the minimum requirements of American National Standard
for Information Sciences—Permanence of Paper for Printed Library Materials, ANSI Z39.48–1984.

Foreword

THE ACS SYMPOSIUM SERIES was first published in 1974 to provide a mechanism for publishing symposia quickly in book form. The purpose of the series is to publish timely, comprehensive books developed from ACS sponsored symposia based on current scientific research. Occasionally, books are developed from symposia sponsored by other organizations when the topic is of keen interest to the chemistry audience.

Before agreeing to publish a book, the proposed table of contents is reviewed for appropriate and comprehensive coverage and for interest to the audience. Some papers may be excluded in order to better focus the book; others may be added to provide comprehensiveness. When appropriate, overview or introductory chapters are added. Drafts of chapters are peer-reviewed prior to final acceptance or rejection, and manuscripts are prepared in camera-ready format.

As a rule, only original research papers and original review papers are included in the volumes. Verbatim reproductions of previously published papers are not accepted.

ACS BOOKS DEPARTMENT

Contents

Poly(β-hydroxyalkanoates)

Biodegradable Polyesters: Synthesis and Characterization

Enzymatic Synthesis

Chemical Synthesis

Polymer Evaluation

Biodegradability and Recycling

Biomedical Application of Polyesters

Preface

The meeting of the Bio/Environmentally Degradable Polymer Society was held for the 7th time in a series in August of 1998. The conference was held in a timely and locally conjunction with the symposium on Biodegradable Polymers held at the 216th American Chemical Society (ACS) meeting, which enabled us to bring together scientists from industrial, academic, and governmental agencies and to discuss the most contemporary environmental biopolymer issues.

In the search for a polyester-based structural material, Carothers at DuPont developed Nylon 66 instead in 1936. Polyesters with molecular weights high enough to be useful as fibers were introduced a few years later. Today, polyesters based on poly(ethylene terephthalate) and ethylene glycol are part of our daily life in form of soda bottles, fibers, and bases for photographic films and magnetic tapes, to mention just a few. Aliphatic polyesters have been recognized for their biocompatibility and their susceptibility to hydrolytic degradation. Whereas current research on structural, aromatic polyesters focuses mainly on processing and engineering issues, aliphatic polyesters gain increasing consideration as biomedical materials. In particular, poly(lactic acid) (PLA) and poly(glycolic acid) (PGA) are the polymers of interest, for implants, drug delivery systems, and scaffolds, because they are biocompatible, degrade within about one month upon implantation, in addition, they are FDA-approved materials. Polymer research constantly develops and probes new routes to the synthesis of polyesters, new catalysts are found, reaction mechanisms are understood in more detail, and efforts focus on tailoring physical properties. In addition to their biocompatibility, PLA, PGA, and copolymers thereof, exhibit mechanical properties, which make them uniquely suited for applications in sutures, staples, screws, clips, fixation rods, and drug delivery systems. High strength, thermoplasticity, and non-toxicity are inherent properties of PLA and PGA, which are complemented by the controllability of the crystallinity, hydrophilicity, and therefore overall degradation rate.

Polyhydroxyalkanoates (PHAs) are polyesters that occur in nature. They are produced by a wide variety of microorganisms as an internal carbon and energy storage, as part of their survival mechanism. Polymer production is triggered when the environment changes to unsuitable living conditions. The storage polymer is intended to sustain the survival of the organism when one or more nutrients become unavailable. Depolymerase enzyme are expressed and degrade this stored polymer. PHAs and in particular poly(β-hydroxybutyrate) (PHB), poly(β-hydroxyvalerate) (PHV), and copolymers thereof have been considered as alternatives to conventional fossil fuel-based bulk polymers. The properties of PHB and PHBV are comparable to polypropylene. Pros and cons of the use of microbially derived commodities are subject to scientific as well as political debates and are reflected in the current volume.

The quest for new materials and environmentally sound procedures drives research to new frontiers, one of which brought natural and synthetic polyesters closer together by using enzymes, natural products, as catalysts for polyester synthesis. This novel approach to environmentally suitable materials makes the terms "synthetic" and "natural polyester" obsolete, creating a new class of materials, polymers "from and through renewable resources". Biocatalytic reactions combine several advantages, their site-specificity is most important for stereoregular reactions and prevent effectively the formation of side-products. Enzymes can be easily recovered after the reaction and can in average go through about 100 cycles of usage. Most importantly, ways to use enzymes in organic solvents have been discovered, opening new avenues for organic (polymer) synthesis. Although the use of enzymes as catalysts in organic reactions is a relatively new branch of chemistry, extraordinary achievements in enzyme catalysis are reported in this publication. Collaborations between enzymologists, engineers, and chemists will carry the field of biocatalytic synthesis to a new level within the near future. This book comprises 21 chapters related to research being pursued on the forefront of biopolyesters and enzymatic catalysis.

The editors thank all contributors to this publication and wish them success in their ongoing pursuit for novel materials and advanced technologies that contribute to better the world around us.

The Symposium on Polymers from Renewable Resources was supported in part by ACS Corporation Associates Task Force for Industrial Programming. The generous contribution is gratefully acknowledged.

CARMEN SCHOLZ
Department of Chemistry
University of Alabama in Huntsville
301 Sparkman Drive, MSB 333
Huntsville, AL 35899

RICHARD A. GROSS
Department of Chemical Engineering, Chemistry and Materials Science
Polytechnic University
Six Metrotech Center
Brooklyn, NY 11201

Chapter 1

Biopolyesters and Biocatalysis Introduction

Carmen Scholz[1] and Richard A. Gross[2]

[1]Department of Chemistry, University of Alabama in Huntsville,
301 Sparkman Drive, MSB 333, Huntsville, AL 35899
[2]NSF Center for Biocatalysis and Bioprocessing of Macromolecules,
Department of Chemical Engineering, Chemistry and Materials Science,
Polytechnic University, Six Metrotech Center, Brooklyn, NY 11201

Recently, there has been great interest by both academic and industrial scientists to develop polyesters that provide an environmental benefit. Such polyesters, denoted herein as 'biopolyesters', might be synthesized from readily renewable resources, by methods such as biocatalysis that by-pass non-desirable toxic chemicals, and that degrade and are mineralized after disposal. Most of the polyesters used today are obtained by classical chemical synthetic techniques. As described in this book, a new arsenal of biocatalytic strategies is under study that provides advantages including all aqueous processes, enantio- and regioselective transformations, the use of non-toxic catalysts, one-pot multistep transformations and efficient conversions of renewable resources to value-added products.

Background

The term "Biopolyester" can be understood in several different ways. Biopolyesters can be interpreted as polyesters of strictly biological origin. One can interpret biopolyesters also as polyesters that have been synthesized by biological means, for instance by enzyme-catalyzed polymerization reactions. Moreover, there are hybrids between these two strict definitions of biopolyesters. For example, monomer synthesis for poly(lactic acid), (PLA), is by a biological process in which lactic acid is produced microbially by the fermentation of a renewable, polysaccharide-based resource, mostly corn. Lactic acid is subsequently polymerized chemically into PLA by a condensation reaction.

Polymers preferred by nature seem to be polyamides, in biological terms proteins, and polysaccharides. Proteins fulfill in biological systems catalytic functions (enzymes), regulatory functions (peptides), and serve as transport vehicles. In addition, proteins form structural and storage units and have defensive functions. Polysaccharides are the other large group of natural polymers. They exist in the form of cellulose, starch, alginate, glycogen and many others. All of the above mentioned polysaccharides fulfill structural and storage functions. The most significant, life sustaining biological polymers, DNA and RNA, are polynucleotides. Using chemical nomenclature, polynucleotides are poly(phosphodiesters). Their backbone is composed of ribose and phosphoric acid, which are linked by phosphodiester bonds. This ribose-phosphate-ester linkage is the only covalent bond that connects the four nucleic bases in their specific order. DNA and RNA are recognized for their extraordinary function attributes, which are based upon the nucleosides and their order of arrangement. Therefore, they are classified as polynucleotides rather than polyesters.

Truly naturally produced, high molecular weight biopolyesters exist only in the form of poly(hydroxyalkanoates), (PHAs). PHAs, just like natural rubber, hold an "outsider" position in nature. PHAs are polyesters while rubber is a natural polyolefin, both of which seem to be the product of one of nature's exotic moods, down a playful alley off the common track. Lemoigne[1] first discovered PHAs in 1925 while he was examining *Bacillus megaterium*. PHAs then fell into oblivion for the next couple of decades. The world was taken by storm by synthetic polymers. Staudinger[2] had elucidated and forcefully presented the physical structure of polymers, showing that polymers are covalently linked molecules rather than agglomerates of small molecules. Polymer research gained momentum over the next several years, driven by scientific inquisitiveness and the quest for new materials as well as by the machinery of the Second World War. Plastics can be made for all different kinds of applications with specific, almost made-to-measure properties. There seemed to be no limit to the development of new polymers with properties mimicking the strength of steel, the softness of cotton and everything in between. Polyisoprene superceded natural rubber, giving the same kind of boost to the automobile industry that Nylon 66 gave to the fashion industry. Natural polymers were thrust into the background. The euphoria over synthetic polymers started to wane as a result of the oil crisis in the 1970s. The exhaustibility of fossil fuels was suddenly recognized. Energy conservation, recycling, reusability and the search for new alternatives and renewable resources became significant issues. In addition, environmental concerns and a developing sense of environmental responsibility evolved at about the same time, demanding environmentally sound production processes, that would reduce pollution, improve waste management, and provide biodegradable consumer products. In short "green" Zeitgeist now took the world.

In Europe, where population density is high and land area is much more utilized than in North America, the "Green Movement" developed faster. More focused, "green" alternative groups developed politically and rather quickly into reputable "green parties". Several of these "green parties" share governmental power throughout Europe today.

Therefore, it is not surprising that the first commercial facility for the production of fully biodegradable microbially produced polyesters was established by ICI[3] in England. Poly(β-hydroxybutyrate-co-valerate), (PHBV), was produced from glucose and propionic acid by microbial fermentation of *Alcaligenes eutrophus* (now renamed *Ralstonia eutropha*) and marketed under the tradename Biopol[TM]. One of the first commercial products made from PHBV were shampoo bottles marketed by Wella AG, Germany. Biodegradability is viewed by some as one of the most important assets of biopolyesters. Upon disposal in a biologically active environment, biodegradable biopolyesters can serve as a food source for numerous microorganisms, including fungi, bacteria, and algae. Degradation rates of PHBV are generally rapid in a wide variety of disposal environments including soil, compost and marine ecosystems. High production costs and/or an inability to achieve suitable performance criteria have thus far been obstacles to the commercialization of many biodegradable polymeric products. Thus, despite the biodegradability of some biopolyesters, this property attribute has not been sufficient to allow these products to gain a large share of the plastics market. Even though customers in European countries are more concerned with environmental issues and are, in general, more willing to pay a higher price for an environmentally sound product, commercial success is still governed by a need to reach a suitable cost-performance profile. As for the case of the shampoo bottles; the bottles were more expensive than the shampoo they carried!

Research into biopolyesters, and biocatalytic routes to these polymers, is a broad field that often requires the involvement of interdisciplinary research teams. For example, biologists, chemists and polymer scientists need to communicate and conduct research in a coordinated manner to ultimately achieve a positive outcome. This characteristic of the field will be reflected in the research described in the following chapters. The remainder of this chapter will provide a summary of prominent biopolyesters that are currently in use or are the subject of much research.

Structure of Biopolyesters

All polyesters, of natural and synthetic origin, are characterized by the following common formula, see Figure 1.

Many aliphatic polyesters are obtained by lactone ring opening polymerization, see below. The most prominent representative is polycaprolactone (PCL), (R= $(CH_2)_5$), which is commercially synthesized by ring-opening polymerization of ϵ-caprolcatone, see Figure 4, using metal oxides as catalyst.[4] Some of the earliest reports on the biodegradation of aliphatic polyesters were performed by using PCL.[5,6] Based on its biodegradability, PCL finds primarily applications as blown films for food and yard compost bags and matrix systems for controlled and slow release of pesticides, herbicides and fertilizers. Based on its low melting point (about 60 °C), excellent melt formability and high rigidity in the solid state, PCL has been successfully introduced as plaster replacement in orthopedic and orthotic applications. Due to its low glass transition temperature PCL is often used as additive for other, more brittle polymers. Even though the formation of miscible blends is rather rare in polymer science, PCL is miscible with a variety of polymers. For example, in blends of poly(vinyl chloride) with PCL, the latter acts as plasticizer, thus generating a processable material by reducing the modulus of poly(vinyl chloride).[7] It is equally important that PCL adds biodegradability to the material. Upon disposal in a landfill or compost facility PCL will biodegrade leaving behind the now porous polyolefin bulk material. Since plasticizers can make up to 30 weight% of the material, the use of biodegradable additives contributes significantly to a cleaner environment.

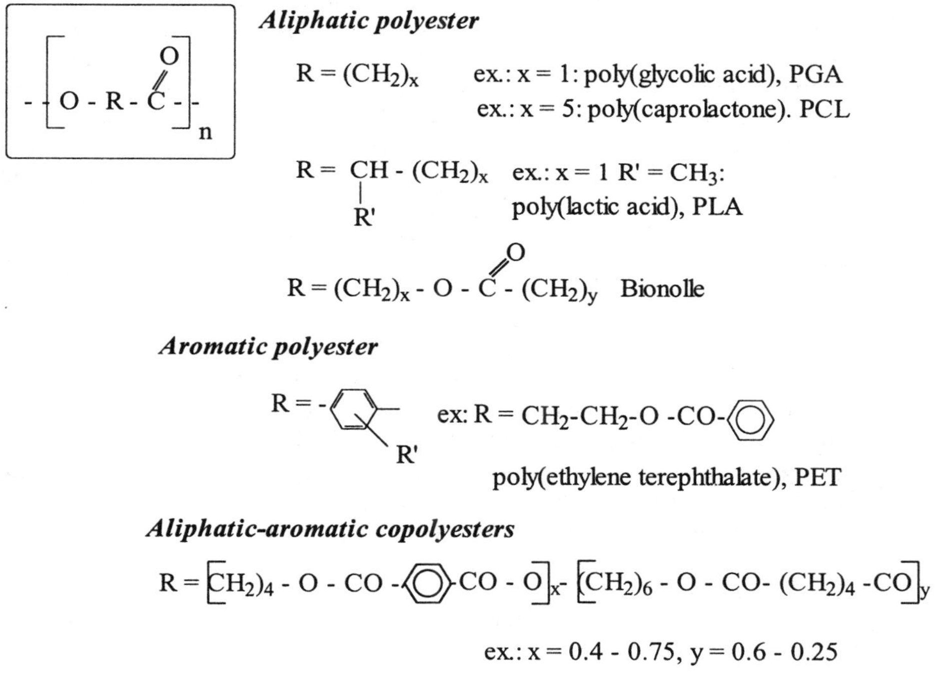

Aliphatic polyester

$R = (CH_2)_x$ ex.: $x = 1$: poly(glycolic acid), PGA
ex.: $x = 5$: poly(caprolactone). PCL

$R = CH - (CH_2)_x$ ex.: $x = 1$ $R' = CH_3$:
$\quad\quad |$
$\quad\quad R'$ poly(lactic acid), PLA

$R = (CH_2)_x - O - \overset{\displaystyle O}{C} - (CH_2)_y$ Bionolle

Aromatic polyester

$R = -$⬡$-$ ex: $R = CH_2-CH_2-O -CO-$⬡
$\quad\quad |$
$\quad\quad R'$

poly(ethylene terephthalate), PET

Aliphatic-aromatic copolyesters

$R = [CH_2)_4 - O - CO -$⬡$CO - O]_x - [CH_2)_6 - O - CO- (CH_2)_4 -CO]_y$

ex.: $x = 0.4 - 0.75$, $y = 0.6 - 0.25$

Figure 1: General Formula of Polyesters: Depending on the structure of R, polyesters are divided into aliphatic and aromatic polyesters

Bionolle is a family of aliphatic polyesters that has been developed and is now commercialized by Showa High Polymer Co., Ltd. in Japan. Bionolle polymers are structurally different from PCL, in that they are the product of a condensation polymerization of an aliphatic diol with a dicarboxylic acid. Using a variety of different diols and dicarboxylic acids leads to the possibility of fine-tuning the physical properties of the material. Isocyanate moieties are commonly added for structural rigidity and toughness. Bionolle is the Japanese counterpart to PCL. It is also biodegradable in soil, active sludge and in compost facilities. As for PCL, biodegradation results in the formation of carbon dioxide, water and biomass and no intermediate metabolism products are released that could be harmful to the ecosystem. Hence, Bionolle is used in similar types of application as described above for PCL.

Poly(lactic acid), PLA, and poly(glycolic acid), (PGA), are aliphatic polyesters that readily degrade by chemically-induced hydrolysis under physiological conditions. Therefore, PLA, PGA and their respective copolymers have found broad application as bioresorbable sutures, implants and drug delivery systems.[8,9] They are synthesized from their respective cyclic dimers, usually by ring-opening polymerization, see Figure 2.

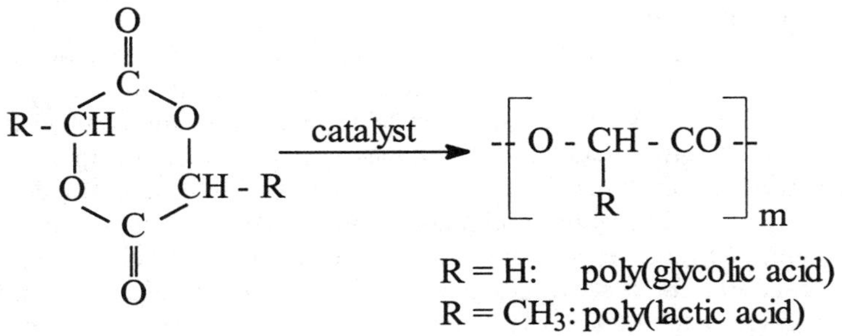

$$R = H: \quad \text{poly(glycolic acid)}$$
$$R = CH_3: \text{poly(lactic acid)}$$

Figure 2: Ring opening polymerization of lactides

Lactic acid is obtained by the fermentation of engineered microbes of the genus *Lactobacilli*. These microorganisms are highly efficient sources of lactic acid. *Lactobacilli* can be subdivided into strains that produce either the L(+) isomer or the D(-) isomer. Fermentation has been optimized to achieve high yields in which batch processes give an average 2 g/L lactic acid per hour. A variety of different sugars are used as carbon sources in the fermentation process. These sugars are either specifically prepared enzymatically from starch for lactic acid production, or are by-products from fruit processing. In addition, lactose, a by-product in the cheese industry, can also be used as a carbon source for lactic acid production. In all cases sugars from renewable resources are transformed into a value-added product by an enzymatic whole-cell catalysis process. The subsequent steps in the production of PLA are based on synthetic chemistry and involve the formation of the dimer by a self-condensation reaction that yields a low molecular weight prepolymer. Depolymerization of the prepolymer yields lactide, which is then polymerized through ring-opening polymerization. The prepolymer can also be polymerized into high molecular weight PLA by the action of chain coupling agents.

In contrast to lactic acid, glycolic acid is produced on an industrial scale by a chemical process. Glycolic acid is present in small amounts in a wide variety of fruits and vegetables. It accumulates during photosynthesis in a side path of the Krebs cycle. So far, economically viable methods to produce glycolic acid in photosynthetic biological systems do not exist. On an industrial scale, carbon monoxide, formaldehyde and water are reacted at elevated temperature and pressure to yield glycolic acid. The acid is thermally converted into its dimer, which eventually is polymerized by a ring-opening polymerization.

More recently, aliphatic-aromatic copolyesters gained much commercial consideration. These copolyesters are synthesized by a condensation polymerization from butanediol and terephthalic acid, as the main components. To tailor the processability and to enhance biodegradation, these copolymers are further modified by copolymerizing with linear dicarboxylic acids (e.g. adipic acid) and glycol components with more than four methylene groups (e.g. hexanediol). One example for an aliphatic-aromatic copolyester is depicted in Figure 1. These copolymers were initially found useful as adhesives.[10] To enhance the environmentally benevolent aspect of these materials and to broaden the range of their use, aliphatic-aromatic copolyesters blended with cellulose esters have been processed into useful fibers, films and molded objects.[11,12] Biodegradation of aliphatic-aromatic copolyesters has been demonstrated in compost facilities.[13] *Actynomycetes*, which are gram-positive bacteria, have been identified as the major microbial species capable of degrading these copolyesters. The formation of aliphatic-aromatic copolyesters using enzymes as catalysts receives currently much consideration. Under tailored conditions, the use of lipases as catalysts for the synthesis of aliphatic-aromatic copolyesters led to polymers with an extremely low polydispersity. Lipases have also been successfully employed as catalysts in studies demonstrating the feasibility of conducting the polymerization in supercritical fluids.

To the best of our knowledge, the only high molecular weight biopolyesters that are currently produced from renewable resources by a whole cell biocatalytic process are poly(hydroxyalkanoates)[14], see Figure 3. PHAs are produced by a wide variety of bacteria and serve as a carbon and energy storage site for microorganisms. Under optimum fermentation conditions more than 90% of the cell dry weight may consist of PHAs. These biopolyesters are produced by the concerted action of three enzymes that include β-ketoacyl-CoA thiolase, acetoacetyl CoA reductase and P(3HB) polymerase. The polymerase functions to convert bioproduced monomers to high molecular weight polymers. Biopolyester synthesis by these organisms is triggered when the environment of the microorganism changes to unsuitable growth conditions, that is the lack of one or several nutrients or oxygen depletion for aerobic species.

Most PHA-accumulating bacteria produce poly(β-hydroxybutyrate), PHB, the PHA with a methyl side chain repeat unit, see Figure 3. Microbial PHB has a very high molecular weight ranging between 200, 000 and 500, 000 Dalton. Numerous depolymerase enzymes exist in a wide variety of environments that recognize the unique structure of these polymers. Specifically, the polymers are isotactic by virtue of having enantiopure β-substituted β-hydroxylalkanoate repeat units in the (R)-configuration. Thus, these polymers are readily degradable upon disposal as long as they are placed into environments with water and a diverse natural microflora.

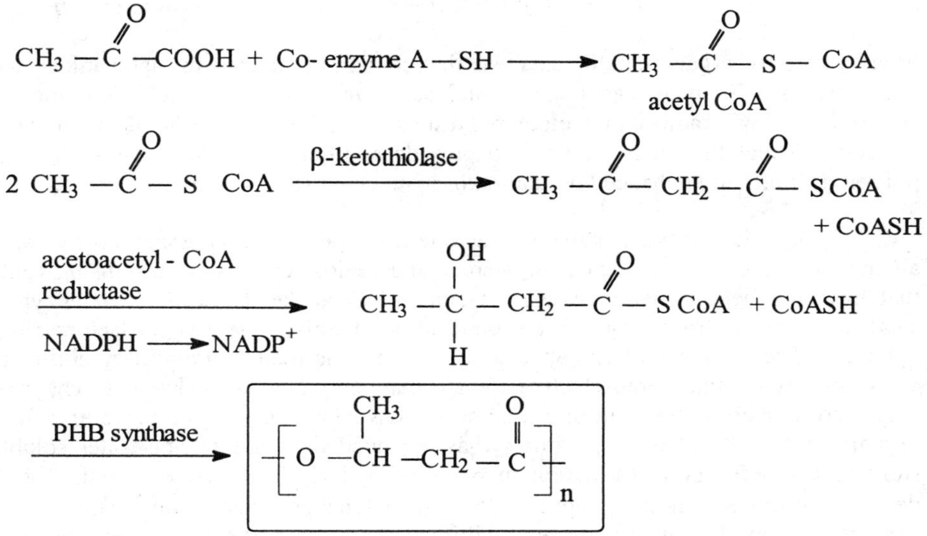

$$CH_3 - \overset{\overset{\displaystyle O}{\|}}{C} - COOH \ + \ Co\text{- enzyme A} - SH \longrightarrow CH_3 - \overset{\overset{\displaystyle O}{\|}}{C} - S - CoA$$

acetyl CoA

$$2\ CH_3 - \overset{\overset{\displaystyle O}{\|}}{C} - S\ \ CoA \ \xrightarrow{\ \beta\text{-ketothiolase}\ } \ CH_3 - \overset{\overset{\displaystyle O}{\|}}{C} - CH_2 - \overset{\overset{\displaystyle O}{\|}}{C} - S\ CoA$$

+ CoASH

acetoacetyl - CoA
reductase

NADPH $\longrightarrow$ NADP$^+$

$$CH_3 - \overset{\overset{\displaystyle OH}{|}}{\underset{\underset{\displaystyle H}{|}}{C}} - CH_2 - \overset{\overset{\displaystyle O}{\|}}{C} - S\ CoA \ \ + CoASH$$

PHB synthase $\longrightarrow$

$$\left[O - \overset{\overset{\displaystyle CH_3}{|}}{CH} - CH_2 - \overset{\overset{\displaystyle O}{\|}}{C} \right]_n$$

Figure 3: Schematic presentation of the enzymatic synthesis of PHB

PHAs can also be obtained by ring-opening polymerization of their respective lactones as is typical for the synthesis of a variety of polyesters, e.g. PCL, see Figure 4. The ring-opening polymerization of β-butyrolactone results in the formation of PHB synthetic analogs. Judicious selection of the catalyst can drive the reaction to yield a highly isotactic polymer. However, without the tedious synthesis of the enantiopure monomer, chemical synthesis fails to provide a complete stereoregular product that has only the natural configuration. It is noteworthy that the synthetic route does provide options for stereochemical variation. The synthesis of stereocopolymers, provides an opportunity to 'tailor' the functional properties of these products. Furthermore, synthetic routes have led to the synthesis of a predominantly syndiotactic PHB that has very different properties than its natural analogue.

Molecular weight and stereochemical control of polyesters synthesized by lactone-ring-opening polymerization is dependent on the catalyst used. There exists a

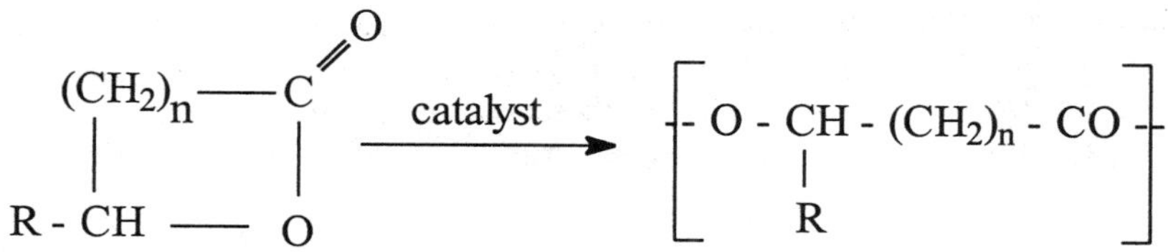

$$(CH_2)_n \!-\! C \overset{O}{\diagup} \quad \xrightarrow{\text{catalyst}} \quad \left[-O-CH-(CH_2)_n-CO- \right]$$

Figure 4: Ring opening polymerization of lactones; for example:
n = 1, R = CH₃: PHB, n = 4, R = H: poly(caprolactone) (PCL)

broad variety of organometallic catalysts that contain tin, aluminum, lanthanides and other metals. In many cases these catalysts have proven to be efficient for the synthesis of high molecular weight polyesters. Work continues to develop these catalysts so that they may be used to provide additional benefits such as 'living' polymerizations and enhanced regulation of main chain stereochemistry.

There is a desire to carry out lactone ring-opening polymerizations by using all natural catalysts, with enhanced stereo- and regioselectivity. A promising route that has only been under investigation since 1993 is the use of *in-vitro* enzyme-catalysis. Certain lipases have been found to be effective catalysts for lactone ring-opening. The hallmark of enzymes such as lipases is their extraordinary ability to carry out regio- and stereoselective transformations. New knowledge in enzyme-catalyzed transformations in non-traditional environments (e.g. reverse micelles, organic media, elevated temperatures) has extended their use to non-water soluble substrates and for bond formation in place of hydrolysis. Chapters in this book describe recent advances in lipase-catalyzed lactone polymerizations. This is one example of how the barrier between traditional chemistry and biology was overcome and how an interdisciplinary research field arose from two classical fields of science.

Biodegradability

The ester bonds of polyesters lead to the susceptibility of these chains to hydrolytic breakdown.[15] For example, in acidic environments, protonation of the carbonyl group catalyzes the cleavage of Carbon - Oxygen bonds along polyester chains, see Figure 5. Hydrolysis of PHAs, PLA, PGA or PCL results in the formation of hydroxy acid products of variable chain length. The rate of chain cleavage events by hydrolysis depends on many factors including the structure of pending groups (denoted as R) as well as the structure of main chain groups that are between ester linkages. These structural entities impact the extent of water adsorption into materials, chain stiffness, and the formation of crystalline domains. All of these factors are important determinants in the rate that biopolyesters will degrade by chemical-mediated hydrolytic reactions. Intermolecular π-electron interactions can lead to high degrees of crystallinity in the corresponding material. For example, PET (Fig. 1) is capable of forming strong intermolecular interactions, which results in high crystallinity and low accessibility of ester groups to water. For these reasons, PET can remain in contact with aqueous solutions in soda bottles without significant hydrolysis.

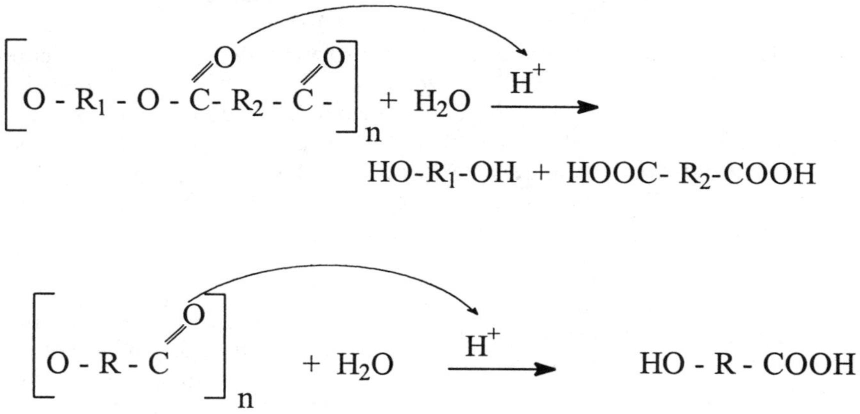

$$\left[O - R_1 - O - \overset{\displaystyle O}{\overset{\displaystyle \|}{C}} - R_2 - \overset{\displaystyle O}{\overset{\displaystyle \|}{C}} - \right]_n + H_2O \xrightarrow{\quad H^+ \quad}$$

$$HO\text{-}R_1\text{-}OH + HOOC\text{-}R_2\text{-}COOH$$

$$\left[O - R - \overset{\displaystyle O}{\overset{\displaystyle \|}{C}} \right]_n + H_2O \xrightarrow{\quad H^+ \quad} HO - R - COOH$$

Figure 5: Hydrolysis of polyesters

In contrast, polyesters such as PLA and PGA are highly susceptible to hydrolytic degradation. This can be explained by their small R-groups (methyl or hydrogen, respectively) and the fact that ester groups occur at short distances along the chain. Even though PLA and PGA can be highly crystalline, the flexibility of chains, high density of ester-groups along the chain and the absorption of water into the interior of the corresponding materials facilitates their hydrolytic breakdown. The degradation of polyesters proceeds either by hydrolysis throughout the bulk or by enzymatic degradation facilitated by endo- and exo-enzymes. The latter starts from the surface of the material. Hydrolytic degradation necessitates a certain degree of hydrophilicity of the polymer, so that water molecules can access the bulk of the sample. Materials characterized by a strong hydrophobic character and/or a high degree of crystallinity will generally show good tolerance to hydrolytic breakdown (e.g. PET). Enzymes attack a polymeric material from the surface, degrading first the amorphous, easily accessible regions, see Figure 6. Therefore, the initial phase of an enzymatic degradation is often characterized by an apparent increase in crystallinity.[16] Exo-enzymes attack the polymer chain from the terminus, removing C_2 units. In contrast, endo-enzymes cleave interior esters along the chain at random. Under a strictly chemical viewpoint, biodegradation is a hydrolysis reaction catalyzed by enzymes. Chemical hydrolysis under acidic conditions normally occurs by first protonating the carbonyl oxygen of a C=O bond, see Figure 5. The chemical hydrolysis of ester groups along polylactic acid chains proceeds throughout the bulk of these materials. This is in sharp contrast to enzymatic degradation mechanisms that occur by surface erosion. In addition, hydrolytic degradation will also depend on the geometric dimensions of the materials. PGA and PLA are well established polymers used in biomedical applications. In general, for PGA and PLA implants exposed to the physiological conditions of our bodies, complete degradation of PGA does not exceed 12 months and PLA degrades within 24 months.[17] Other polyesters have however also been studied for the use as temporarily or permanently implantable biomaterials. Even highly hydrophobic and therefore strongly thrombogenic materials, such as PET, have been used as implants because of the lack of more suited materials. Knitted PET fabric is used in prosthetic valves to cover the core of the sewing ring as well as other surfaces where coverage by fibrotic or endothelial tissue is desirable. Associated thrombosis immediately after implantation was controlled through anticoagulants. Polyesters that are biocompatible, and that decompose *in-vivo* on demand, continue to be of great interest. Besides the well-established use as suture materials, PLA and blockcopolymers thereof containing hydrophilic segments are currently under investigation for the use in drug delivery systems.

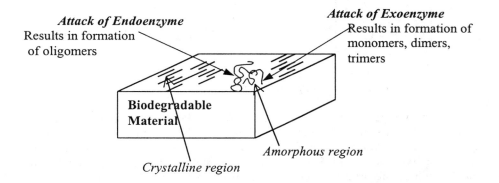

Figure 6: Biodegradation by surface of biodegradable materials, action of exo- and endo-enzymes

Conclusion

Increased consideration of the environment has lead to a redirection of activities in the development of polyesters. In addition to a desire to retain functional attributes, researchers are taking a close look at opportunities to synthesize polyesters from renewable resources, ways that circumvent the use or accumulation of toxic reagents, and that offer an option of biodegradability upon disposal. One outcome of these efforts is to look towards biological production routes. For example, the microbial synthesis of PHAs takes place in a one-pot fermentation process, in water, at approximately room temperature. Inputs into the fermentation are ordinarily readily renewable feedstocks. The output is completely degradable polyesters along with wastes such as biomass that can function as a safe fertilizer. In the case of poly(lactic acid), the monomer is produced by the conversion of agricultural feedstocks by fermentation. This monomer is converted to polymer by a solventless all chemical process. The corresponding polymers are hydrolytically degradable and, microbes mineralize products of hydrolytic degradation. More recently, *in-vitro* lipase-catalysis has proven useful for the synthesis of polyesters. Advantages of the use of lipase-catalysts are that they circumvent the use of heavy metal catalysts, provide extraordinary levels of stereo- and regioselectivity, reduce the need for energy input (lower temperature reactions) and represent a family of safe all natural catalysts (see Kaplan and Gross in this book). Because of the ability to 'tailor' the physical properties of biopolyesters as well as to achieve biocompatibility and controlled biosorption rates, these polymers have excellent potential for increased development and use in a wide range of biomedical applications (see Borden et al. in this book).

References
1   Lemoigne, M. *CR Acad. Sci.* **1925**, 180, 1539; Lemoigne, M. *Ann. Inst. Pasteur* **1925**, 39, 144; Lemoigne, M. *Bull. Soc. Chim. Biol.* **1926**, 8, 770
2   Staudinger, H. *"Die Hochmolecularen Organischen Verbindungen"* Springer Verlag Berlin, **1932** (reprinted 1960)
3   Howells, E.R. *Chem. Ind.* **1982**, 15, 508
4   Dubois, P.; Barakat, I.; Jerome, R.; Teyssie, P. *Macromolecules* **1993**, 26, 4407
5   Benedict, C.V.; Cameron, J.A.; Huant, S.J.; *J. Appl. Polym. Sci.* **1983**, 28, 335
6   Goldberg, D. (review) *J.Environm. Polym. Degrad.* **1995**, 3, 61
7   Hubbell, D.S.: Cooper, S.L. *J. Appl.Polym. Sci.* **1977**, 21, 3035
8   Gilding, D.K.; Reep, A.M. *Polymer* **1979**, 20 459
9   Barrows, T.H. *Clin. Mat.* **1986**, 1, 233
10 Sublett, B.J. US Patent 4,419,507 **1983**
11 Buchanan, C.M.; Gardner, R.M.; Wood, M.D.; White, A.W. Gedon, S.C.; Barlow, Jr., F.D. US Patent 5,292,783 **1994**
12 Buchanan, C.M.; Gardner, R.M.; White, A.W. US Patent 5,446,079 **1995**
13 Kleeberg, I.; Hetz, C.; Kroppenstedt, R.M.; Müller, R.-J.; Deckwer, W.D. *J. Appl. Environm. Microbiol.* **1998**, 64, 1731
14 Doi, Y. "Microbial Polyesters" VCH Publishers, Inc. **1990**
15 Swift, G. *Polym. Mater. Sci. Eng.* **1992**, 66, 403
16 Nishida, H.; Tokiwa, Y. *Polym. Mater. Sci. Eng.* **1992**, 66, 137
17 Kohn, J.; Langer, R. in "Biomaterials Science" *An Introduction to Materials in Medicine* (ed.: B.D. Ratner, A.S. Hoffman, F.J. Schoen, J.E. Lemons) New York Academic Press **1996**, 64

Poly(β-hydroxyalkanoates)

Chapter 2

Interesting Carbon Sources for Biotechnological Production of Biodegradable Polyesters: The Use of Rape Seed Oil Methyl Ester (Biodiesel)

A. Steinbüchel, I. Voß, and V. Gorenflo

Institut für Mikrobiologie, Westfälische Wilhelms-Universität Münster, Corrensstraße 3, D–48143 Münster, Germany

In the past, several interesting carbon sources were used for the synthesis and biotechnological production of biodegradable polyhydroxyalkanoic acids (PHA). A short overview will be provided on our former studies in this field. Presently, several Gram-positive and Gram-negative bacteria were analyzed for their ability to use rape seed oil methyl esters (biodiesel) as a sole carbon source for growth and biosynthesis of PHA. Strains of *Nocardia corallina, Ralstonia eutropha* and *Pseudomonas aeruginosa* were the most suitable candidates; they accumulated poly(3-hydroxybutyric acid), a copolyester of 3-hydroxybutyric acid and 3-hydroxyvaleric acid and a copolyester consisting of various 3-hydroxyalkanoic acids of medium-chain-length, respectively, when cultivated in a mineral salts medium that contained biodiesel as sole carbon source. *R. eutropha,* which grew on biodiesel at a doubling time of 2.2 h, was cultivated at a 30 L scale in a fed-batch process. Approximately 185 g biomass was obtained within 45 h from 280 g biodiesel, and the cells contained 68% (w/w) poly(3-hydroxybutyric acid) so that 125 g of the polyester could be isolated.

1 Introduction

Poly(hydroxyalkanoic acids), PHA, represent a complex class of naturally occurring biodegradable thermoplastics and elastomers, and approximately 120 different hydroxyalkanoic acids are currently known as constituents of these microbial polyesters (*1-4*). These polyesters can be synthesized *in vivo* and *in vitro* resulting in the formation of water insoluble granules. Only some of the known PHAs can be obtained from simple carbon sources that are cheap and abundantly available. Most PHAs can only be produced from precursor substrates which are structurally related to the constituents that are to be incorporated into the polyesters (*5*).

In general, studies on the biosynthesis of coenzyme A thioesters of hydroxyalkanoic acids in bacteria and on the properties of PHA synthases, which use the thioesters as substrates for the polymerization reaction, will contribute to engineer metabolic pathways that allow the synthesis and production of as many PHAs as possible from simple unrelated carbon sources. Preferred substrates will be those which are available in large quantities from agricultural crops, from abundant fossil carbon sources such as lignite (*6*) or from cheap waste materials (*5*). The final aim should be the production of such PHAs also from carbon dioxide. Metabolic engineering in bacteria may contribute to prove the concepts for corresponding studies in transgenic plants (*7*).

Approximately 12 years ago, our laboratory started to study the physiological, biochemical and molecular basis of PHA biosynthesis in bacteria. This research focussed mainly on three aspects: (i) After the genes for the poly(3-hydroxybutyric acid) biosynthesis pathway of *Ralstonia eutropha* were cloned (*8*), we also studied the PHA biosynthesis genes in many other bacteria. Meanwhile, the structural genes for PHA synthases and for enzymes providing the substrates for this key enzyme of PHA biosynthesis were cloned from more than 38 different bacteria in our and several other laboratories. A recent review summarizes the results of these studies and provides an alignment of the primary structures of 30 different PHA synthases (*9*). (ii) The second focus was the analysis of pathways which allow the synthesis of coenzyme A thioesters of hydroxyalkanoic acids from intermediates of central pathways like the amino acid metabolism, the citric acid cycle and the fatty acid metabolism. For example, recently the missing link between the fatty acid *de novo* synthesis and PHA biosynthesis, which is a 3-hydroxyacyl-acyl carrier protein-coenzyme A transferase was identified (*10*). A recent review also summarizes this aspect of PHA research, which is studied in several laboratories and which is important, since knowledge of such pathways will allow a more efficient production of PHAs from simple carbon sources and in particular renewable resources (*5*). (iii) The third focus is the use of precursor substrates for the production of novel polyesters containing hitherto not described constituents. These studies have provided many new polyesters, and the material properties of an increasing number of these polymers are currently investigated.

Two recent examples for the synthesis of unusual PHAs are copolyesters that contain 2-methyl-3-hydroxybutyric acid from tiglic acid and 2,2-dimethyl-

3-hydroxypropionic acid from 3-hydroxypivalic acid. These studies provided examples of how PHAs are synthesized containing more than one alkyl side chain on carbon atoms that contribute to the backbone of the polyester (*11, 12*).

PHAs containing 4-hydroxyalkanoic acids as constituents exhibit interesting material properties that are quite different from polyesters consisting of 3-hydroxyalkanoic acids. We recently established a process which allows the production of poly(4-hydroxybutyric acid) homopolyester from 4-hydroxybutyric acid employing a recombinant strain of *Escherichia coli* (*13*). In addition, we investigated the production of 4-hydroxyvaleric acid containing copolyesters from levulinic acid. For this, we constructed recombinant strains of *Ralstonia* and *Pseudomonas* and by optimizing the cultivation conditions and introducing mutations in the levulinic acid catabolic pathway, we were able to obtain polyesters with an increased content of 4-hydroxyvaleric acid (*14, 15, C. Brämer unpublished data*).

In addition, we are now able to produce PHAs consisting of medium-chain-length 3-hydroxyalkanoic acids such as 3-hydroxyoctanoic acid or 3-hydroxydecanoic acid in recombinant strains of *E. coli* that expressed the PHA synthase genes from *Pseudomonas aeruginosa* (*16, 17*). Recently, we identified in *P. putida* the gene which encodes for a metabolic link between fatty acid *de novo* synthesis and PHA biosynthesis. This enzyme – 3-hydroxyacyl-acyl carrier protein-coenzyme A transferase – contributes to the synthesis of PHA from simple carbon sources such as e. g. glucose (*10*).

In this study, we investigated whether rape seed oil methyl ester (biodiesel), which is produced at an increasing scale as fuel for automobiles from renewable resources as an alternative to fossil diesel (*18*), can be used as a carbon source for the synthesis and biotechnological production of PHAs.

2 Materials and Methods

2.1 Bacterial Strains, Media and Growth Conditions.

Bacterial strains used in this study are listed in Table 1. The cells were cultivated for 3 days at 30°C under aerobic conditions in nutrient broth (NB) or in mineral salts medium (MSM) according to Schlegel et al. (*19*), containing 0.05 or 0.1% (w/v) NH_4Cl. Rape seed oil methyl ester (biodiesel), that was produced by the Oelmühle Leer (Connemann GmbH & Co.), was added to the MSM as sole carbon source as it was obtained from a local gasoline station (Raiffeisen Genossenschaft). To obtain solid media, 1.5% (w/v) agar was added.

Cultivations in liquid media were done in 100 to 2000 mL Erlenmeyer flasks equipped with baffles and containing 10 to 15% (v/v) medium. The flasks were agitated on a horizontal rotary shaker. For cultivation at a 30 L scale, a Biostat D30 stainless steel stirred tank reactor (Fa. Braun Biotec, Melsungen, Germany) equipped with a DCU control system and 3 rotary blades was used. The culture broth was aerated at a rate of 1 vvm, and MSM containing 0.1% (w/v) NH_4Cl was used. Cells were harvested with a CEPA Z41 continuous centrifuge (Carl Padberg, Lahr, Germany).

2.2 Isolation of PHA from lyophilized Cells.

PHA was isolated from lyophilized cells by extraction with chloroform in a Soxhlet apparatus. The polyester was precipitated from the chloroform by the addition to 10 vol. ethanol. The precipitate was separated from the solvents by filtration. Remaining solvents were removed by exposure of the polyester to a stream of air. Precipitation was repeated two times to obtain highly purified PHA samples.

2.3 Analysis of PHA.

The polyester content of the bacteria was determined by methanolysis of 5-7 mg lyophilized cells in the presence of 15% (v/v) sulfuric acid suspended in methanol, according to Brandl *et al.* (*20*). The resulting methyl esters of the constituent hydroxyalkanoic acids were characterized by gas chromatography in a Model 8420 gas chromatograph (Fa. Perkin-Elmer, Überlingen, Germany), equipped with a flame ionization detector and a stabilwax column (60 m, 0.32 mm ID, 0.5 microns, Fa. Restek GmbH, Bad Soden, Germany) according to Brandl *et al.* (*21*) and as described in detail by Timm *et al.* (*22*).

2.4 Analysis of Methanol.

For the detection of methanol in culture supernatants, the cultivated cells were settled by centrifugation, and aliquots of the supernatants were analyzed by gas chromatography employing a Model GC-9A gas chromatograph (Shimadzu, Kyoto, Japan) equipped with a flame ionization detector and a Poropak QS column. The initial column temperature was 120°C, the temperature gradient was 20°C/min and the final temperature was adjusted to 200°C, the temperatures of the detector and the injector were 250°C. Helium was used as a carrier gas.

2.5 Analysis of Biodiesel.

For the detection of biodiesel lyophilized aliquots of supernatants were subjected to methanolysis, and the resulting methyl esters were analyzed by gas chromatography (see section 2.3.). Fatty acids (C16:0, C18:0, C18:1, C18:3) were used as reference standards and the concentration of biodiesel in the supernatant was calculated by referring to the composition given by the manufacturer.

3 Results

3.1 Growth of PHA accumulating Bacteria on Biodiesel.

Although it is known that biodiesel is degraded in the environment (23) and that methyl esters of fatty acids are easily hydrolyzed by lipases or esterases, bacteria, which are of interest for the production of PHA, have to our knowledge so far not been cultivated on biodiesel. Therefore, we investigated whether such bacteria are able to grow on biodiesel if it was supplied as the sole carbon source in a mineral salts medium (Table 1).

Table 1: Bacterial Strains used in this Study and Utilization of Biodiesel as sole Carbon Source

Bacterial strains (Source or reference)	Growth
Rhodococcus ruber (NCIMB 40126, (24))	++
Rhodococcus opacus PD630 (DSMZ 44193, (25))	+
Rhodococcus opacus MR22 (DSMZ 3346)	+
Nocardia corallina N°724 (26)	++
Ralstonia eutropha H16 (DSMZ 428)	++
Burkholderia cepacia (DSMZ 50181)	++
Pseudomonas aeruginosa PAO1 (DSMZ 1707)	++
Pseudomonas oleovorans (ATCC 29347)	+
Pseudomonas putida KT 2440 (27)	-
Paracoccus denitrificans "Morris" (DSMZ 413)	++

Abbreviations and symbols: ATCC, American Type Culture Collection; DSMZ, Deutsche Sammlung von Mikroorganismen und Zellkulturen; NCIMB, National Collection of Industrial and Marine Bacteria; -, no growth; +, good growth; ++, very good growth.

Several Gram-positive bacteria belonging to the genera *Rhodococcus* and *Nocardia* as well as several Gram-negative bacteria belonging to the genera *Ralstonia*, *Burkholderia*, *Pseudomonas* and *Paracoccus* grew in liquid MSM containing biodiesel as carbon source. *P. putida* KT 2440 was the only strain investigated that did not grow on this carbon source.

Three bacteria were selected for further studies, and the growth was investigated in liquid MSM in more detail (Fig. 1). Whereas *P. aeruginosa* PAO1 exhibited a relatively long lag-phase and a low optical density at the end of the growth phase, *N. corallina* N°724 and *R. eutropha* H16 grew much faster and to higher cell densities, the doubling times were 2.0 h and 2.2 h, respectively. From these data *N. corallina* N°724 seemed to be the most suitable candidate for PHA production on

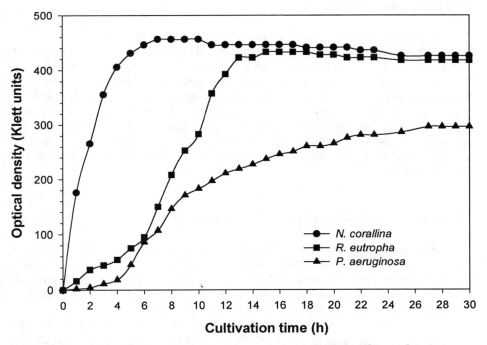

Figure 1. Growth of three representative bacteria investigated in this study on biodiesel. 50 mL MSM containing 0.2% (v/v) biodiesel in a 300 mL Erlenmeyer flask were inoculated with 3 mL of an overnight NB culture. Growth was recorded in a Klett-Summerson colorimeter equipped with a filter of 520-580 nm.

biodiesel (see Fig. 1), however, the amount of PHA was low at the end of the cultivation period.

When the cells obtained from the cultivation experiment shown in figure 1 were analyzed for the PHA content, we found that the cells of *R. eutropha* H16 contained 8.6% poly(3HB), whereas the cells of *P. aeruginosa* PAO1 had accumulated no determinable amounts of PHA.

3.2 Consumption of Methanol.

We also analyzed the cell free supernatants of the culture broths of various bacteria that were cultivated on biodiesel as sole carbon source. These analysis showed that only *P. denitrificans* was able to use the methanol released from the biodiesel completely since no methanol was detected in the supernatant. The supernatants of the various Gram-positive bacteria investigated in this study contained little methanol, whereas those of *R. eutropha* H16, *B. cepacia* and *P. aeruginosa* PAO1 contained the theoretical maximum concentration of methanol that was expected from the available biodiesel if it was not consumed. Further studies will be necessary to address the metabolism of methanol in these bacteria in the presence and absence of cosubstrates.

3.3 Formation of PHA from Biodiesel by selected Bacteria.

Subsequently, we did several fed-batch cultivation experiments in Erlenmeyer flask at the 500 mL scale in order to investigate the formation of PHAs in various bacteria in more detail and also to isolate the polyesters from the cells. In addition, we wanted to obtain information on whether methanol, which is the other hydrolysis product of biodiesel is also used as a carbon source by these bacteria.

Cells of *R. eutropha* H16, *P. aeruginosa* PAO1, *N. corallina* N°724 and *B. cepacia* were cultivated in 500 to 2000 mL MSM containing 0.05% NH_4Cl and 0.2% (v/v) biodiesel. Biodiesel was fed to the cultures after 12, 24, 36 and 48 h at an amount each corresponding to 0.2% (v/v). After approximately 60 h of cultivation the cells were harvested, analyzed, and the polyesters were isolated.

The data compiled in Table 2 show that the cell yields of all four strains were high, with between 44 to 74% of the biodiesel converted into biomass. *R. eutropha* H16 cells possessed the highest PHA content where poly(3HB) contributed 70% (w/w) to the cellular dry mass. In contrast, cells of *P. aeruginosa* PAO1 and *B. cepacia* contained 33 or 28 % poly(3HA$_{MCL}$), respectively. These copolyesters were composed of various 3-hydroxyalkanoates of medium chain length with 3-hyroxydecanoate or poly(3HB-*co*-3H4Pe) as the main constituent, respectively. Cells of *N. corallina* N°724 accumulated a polyester consisting of poly(3HB-*co*-3HV) amounting only up to 6% of the cellular dry mass.

Table 2: Formation of PHAs from Biodiesel

Strain	Scale	Bio-diesel	Harvested Cells	Isolated PHA	Extracted Cell Material	Lost Material
	[mL]	[g]	[g]	[g]	[g]	[g]
R. eutropha	1000	8.9	7.4	5.20	1.61	0.59
			100%	70%	22%	8%
P. aeruginosa	1000	8.9	2.43	0.81	1.45	0.17
			100%	33%	60%	7%
N. corallina	500	4.45	2.20	0.13	1.68	0.39
			100%	6%	76%	18%
B. cepacia	2000	17.8	11.5	3.20	7.79	0.60
			100%	28%	67%	5%

When cells of *R. opacus* PD 630, *R. opacus* MR 22 and *R. ruber* were collected, they formed a slimy phase at the bottom of the tube that could not be freeze-dried to a dry powdery material like the others. Gas chromatographic analysis of this slimy phase revealed biodiesel as the main constituent, containing only small amounts of 3-hydroxyalkanoic acids. This demonstrated that components of the biodiesel were bound to the cells. To obtain reliable data about the composition of PHA accumulated in the cells, it is necessary to isolate the polyesters and separate them from all the other constituents.

3.4 Cultivation of *R. eutropha* H16 on Biodiesel at a 30 L Scale.

From the previous experiments it became clear that *R. eutropha* H16 was the most suitable microorganism for the conversion of biodiesel to PHA. Therefore, we cultivated this strain at a 30 L scale in MSM, containing 0.1% NH_4Cl in a stirred and aerated tank bioreactor (see Fig. 2). The cultivation was done as a fed-batch process with additional biodiesel fed when the carbon source became exhausted. Within 44 h the cells grew to an optical density of approximately 25 units at a doubling time of 2.2 h, and the accumulated poly(3HB) homopolyester contributed to almost 70% of the biomass as revealed by the analysis of whole cells.

Cells were harvested, washed with 0.9% (w/v) NaCl-solution and freeze-dried. From 185 g lyophilized cells, 125 g poly(3HB) were obtained by chloroform extraction and precipitation with ethanol, which corresponds to a polyester content of the cells of 68%.

4 Discussion

Biodiesel is produced from rape seed oil by chemical transesterification of the fatty acids from glycerol to methanol. Oleic acid (63%) and linoleic acid (20%) are

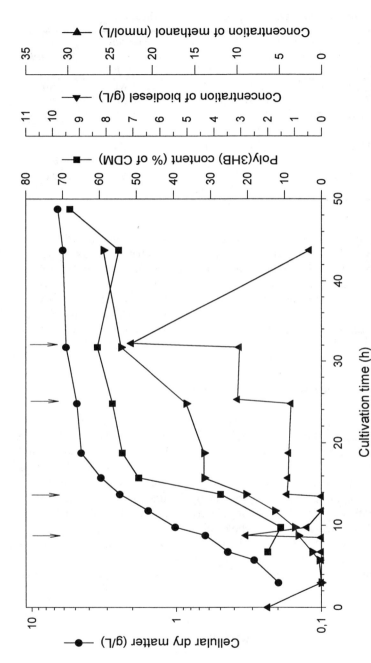

Figure 2. Cultivation of R. eutropha at a 30 L scale. 30 L MSM containing 0.2% (v/v) biodiesel were inoculated with four 500 mL MSM-biodiesel precultures and cultivated in a Biostat D30 stirred tank reactor. After 7, 13, and 25 h cultivation 50 mL and after 32 h 100 mL biodiesel were fed (see arrows). The temperature was kept at 30°C, the aeration rate was adjusted to 1 vvm. The pH was not controlled and decreased from 7.2 to 6.5 during the cultivation.

the major fatty acids of the oil. In this study it was shown that several PHA accumulating bacteria are able to utilize biodiesel as sole carbon source for colony growth an agar plates and also for cell growth in liquid media. This indicates that those bacteria excreted esterases which hydrolyze the methyl esters of fatty acids obtained from the rape seed oil to the corresponding fatty acids and to methanol. The fatty acids are carbon sources that are readily utilized by these bacteria via the β-oxidation pathway as known from many previous studies. *P. denitrificans* also consumed the methanol completely. *P. denitrificans* is very well known to utilize methanol as sole carbon source for growth and has been studied in great detail with respect to the production of PHAs from methanol (*28*). Some bacteria such as *R. eutropha* H16 and *B. cepacia* did not assimilate the methanol, which remained in the medium. From *R. eutropha* it is known that it can not grow on methanol (*29*). All other bacteria assimilated methanol only partially.

It was demonstrated that different polyesters can be obtained from biodiesel. Poly(3HB), poly(3HB-*co*-3HV) and poly(3HA$_{MCL}$) were produced employing *R. eutropha* H16, *N. corallina* N°724 or *P. aeruginosa* PAO1, respectively. Fermentation of *R. eutropha* H16 at the 30 L scale demonstrated that biodiesel may be a suitable carbon source for technical production of poly(3HB) since growth on this carbon source is relatively good and the polyester content of the cells is rather high. This is a good basis and prerequisite to develop a high cell density process for the production of poly(3HB) from biodiesel with *R. eutropha*.

References

1. Hocking, P. J.; Marchessault, R. H. *Chemistry and technology of biodegradable polymers;* 1 st edn.; Griffin, G. J. L. (Ed); Chapman & Hall, London, 1994, Biopolyesters, pp 48-96.
2. Anderson, A. J.; Dawes, E. A. *Microb. Rev.* **1990,** *54,* 450-472.
3. Steinbüchel, A. *Biotechnology*; Rehm, H. J.; Reed, G.; Pühler, A.; Stadler, P.: VCH Weinheim, Germany, 1998, Vol. 6, pp. 403-464.
4. Steinbüchel, A.; Valentin, H. E. *FEMS Microbiol Lett.* **1995,** *128,* 219-228.
5. Steinbüchel, A.; Füchtenbusch, B. *Trends in Biotechnol.* **1998,** in press.
6. Füchtenbusch, B.; Steinbüchel, A. *Appl. Microbiol. Biotechnol.* **1998,** in press.
7. Poirier, Y.; Nawrath, C.; Somerville, C. *Bio/Technology.* **1995,** *13,* 142-150.
8. Schubert, P.; Steinbüchel, A.; Schlegel, H.G. *J. Bacteriol.* **1988,** *170,* 5837-5847.
9. Rehm, B.H.A.; Steinbüchel, A. *Int. J. Biol. Mocr.* **1999,** in press.
10. Rehm, B. H. A.; Krüger, N.; Steinbüchel, A. *J. Biol. Chem.* **1998,** *273,* 24044-24051.
11. Füchtenbusch, B.; Fabritius, D.; Steinbüchel, A. *FEMS Microbiol Lett.* **1996,** *138,* 153-160.
12. Füchtenbusch, B.; Fabritius, D.; Wältermann, M.; Steinbüchel, A. *FEMS Microbiol Lett.* **1998,** *159,* 85-92.
13. Hein, S.; Söhling, B.; Gottschalk, G.; Steinbüchel, A. *FEMS Microbiol Lett.* **1997,** *153,* 411-418.

14. Steinbüchel, A.; Gorenflo, V. *Macromolecular Symposia,* **1997,** *123,* 61-66.
15. Schmack, G.; Gorenflo, V.; Steinbüchel, A. *Macromolecules,* **1998,** *31,* 644-649.
16. Langenbach, S.; Rehm, B. H. A.; Steinbüchel, A. *FEMS Microbiol Lett.* **1997,** *150,* 303-309.
17. Qingsheng, Q.; Rehm, B. H. A.; Steinbüchel, A. *FEMS Microbiol Lett.* **1997,** *157,* 155-162.
18. Connemann, J. *Fat. Sci. Technol.* **1994,** *96;* 536-548.
19. Schlegel, H. G.; Kaltwasser, H.; Gottschalk, G. *Arch. Microbiol.* **1961,** *38,* 209-222.
20. Brandl, H.; Knee, E. J.; Fuller, R. C.; Gross, R. A.; Lenz, R. W. *Int. J. Biol. Macromol.* **1989** *11,* 49.
21. Brandl, H.; Gross, R. A.; Lenz, R. W.; Fuller, R. C. *Appl. Environ. Microbiol.* **1977,** *54.*
22. Timm, A.; Byrom, D.; Steinbüchel A. *Appl. Microbiol. Biotechnol.* **1990,** *33,* 296.
23. Rodinger, W. International conference on standardization and analysis of biodiesel. Vienna, 6th and 7th November 1995.
24. Haywood G. W.; Anderson, A. J.; Williams, D. R.; Dawes, E.A.; Ewing, D. F. *Int. J. Biol. Macromol.* **1991,** *13,* 83-88.
25. Alvarez, H. M.; Mayer, F.; Fabritius, D.; Steinbüchel, A. *Arch. Microbiol.* **1996,** *165,* 377-386.
26. Valentin, H. E.; Dennis, D. *Appl. Environ. Microbiol.* **1996,** *62,* 372-379.
27. Worsey, M. J.; Williams, P. A. *J. Bacteriol.* **1975,** *124,* 7-13.
28. Yamane, T. *FEMS Microbiol. Rev.* **1992,** *103,* 257-264.
29. Steinbüchel, A.; Fründ, C.; Jendrossek, D.; Schlegel, H.G. *Arch. Microbiol.* **1987,** *148,* 178-186.

Chapter 3

Biosynthesis and Properties of Medium-Chain-Length Polyhydroxyalkanoates from *Pseudomonas resinovorans*

Richard D. Ashby, Daniel K. Y. Solaiman, and Thomas A. Foglia

Eastern Regional Research Center, Agricultural Research Service, U.S. Department of Agriculture, 600 East Mermaid Lane, Wyndmoor, PA 19038

In an effort to create new polymers, and provide additional outlets for animal fat and vegetable oil commodities, several triacylglycerols (TAGs) were studied as substrates for the bacterial production of medium-chain-length poly(hydroxyalkanoate) polymers (*mcl*-PHA) by *Pseudomonas resinovorans*. Polymer yields ranged from 40% to 60% of the cell dry weight depending on the substrate. Each of the PHA polymers was characterized with respect to molar mass (gel permeation chromatography), thermal properties (calorimetry), and repeat-unit composition (gas chromatography/mass spectrometry). Repeat-unit chain lengths ranged from C4 to C14 with varying amounts of side-chain unsaturation (double bonds). For example, PHA from coconut oil consisted almost entirely of saturated monomers, whereas PHA from soybean oil contained higher levels of unsaturated side-chains. Variation in repeat-unit composition resulted in polymers with properties ranging from elastomeric to adhesive-like. The presence of olefinic groups in the polymer side-chains allowed for radiation crosslinking which enhanced the tensile strength, flexibility, and Young's modulus of the polymer films.

Poly(hydroxyalkanoates) (PHA) are a class of naturally occurring, optically active polyesters that accumulate in numerous bacteria as carbon and energy storage materials *(1-3)*. In most cases the polymers contain β-linked repeat units and are compositionally distinct based on side-chain structure *(4)*. However, some exceptions exist where the linkages are through the δ position of the monomers (*e.g.*, 4-hydroxybutyrate, 4HB, 4-hydroxyvalerate, 4HV, 4-hydroxyhexanoate, 4HH) *(5-7)*, or through the γ position (5-hydroxyvalerate, 5HV) *(8)*. To date, approximately 100 different PHA structural analogs exist based on the large variety of PHA producers and the broad substrate specificity of the polymerization systems. Because they are viewed as "environmentally friendly," many PHAs are being studied as potential replacements for synthetic plastics in several applications. Triacylglycerols (TAGs) are non-traditional feedstocks for *mcl*-PHA production. By using agricultural products like fats and oils, new polymers may result with unusual repeat unit compositions and properties.

It is known that several bacterial species produce medium-chain-length poly(hydroxyalkanoate) (*mcl*-PHA) polymers from fatty acids *(9-11)*. Recently intact TAGs have also been considered as substrates for PHA production *(10,12-14)*. Organisms that metabolize TAGs to PHA polymer presumably have to meet two requirements: (i) production of a lipase enzyme that hydrolyzes the TAG to liberate the long-chain fatty acids and (ii) production of PHA from long-chain fatty acids. Lipase production is a common feature among *Pseudomonas* species *(15,16)*, and likewise many pseudomonads produce *mcl*-PHA from a variety of carbon sources including long-chain fatty acids *(11,17,18)*. Therefore, under certain growth conditions, selected bacterial strains have the ability to produce PHA polymers from TAGs. In addition, PHA biosynthesis from TAGs may provide new polymers with repeat-unit compositions containing saturated and unsaturated side-chains resulting in materials with elastomeric properties.

One way to increase the strength of a polymer film is to crosslink the matrix. This can be accomplished by the use of chemical additives (peroxides, sulfur vulcanization) *(19,20)* or physical treatments (radiation) *(21)*. Generally, the crosslink density is dependent upon the method used, and the number of functional groups present. Radiation provides a quick, and relatively clean method of crosslinking without the addition or formation of contaminants in the polymer matrix. The interaction of radiation with polymeric materials generally results in three-dimensional network structures with improved tensile properties *(22)*. Because of this, three of the *mcl*-PHAs were selected (based on the number of olefinic groups present in their side-chains) for irradiation studies to determine the effects of olefinic content on radiation induced crosslink efficiency, and tensile properties. PHA films produced from coconut oil (PHA-C; low olefinic concentration), tallow (PHA-T; intermediate olefinic concentration), and soybean oil (PHA-So; high olefinic concentration) were solution cast and irradiated using a γ-emitter (cesium-137). The films were characterized by tensile testing before and after radiation treatment to compare the effect of γ-irradiation on the polymer tensile properties.

Experimental

Materials

All simple salts were obtained from Sigma Chemical Company (St. Louis, MO) and used as received. Each TAG was obtained as a commodity material as follows: coconut oil (C) and olive oil (O) (Sigma Chemical Company), tallow (T) (Miniat Inc., Chicago, IL), lard (L) (Holsum Foods, Waukesha, WI), butter oil (B) (ERRC dairy pilot processing plant), high oleic acid sunflower oil (Su) (SVO Enterprises, Eastlake, OH), and soybean oil (So) (purchased from the supermarket under the trade name Wesson oil). All organic solvents were HPLC grade (Burdick and Jackson, Muskegon, MI).

Strain Information and Polymer Synthesis

Pseudomonas resinovorans NRRL B-2649 was obtained from the National Center for Agricultural Utilization Research, Agricultural Research Service, United States Department of Agriculture, Peoria, IL, and used as the producer strain in all fermentations. Stock cultures were prepared and stored as described previously *(14)*.

Shake flask experiments were conducted at 500 mL volumes (in 1 liter Erlenmeyer flasks) under batch culture conditions in Medium E* (for medium composition see reference *9*). Each TAG substrate was heated at 60°C for 15 minutes and added to sterile Medium E* at a concentration of 1% (w/v). The flasks were inoculated with a 0.1% inoculum from a thawed cryovial. Bacterial growth and polymer production were carried out at 30°C with shaking at 250 rpm, for 72 hours in an orbital shaker-incubator. At 72 hours the cells were pelleted and washed twice in deionized water by centrifugation (8000 × g, 20 minutes, 4°C). Unused fats (solids) were removed from the cultures prior to centrifugation by filtration through cheesecloth. Unused oils concentrated at the air/liquid interface upon centrifugation and were removed either with the supernatant or by wiping the walls of the centrifuge bottles with paper towels. The cell pellets were then lyophilized (~24 hours) to a constant weight.

Electron Microscopy

Scanning Electron Microscopy (SEM)
A solution of 1% glutaraldehyde-0.1 M imidazole-HCl buffer (pH 6.8) was flooded onto the surface of a 48 hour medium E* agar plate inoculated with *P. resinovorans* and containing emulsified olive oil as the carbon substrate (the medium was modified from that described in reference *23*). The cultures were incubated at room temperature for 2 hours and stored in sealed containers at 4°C.

Selected areas of the cultures were excised, washed by immersion in imidazole buffer for 30 minutes, and immersed in 2% osmium tetroxide-0.1 M imidazole buffer for two hours. Samples were then washed in distilled water, dehydrated in a graded series of ethanol solutions and critical point dried from liquid carbon dioxide. Finally, the dried samples were mounted on aluminum specimen stubs with colloidal silver adhesive and coated with a thin layer of gold by DC sputtering at low voltage (Model LVC-76, Plasms Sciences, Lorton, VA). Photographic images of the surfaces of sample cultures were made using a JSM-840A scanning electron microscope (JEOL USA, Peabody, MA) operated in the secondary electron imaging mode.

Transmission Electron Microscopy (TEM)
One hundred microliters of 10% glutaraldehyde was added directly to 900 μL of a 72 hour broth culture of *P. resinovorans* and rapidly mixed in a 1.5 mL tube. The mixture was then spun in a Model 5413 Eppendorf centrifuge for 5 minutes. After standing at room temperature for 2 hours, the cell pellets were washed with 0.1 M imidazole buffer (pH 6.8) for 30 minutes and immersed in 2% osmium tetroxide-0.1 M imidazole buffer for 2 hours. Pellets were washed in distilled water, dehydrated in a series of ethanol solutions, soaked in propylene oxide and infiltrated with a 1:1 (v/v) mixture of propylene oxide and an epoxy resin mixture overnight. Pellets were next embedded in fresh epoxy resin mixture and cured for 48 hours at 55°C. Thin sections of the embedded samples were cut with diamond knives and stained with solutions of 2% uranyl acetate and lead citrate. Photographic images were made using a Model CM12 scanning-transmission electron microscope (Philips Electronics, Mahwah, NJ) operated in the bright field imaging mode.

PHA Purification and Isolation

Supercritical fluid extraction (SFE) was used to remove residual TAG material prior to PHA isolation. The dried bacterial cells (5 g per extraction) were packed tightly into a 24 mL stainless steel extraction vessel (rated at 10,000 psi). Polypropylene wool was packed in front and behind the bacterial cells to protect the vessel frits (2 μm). The extraction vessel was placed in the SFE apparatus oven (Applied Separations, Allentown, PA). The oven was adjusted to 60°C and the restrictor valve temperature was set at 100°C. When the vessel temperature reached 60°C carbon dioxide at 5000 psi was allowed to flow through the extraction vessel at 1.5 L/minute for 1 hour. The pressure was then raised to 7000 psi and carbon dioxide flow continued for 1 hour at 1.5 L/minute In the final hour, carbon dioxide flowed at 1.5 L/minute at a pressure of 9000 psi. At the end of 3 hours, the vessel temperature and pressure were reduced to room temperature and atmospheric pressure and the bacterial cells removed from the vessel for polymer isolation.

PHA isolation was accomplished by stirring SFE extracted cells (approximately 1 g) for 24 hours in chloroform (100 mL) at 30°C. The insoluble cellular material was removed by vacuum filtration through Whatman #1 filter paper (Whatman International Ltd., Maidstone, England) and the solvent was concentrated (to approximately 5 mL) in a rotary evaporator. Each *mcl*-PHA was precipitated by dropwise addition of the polymer concentrate into excess cold methanol and the precipitate dried *in vacuo* for 24 hours at 25°C and 1.0 mm Hg.

Instrumental Procedures

PHA repeat unit compositions were determined by gas chromatography (GC) and GC/mass spectrometry (GC/MS) of the β-hydroxymethyl esters (β-HMEs). Samples were prepared according to the procedures of Brandl et al. *(9)* and analyzed by GC and GC/MS using conditions reported previously *(10)*.

Molar mass averages were determined by gel permeation chromatography (GPC) after the procedure of Cromwick et al. *(10)*. Polystyrene standards (Polysciences Corp., Warrington, PA) were used to generate a calibration curve from which PHA molar masses were determined without further corrections. Chloroform was used as the eluent at a flow rate of 1 mL/min. The sample concentration and injection volume were 0.5% (w/v) and 200 μL, respectively. The thermal properties and the carbon (^{13}C) nuclear magnetic resonance (NMR) spectra of each *mcl*-PHA were obtained as described elsewhere *(14)*.

Film Preparation and Irradiation

Films were cast from solutions of PHA-T, PHA-C, or PHA-So. Solutions were prepared by dissolving 1.5 g of purified PHA in 15 mL of chloroform. Films were cast in glass petri dishes (100 mm × 15 mm), and the solvent evaporated under a nitrogen atmosphere. The resulting films were approximately 0.1 mm thick, and were stored desiccated in the dark under a nitrogen purge prior to irradiation. The films in the glass petri dishes were irradiated with 50 kGy of radiation (energy was based on an 8.5 h exposure) using a cesium-137 source at 20°C in a nitrogen atmosphere. The irradiator and dosimetry were described previously *(24)*.

Sol/Gel Analysis

Sol/gel tests were performed on both non-irradiated and irradiated film samples according to the method described previously *(25)*.

Tensile Property Measurement

Tensile properties of the PHA films were measured at 23°C and 50% relative humidity with a gauge length of 25 mm. An Instron tensile testing machine, model 1122 (Canton, MA), was used throughout this work. The cross-head was maintained at a constant speed of 50 mm/minute Measurements included tensile strength, elongation to break and Young's modulus. Tensile strength and elongation to break were defined as the ultimate stress and strain, respectively. Young's modulus is a physical quantity representing the stiffness of a material. It was determined by measuring the slope of a line tangent to the initial stress-strain curve. All the data were calculated and collected through the use of Instron series IX automated materials testing system version V.

Results and Discussion

mcl-PHA Production and Characterization

While many species of *Pseudomonas* have the ability to produce *mcl*-PHA from fatty acids, only two (*P. resinovorans* and *P. aeruginosa*) have been documented to produce *mcl*-PHA from TAGs *(10, 13)*. Cromwick et al. *(10)* explored the reason why *P. resinovorans* grew and produced *mcl*-PHA from TAGs while the other known PHA producers tested did not. By using a fluorescent assay developed by Moreau *(26)*, it was determined that *P. resinovorans* was the only organism tested that exhibited extracellular esterase (lipase) activity in the presence of TAG substrates. This enzyme activity liberates the fatty acids from the glycerol backbone of the TAG, thus providing a substrate that could be used in *mcl*-PHA synthesis. The inability of the other PHA-producing pseudomonads to grow and produce PHA from TAGs suggests that the esterase (lipase) activity is necessary for PHA biosynthesis by *P. resinovorans*.

Each animal fat and vegetable oil is unique in its fatty acid composition. For this reason, seven different TAGs (tallow, lard, butter oil, olive oil, high oleic acid sunflower oil, coconut oil, and soybean oil) were screened as potential substrates for *mcl*-PHA production. Scanning electron microscopy (SEM) showed that *P. resinovorans* attaches itself to the fat or oil droplets, thus providing maximal access to the substrate (Fig. 1a). In each case, *P. resinovorans* produced an *mcl*-PHA polymer, which was evident by the presence of one or more PHA granules per bacterial cell (Fig. 1b). The cells were harvested by centrifugation and the cellular biomass and polymer yield determined gravimetrically. The PHA content was calculated as a percentage of the cell dry weight (CDW) (Table I).

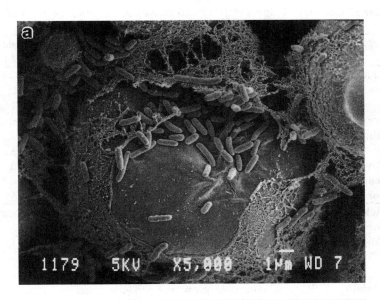

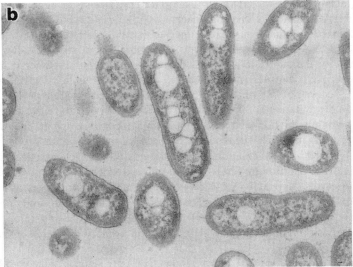

Figure 1. Electron micrographs showing growth and polymer production by P. resinovorans *on TAG substrates: a) SEM of bacterial cells attached to a droplet of olive oil (the length of the average bacterial cell is approximately 1 μm). b) TEM of* P. resinovorans *grown on tallow showing granule production within the cells (scale: 22 mm = 1 μm).*

Table I. Cell Dry Weights (CDW) and *mcl*-PHA Polymer Content of
P. resinovorans Grown on TAG Substrates

Substrate	Crude cell yield (g/L)	PHA content (% CDW)	PHA yield[a] (g/L)
Animal fats			
Tallow	2.4	43	1.0
Lard	3.8	55	2.1
Butter oil	3.7	53	2.0
Vegetable oils			
Olive	4.0	58	2.3
Sunflower (HO)[b]	4.0	57	2.3
Soybean	4.1	59	2.4
Coconut	4.1	61	2.5

[a]PHA yield is calculated by multiplying the cell yield (g/L) by the PHA content (% dry weight) of the cells.
[b]HO = "high oleic."

The ability of *P. resinovorans* to grow and produce equal cell masses and polymer yields on coconut oil (highly saturated TAG) and soybean oil (highly unsaturated TAG) indicates that the concentration of *cis* double bonds has little effect on substrate metabolism. This suggests that the discrepancy between cell growth and PHA yields on animal fats *vs.* vegetable oils is primarily due to the physical state of the substrate at 30°C (incubation temperature) rather than fatty acid composition. The titer (melting range) of each vegetable oil tested is less than 25°C. Because of this, these oils are easily dispersed throughout the culture media by simple agitation at 30°C. In contrast, lard and butter oil have titers near 33°C, only slightly higher than the incubation temperature, while the titer of tallow is greater than 40°C. The result is that some of the animal fat TAGs remain solid throughout the fermentation, which retards both bacterial growth and polymer yield. In addition to TAGs, animal fats and vegetable oils contain small amounts of sterols (approx. 0.1%-0.5%) and tocopherols (primarily α and γ). Though the utilization of these compounds for cell growth and polymer production cannot be eliminated, high cell and polymer yields (along with polymer composition) suggest that TAGs were the primary carbon sources used by the bacterium.

The composition of *mcl*-PHAs from long chain fatty acids is controlled by the specificity of the PHA-synthesizing system, the structures of the fatty acids, and the degradation pathway for long chain fatty acids *(11)*. It has been reported that *P. oleovorans* grown on alkanoic acids produces *mcl*-PHAs that reflect the chain length of the substrate *(27)*. For example, when grown on hexanoate or octanoate the *mcl*-PHA was composed predominantly of 3-hydroxyhexanoate and 3-hydroxyoctanoate monomers, respectively. When grown on alkanoates with chain lengths greater than C_{10}, it was observed that C_8 and C_{10} were the

Table II. Repeat-Unit Composition of Poly(Hydroxyalkanoate) Polymers Isolated from *P. resinovorans* Cultures Grown on TAG Substrates

Isolate	β-hydroxymethyl ester (%)[a]									
	$C_{4:0}$	$C_{6:0}$	$C_{8:0}$	$C_{10:0}$	$C_{12:0}$	$C_{12:1}$	$C_{14:0}$	$C_{14:1}$	$C_{14:2}$	$C_{14:3}$
PHA-T	Tr[b]	3	15	46	17	4	6	9	n.d.[c]	n.d.
PHA-L	Tr	7	26	34	14	4	4	8	3	n.d.
PHA-B	Tr	9	31	35	15	Tr	4	5	n.d.	n.d.
PHA-O	1	8	29	33	14	1	3	10	1	n.d.
PHA-Su	2	5	22	35	14	3	3	13	3	n.d.
PHA-So	Tr	4	18	32	8	14	4	9	10	Tr
PHA-C	Tr	7	33	40	16	1	3	Tr	n.d.	n.d.

[a]Average relative percent (n=5) as determined by GC of the β-HMEs obtained by acid hydrolysis and methylation of each *mcl*-PHA polymer.
[b]Tr (trace) = < 1%.
[c]n.d. = not detected.

predominant monomers in each *mcl*-PHA. Because each TAG contains different ratios of saturated and unsaturated fatty acids, the composition of the TAG based *mcl*-PHAs should vary to reflect the substrates and the enzymatic makeup of the organism (particularly those involved in β-oxidation of unsaturated fatty acids). The repeat-unit composition of each polymer was determined by GC and GC/MS of the β-hydroxymethyl esters (β-HME) and are listed in Table II.

The electron-impact (EI) mass spectrum of β-hydroxydecanoate methyl ester is shown in Figure 2. Each β-hydroxyalkanoate methyl ester contained an ion fragment at $m/z = 103$ due to ions formed by cleavage between carbons 3 and 4. The chain length of each methyl ester was based on GC retention time and confirmed by mass spectrometry. Lack of silylation results in a loss of water, and methanol in these molecules. Therefore, molecular ions (M^+) were determined based on $m/z = M^+ - 18$ (loss of water) and $m/z = M^+ - (18 + 32)$ (loss of water + methanol) for each β-HME. Regardless of the substrate, the predominant repeat units in each *mcl*-PHA were β-hydroxyoctanoate and β-hydroxydecanoate. This is likely due to enhanced enzymatic specificity of the PHA synthase for 3–hydroxyoctanoyl-CoA and 3-hydroxydecanoyl-CoA. Each polymer contained a unique repeat-unit composition that ranged in carbon chain length from C_4 to C_{14} with some mono-, di-, and tri-unsaturation in the C_{12} and C_{14} alkyl side chains. Unlike the other *mcl*-PHAs, PHA-T had a C_8:C_{10} ratio of about 3:1 whereas the others had ratios of 1.4 ($\pm$ 0.3):1. Tallow is composed of both a liquid and a solid fraction at room temperature. The variation in the C_8:C_{10} ratio between PHA-T and the other *mcl*-PHAs may be the result of selective utilization of the liquid

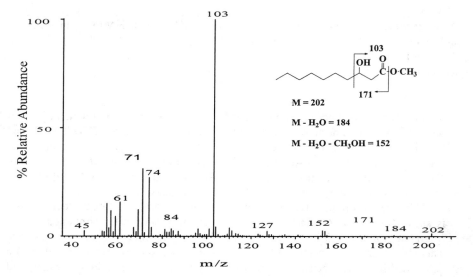

Figure 2. Electron-impact mass spectrum of β-hydroxydecanoate methyl ester. The presence of an ion fragment with m/z = 103 is indicative of a β-hydroxyalkanoate methyl ester. (SOURCE: Reproduced with permission from ref. 14.)

portion of the tallow rather than the more insoluble fraction. PHA-C contained very low levels of unsaturated monomers. This was undoubtedly due to the low levels of unsaturation in the substrate itself (oleic acid, 4.6%; linoleic acid, 0.9%). The PHA-So repeat-unit composition was more complex than the other PHAs. It contained a much larger fraction of di-unsaturated C_{14} monomers (and a small $C_{14:3}$ fraction) and, with the exception of PHA-Su, had more than twice as many unsaturated side chains as the other *mcl*-PHAs.

The PHAs also were studied by ^{13}C-NMR (50 MHz). Figure 3 shows typical spectra of *mcl*-PHAs, in particular PHA-C, PHA-T, and PHA-So. The chemical shift assignments are based on those reported previously *(18, 28, 29)*. The chemical shifts of the peaks in the 0 - 120 ppm region are identical apart from slight differences in their relative intensities. From the chemical shift intensities between 120 and 140 ppm (olefinic carbons) it was confirmed that PHA-C contained very few unsaturated side chains, whereas the side chains from PHA-So were highly unsaturated.

As the concentration of olefinic carbons increased, the polymers became more amorphous. This was verified by measurement of the thermal properties of each polymer by calorimetry (Table III). With the exception of PHA-Su and PHA-So, each *mcl*-PHA was elastomeric at room temperature. PHA-C had the highest glass transition temperature (T_g), melting temperature (T_m), and enthalpy of fusion (ΔH_m) of the polymers. While it seems likely that these increases were primarily due to a more ordered packing arrangement as the degree of side-chain unsaturation decreased, the increased molar masses of the *mcl*-PHAs containing more saturated monomers also may have helped to increase the T_g and T_m of the polymers.

Table III. Molar Masses[a] and Thermal Properties of TAG-Derived *mcl*-PHA Polymers Produced by *P. resinovorans*

Isolate	M_n ($\times 10^3$)	M_w ($\times 10^3$)	M_w/M_n	T_g (^oC)	T_m (^oC)	ΔH_m (J/g)
PHA-T	93	269	2.89	-46	42	11.7
PHA-L	84	220	2.62	-46	42	9.8
PHA-B	101	298	2.95	-43	43	12.0
PHA-O	82	214	2.61	-46	42	10.1
PHA-Su[b]	61	149	2.44	-47	-	-
PHA-So[b]	57	121	2.13	-47	-	-
PHA-C	133	449	3.38	-40	45	15.4

[a]All molar masses were determined by gel-permeation chromatography.
[b]The *mcl*-PHA polymers from high oleic sunflower oil and soybean oil were amorphous and showed no melting transition.

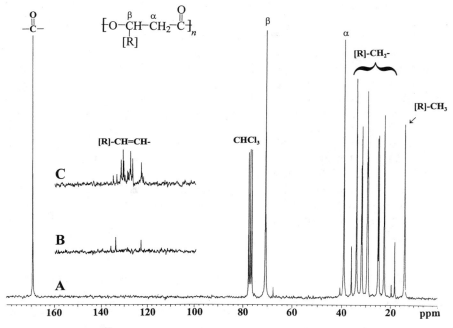

Figure 3. The ^{13}C-NMR spectrum of mcl-*PHA polymers isolated from* P. resinovorans *grown on a) coconut oil (PHA-C), b) tallow (PHA-T), and c) soybean oil (PHA-So). (SOURCE: Reproduced with permission from ref. 14.)*

At the other extreme, PHA-So and PHA-Su had T_g's of -47°C and T_m's and ΔH_m's that were not measurable. This indicated that these polymers were completely amorphous and, in fact, they were almost liquid-like at room temperature. In contrast to PHA-C, for example, this was probably due to the smaller molar masses and higher concentrations of double bonds in the side-chains, which resulted in a lack of packing uniformity in the polymer.

In addition to composition, molar mass also influences polymer properties. The molar masses of the PHA polymers are given in Table III. It can be seen that an increased molar mass was associated with an increased concentration of saturated monomer side-chains. This suggests that an increased concentration of unsaturated fatty acids in the substrate interrupts the polymerization efficiency of the system. Because equal polymer yields were obtained from both coconut oil and soybean oil, it seems likely that the presence of unsaturated fatty acids, while not inhibiting polymerization, causes a more efficient termination of polymerization and results in a larger number of smaller molar mass PHA-So chains. In addition, unsaturated free fatty acids require additional isomerase and epimerase enzymes for oxidation to produce the CoA intermediates required for polymerization. The increased energy requirement involved in the production of two additional enzymes may slow the polymerization process and result in polymers with smaller relative molar masses. Whatever the predominant mechanism, it may be that both chemistries exist, resulting in a molar mass reduction in *mcl*-PHAs containing higher concentrations of unsaturated side-chains.

Radiation Effects

Radiation affects the properties of polymeric materials according to the type and source of radiation, the nature of the polymer structure, and the mechanism of reaction. It has been well established that crosslinking of a polymeric material requires the presence of functional groups *(30)*. Olefinic groups are used frequently to this end since they can be crosslinked by both chemical *(19, 20)* and irradiation methods *(21, 31, 32, 33)*. As seen from the *mcl*-PHA compositions (Table II), many of the double bonds from the TAG substrate were preserved by the bacterial PHA polymerization process. These groups provide the functionality necessary for crosslinking. Gamma-irradiation of a polymeric material generates radicals that lead to crosslink formation. However, these radicals also cause chain scission, the extent of which is based on the chemical composition of the material. Formally, the two processes, *i.e.*, crosslinking and chain scission, occur simultaneously, making it necessary to achieve some balance between the two processes to enhance the tensile properties of each material.

In the present study three different *mcl*-PHAs were chosen based on their differing olefinic concentrations and γ-irradiated to determine the effects of side-chain unsaturation on radiation induced enhancement of polymer tensile properties. After irradiation, each of the polymer films was subjected to a sol/gel analysis to determine the extent of crosslinking using non-irradiated films as controls. For non-irradiated films, the gel fraction amounted to between 1.0% and 1.4% of the total film weight tested. This indicated that prior to irradiation the films were relatively free of crosslinks. After irradiation, the gel fractions increased to 8%, 60%, and 91% for PHA-C, PHA-T, and PHA-So, respectively. Accordingly, the extent of crosslinking was proportional to the number of olefinic groups present in each *mcl*-PHA polymer side-chain. Tensile properties were measured immediately after irradiation to limit auto-oxidation. In each case, the tensile properties were enhanced by irradiation (Table IV).

The most interesting result occurred with PHA-So films. While the tensile properties of the PHA-T film and PHA-C film increased, the PHA-So films were stiffened from a liquid-like texture to a solid film. This allowed the measurement of the tensile properties of PHA-So, which could not be done prior to irradiation. The tensile strength and percent elongation of the irradiated PHA-So film were 0.7 MPa and 25%, respectively, showing it to be a weaker film than the PHA-T and PHA-C crosslinked films. Young's modulus, which is a measure of the stiffness of a material, also increased in each polymer film after irradiation. Interestingly, PHA-So had the highest modulus after irradiation. These results indicated that irradiation had the greatest effect on the PHA-So films, which can be directly related to the high concentration of olefinic groups present in the PHA-So polymer.

Gel permeation chromatography (GPC) of the soluble (sol) fractions from the sol/gel analyses of the irradiated polymers (Table V) revealed that the molar

Table IV. Tensile Properties of Select Irradiated and Non-Irradiated *mcl*-PHA Polymer Films

Isolate	Radiation Dose (kGy)[a]	Tensile Strength (MPa)	% Elongation	Young's Modulus (MPa)
PHA-C	0	2.5	242	2.0
	50	5.1	360	2.6
PHA-T	0	3.0	320	1.7
	50	4.9	360	3.0
PHA-So[b]	0	-	-	-
	50	0.7	25	3.1

[a]Radiation source was cesium-137.

[b] In the absence of radiation PHA-So was amorphous and resulted in a non-coherent film. (SOURCE: Reproduced with permission from ref. 33.)

Table V. Sol Fraction[a] Molar Masses of Select Irradiated *mcl*-PHA Polymer Films

Isolate (film)	Radiation Dose (kGy)	Molar mass (sol fraction)		
		M_n ($\times 10^3$)	M_w ($\times 10^3$)	M_w/M_n
PHA-C	50	59	205	3.47
PHA-T	50	48	194	4.04
PHA-So	50	33	102	3.09

[a]Sol fractions were obtained from the chloroform soluble portions of the sol/gel analysis after film irradiation. (SOURCE: Reproduced with permission from ref. 33.)

masses had M_n and M_w values that were smaller and more polydisperse than the non-irradiated sol controls (Table III).

This indicated that, in addition to crosslinking, random chain scission also occurred during the irradiation process. The formation of small quantities of large molar mass polymers helped increase polydispersity (M_w/M_n). Specifically, the total amount of large molar mass fraction material varied from 3% (PHA-So) to 18% (PHA-C) with average M_n values well above 6.0×10^6 g/mol. Because irradiation is a balance between crosslinking and chain scission, it is entirely possible that the large molar mass fraction was the result of the radiation induced formation of microgels that are small enough to pass through a 0.45 μm filter prior to GPC analysis. PHA-C is composed of the fewest olefinic carbons. This results in a limited number of crosslink sites compared to chain scission sites. Because of this, a larger concentration of microgels may be produced by PHA-C when compared to PHA-So during irradiation. This combination of microgel formation and chain scission resulted in increased polydispersities for the sol fractions of the irradiated films.

In conclusion, it has been demonstrated that *P. resinovorans* produces *mcl*-PHA polymers from TAG substrates. Each *mcl*-PHA synthesized from TAG substrates reflects the fatty acid make-up of the TAG. Because of this, different polymers can be produced from various TAGs ranging from highly saturated (PHA-C) to highly unsaturated (PHA-So). The presence of olefinic groups in the *mcl*-PHA side-chains allows the manipulation of polymer properties by irradiation. Radiation generally results in polymer crosslinking; some chain scission, however, does occur. A positive balance between crosslinking and chain scission provides a relatively clean method of improving tensile properties of the TAG-derived *mcl*-PHAs. Lastly, optimization of *mcl*-PHA production from TAGs may provide farmers an additional outlet for fat and oil commodities, or provide a method to recycle used oils and greases.

Acknowledgement

The authors thank Rob DiCiccio and Bina Christy for their technical assistance, as well as Dr. Alberto Nunez for performing the GC/MS analyses, Dr. Donald Thayer for performing the polymer irradiation, Dr. Peter Cooke for the electron microscopy, Ms. Janine Brouillette for the ^{13}C-NMR, Dr. Jim Hampson for the supercritical fluid extraction, and Dr. Cheng-Kung Liu for the tensile testing.

References

1. Anderson, A. J.; Dawes, E. A. *Microbiol. Rev.* **1990**, *54*, 450-472.
2. Brandl, H.; Gross, R. A.; Lenz, R. W.; Fuller, R. C. In *Advances in Biochemical Engineering/Biotechnology*; Springer, Berlin, 1990, Vol. 41, pp 77-93.
3. Doi, Y. *Microbial Polyesters*; VCH, New York, 1990.
4. Steinbüchel, A.; Valentin, H. E. *FEMS Microbiol. Lett.* **1995**, *128*, 219-228.
5. Kunioka, M.; Nakamura, Y.; Doi, Y. *Polymer Commun.* **1988**, *29*, 174-176.
6. Valentin, H. E.; Schönebaum, A.; Steinbüchel, A. *Appl. Microbiol. Biotechnol.* **1992**, *36*, 507-514.
7. Valentin, H. E.; Lee, E. Y.; Choi, C. Y.; Steinbüchel, A. *Appl. Microbiol. Biotechnol.* **1994**, *40*, 710-716.
8. Doi, Y.; Tamaki, A.; Kunioka, M.; Soga, K. *Makrom. Chem. Rapid. Commun.* **1987**, *8*, 631-635.
9. Brandl, H.; Gross, R. A.; Lenz, R. W.; Fuller, R. C. *Appl. Environ. Microbiol.* **1988**, *54*, 1977-1982.
10. Cromwick, A-M.; Foglia, T. A.; Lenz, R. W. *Appl. Microbiol. Biotechnol.* **1996**, *46*, 464-469.
11. Eggink, G.; van der Wal, H.; Huijberts, G. N. M.; de Waard, P. *Ind. Crops Products* **1993**, *1*, 157-163.
12. Shimamura, E.; Kasuya, K.; Kobayashi, G.; Shiotani, T.; Shima, Y.; Doi, Y. *Macromolecules* **1994**, *27*, 878-880.
13. Eggink, G.; de Waard, P.; Huijberts, G. N. M. *Can. J. Microbiol.* **1995**, *43[supple 1]*, 14-21.
14. Ashby, R. D.; Foglia, T. A. *Appl. Microbiol. Biotechnol.* **1998**, *49*, 431-437.
15. Gilbert, E. J. *Enzyme Microb. Technol.* **1993**, *15*, 634-645.
16. Jaeger, K-E.; Ransac, S.; Dijkstra, B. W.; Colson, C.; van Heuvel, M.; Misset, O. *FEMS Microbiol Rev.* **1994**, *15*, 29-63.
17. Timm, A.; Steinbüchel, A. *Appl. Environ. Microbiol.* **1990**, *56*, 3360-3367.
18. De Waard, P.; van der Wal, H.; Huijberts, G. N. M.; Eggink, G. *J. Biol. Chem.* **1993**, *268*, 315-319.
19. Gagnon, K. D.; Lenz, R. W.; Farris, R. J.; Fuller, R. C. *Polymer* **1994a**, *35(20)*, 4358-4367.

20. Gagnon, K. D.; Lenz, R. W.; Farris, R. J.; Fuller, R. C. *Polymer* **1994b**, *35(20)*, 4368-4375.

21. De Koning, G. J. M.; van Bilsen, H. M. M.; Lemstra, P. J.; Hazenberg, W.; Witholt, B.; Preusting, H.; van der Galien, J. G.; Schirmer, A.; Jendrossek, D. *Polymer* **1994**, *35(10)*, 2090-2097.

22. Shultz, A. R. *Encyclopedia of Polymer Science and Technology*; Interscience Publishers, New York, 1966, Vol. 4, pp. 398-414.

23. Kouker, G.; Jaeger, K-E. *Appl. Environ. Microbiol.* **1987**, *53(1)*, 211-213.

24. Shieh, J. J.; Jenkins, R. K.; Wierbicki, E. *Radiat. Phys. Chem.* **1985**, *25*, 779-792.

25. Ashby, R. D.; Cromwick, A-M.; Foglia, T. A. *Int. J. Biol. Macromol.* **1998**, *23(1)*, 61-72.

26. Moreau, R. A. *Lipids* **1989**, *24*, 691-698.

27. Huisman, G. W.; de Leeuw, O.; Eggink, G.; Witholt, B. *Appl. Environ. Microbiol.* **1989**, *55(8)*, 1949-1954.

28. Abe H.; Doi, Y.; Fukushima, T.; Eya, H. *Int. J. Biol. Macromol.* **1994**, *16(3)*, 115-119.

29. Huijberts, G. N. M.; de Rijk, T.; de Waard, P.; Eggink, G. *J. Bacteriol.* **1994**, *176*, 1661-1666.

30. Mercx, F. P. M.; Boersma, A.; Damman, S. B. *J. Appl. Polym. Sci.* **1996**, *59*, 2079-2087.

31. Lin, C. H.; Maeda, M.; Blumstein, A. *J. Appl. Polym. Sci.* **1990**, *41*, 1009-1022.

32. Haddleton, D. M.; Creed, D.; Griffin, A. C.; Hoyle, C. E.; Venkataram, K. *Makromol. Chem., Rapid Commun.* **1989**, *10*, 391-396.

33. Ashby, R. D.; Foglia, T. A.; Liu, C-K.; Hampson, J. W. *Biotechnol. Lett.* **1998**, *20*, 1047-1052.

Chapter 4

The Structural Organization of Polyhydroxyalkanoate Inclusion Bodies

L. J. R. Foster

Department of Biotechnology, The University of New South Wales, Sydney, New South Wales 2052, Australia

Polyhydroxyalkanoates are a diverse family of natural polyesters with potential as biodegradable replacements for commercial bulk commodity plastics based on petrochemicals. Despite the abundance of research regarding their production and properties, little is known about the structure of their intracellular organisation. This chapter reviews the structural organisation of polyhydroxyalkanoate inclusion bodies. Investigations regarding the architecture of these inclusion bodies have important implications for the biotechnological production of polyhydroxyalkanoates in microbes and other organisms, for downstream processing and in vitro precision polymerisation.

In 1925, Maurice Lemoigne, a microbiologist at the Lille branch of the Pasteur Institute, gazed down the eyepiece of his microscope and saw whitish, refractile bodies (Figure 1) in samples of *Bacillus megaterium*. His curiosity piqued, Lemoigne spent the next few years isolating and identifying the composition of these inclusion bodies. He determined that they consisted of a novel polyester: poly(-3-hydroxybutyrate) (*1,2*), (PHB-Figure 2). Lemoigne also discovered that these intracellular inclusions were synthesised by *B. megaterium* under conditions of excess environmental carbon and one or more limiting nutrients (*1-4*). It is now known that they serve as a carbon storage mechanism in much a similar fashion as glycogen in mammals. Once isolated, however, PHB possesses characteristics that make it a favourable biodegradable alternative to conventional plastics based on limited resources (*5-7*).

© 2000 American Chemical Society

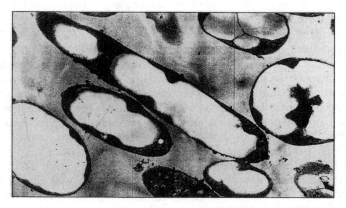

Figure 1:   Electron micrograph of Pseudomonas oleovorans illustrating
refractile intracellular Polyhydroxyalkanoate (PHA) inclusion bodies [x10k].

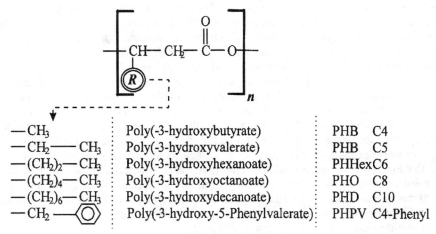

—CH₃	Poly(-3-hydroxybutyrate)	PHB C4
—CH₂——CH₃	Poly(-3-hydroxyvalerate)	PHB C5
—(CH₂)₂—CH₃	Poly(-3-hydroxyhexanoate)	PHHexC6
—(CH₂)₄—CH₃	Poly(-3-hydroxyoctanoate)	PHO C8
—(CH₂)₆—CH₃	Poly(-3-hydroxydecanoate)	PHD C10
—CH₂——⬡	Poly(-3-hydroxy-5-Phenylvalerate)	PHPV C4-Phenyl

Figure 2:   Chemical formula for PHAs and examples of their structural diversity.

Interest in PHB and its material properties grew rather slowly until its commercial potential as a biocompatible medical implantation device was investigated and tested by W.R. Grace & Co. in the 1950's *(8)*. This brief foray in the commercial arena yielded little success, but served to focus the important potential of PHB to the scientific community. Research increased, in the 1980's the Imperial Chemical Company (ICI - now Zeneca) discovered that by feeding *Alcaligenes eutrophus* mixtures of carbon substrates a random copolymer of PHB and poly(-3-hydroxyvalerate), (PHV), could be produced *(9,10)*. Alterations in feed regimens eventually led to the commercialisation of a range of P(HB-*Co*-HV) copolymers under the trade name 'Biopol' *(11, ICI personal Communication 1989)*. Biopol^{TM} is marketed under the appropriate, and appealing, banner; 'Nature's Plastic', its major potential is as a biodegradable replacement to the prevalent bulk commodity plastics which are chemically synthesised from petroleum products and responsible for disturbing amounts of pollution. Unfortunately, present production techniques for PHB and Biopol^{TM} ensure that their cost is non-competitive with current plastics such as polyethylene *(12)*. It can be argued that if conventional plastics were to carry the financial burden of their medically and environmentally safe production and disposal then their price would probably be more in line with that of Biopol^{TM}. The commercial situation is slowly changing as legislation regarding the safe disposal of commodity plastics is introduced and current research lowers the production costs of PHB and Biopol^{TM}.

With the goal of commercialisation in mind, many researchers have taken advantage of recent advances in the field of molecular genetics to improve polymer production *(12,13)*. There has been considerable research to produce transgenic plants capable of synthesising PHB from renewable resources with financially viable yields *(14-16)*. Similarly, considerable research is being conducted into the material properties of PHB and P(HB-*Co*-HV) as well as their biodegradation in various environments of the ecosphere *(17-20)*.

PHB and P(HB-*co*-HV) are, however, but two members of a diverse family of polyesters called polyhydroxyalkanoates, (PHAs). PHAs have the general formula shown in Figure 2, where the 'R' group can be any number of substituents, from relatively short or long alkyl chains to chemically functional groups. A 1995 survey by Steinbüchel and Valentin identified 93 different monomeric components, this number has since increased to over 100 *(21, R..W. Lenz, Personal Communication)*. The material characteristics of such PHAs obviously vary considerably according to their substituents, from brittle plastics to elastomers *(7,8,22,23)*. The stage is set, therefore, for PHAs to assume a leading role as the biodegradable alternative to conventional commodity plastics and much of the research has been to this goal.

Questions from 1925, however, remain unanswered. Relatively little is known about the structure of the PHA inclusion bodies and their interaction with the related enzymatic mechanisms. Apart from the pursuit of academic inquisitiveness, investigations regarding the architecture of these PHA inclusion bodies have important implications for their biotechnological production in microbes and other organisms, for downstream processing and for their *in vitro* precision polymerisation. This chapter reviews the literature concerning the structural organisation of PHA inclusion bodies.

Granules or Inclusion Bodies?

A review of the literature with regards to PHA inclusion bodies shows the majority of publications pre 1990's referring to them as 'granules'. This was most likely due to the perception of these refractile bodies as somewhat crystalline solids, the same properties that PHB exhibits when extracted through solvent from cells. Research by Marchessault and others, however, determined that the *in vivo* PHB does not behave in a crystalline fashion. In the late 1980's research by Sanders and coworkers on the *in vivo* structure of PHB demonstrated that the polyester was not solid, neither was it a liquid or in solution, but most likely an amorphous elastomer *(24,25)*. This led to a period of some confusion as the terms 'granule' and 'inclusion body' were interchangeable. The 'switch' from granule to inclusion body appears to have been firmly cemented in the early 1990's, when Sander's group demonstrated the production of amorphous, biomimetic PHB granules *(26)*. At this point a change in terminology was beneficial to prevent confusion between the native inclusion bodies and their 'artificial' counterparts or crystalline granules. In this chapter, therefore, they are referred to in this fashion. Where quotes with the term 'granule(s)' have been utilised, the word has been replaced with 'inclusion body(ies)' to avoid any confusion. I trust the authors will forgive me this presumption.

Is This a Membrane I see Before Me?

Lemoigne's 1925 observations of the refractile PHB inclusion bodies in *B. megaterium*, his perceptive conclusions and the subsequent research into the material properties of their polymeric component, led to the commercial exploration in the 1950's by W.R. Grace & Co. It wasn't until 1956 however, that investigations concerning the structural organisation of the PHB inclusion bodies was reported *(27)*. In this paper Chapman examined ultra-thin sections of *B. megaterium* and *B. cereus* using electron microscopy and showed the presence of discrete PHB inclusion bodies. In subsequent, unrelated studies examining the internal structure of bacterial cells, Vatter & Wolfe (1958) demonstrated the presence of a membrane system with pigment containing chromatophores in *Rhodospirillum rubrum (28)*. A re-examination of Chapman's work, however, showed no evidence of such a membrane system for PHB inclusion bodies. With these results in mind, Cohen-Bazire & Kunisawa characterised the reserve material of *R. rubrum* synthesised under different cultural conditions and in 1963 reported the absence of a limiting membrane for both glycogen and PHB inclusion bodies *(29)*. Unfortunately, their embedding material; Vestapol, solubilised the PHB and would most likely have destroyed any membrane.

The results of Chapman were contradicted a year later by Merrick and coworkers who examined the appearance of inclusion bodies, containing PHB, isolated from *B. megaterium* and *B. cereus* using the then relatively new technique of carbon replication and electron microscopy *(30)*. Their research showed, quite clearly, that...."On many of the inclusion bodies, fragments of a delicate skin-like structure can be seen. This membranous material, when intact, is apparently wrapped around the inclusion bodies."....These observations varied however, with some of the inclusions possessing rough and irregular surfaces while others appeared smooth (Figure 3a). In some instances small protrusions extending from larger inclusions were observed...."This may represent the early formation of another PHB inclusion body encased within the same membrane."....Furthermore, when the native inclusion bodies were treated with 0.1% sodium lauryl sulphate and subsequently examined it was observed that the membrane coats were disrupted, with torn, raised edges of approximately 150-200 Angstroms (A) thickness (Figure 3b) *(30)*. They also reported that these inclusion bodies varied in size and shape with the majority being spherical between 0.2 and 1.1 microns in diameter and were likely to coalesce, a finding also reported by Chapman. The property to coalesce is most likely due to the preparative drying conditions.

Around the same time, research by Merrick and coworkers regarding the biochemical activities associated with *B. megaterium* and *R. rubrum* inclusion bodies somewhat contradicted the work by Cohen-Bazire & Kunisawa. Their 1965 report investigating the relationship between inclusion body morphology and their associated depolymerase activity, together with their earlier work on the structural organisation of the inclusions illustrated that...."..the inclusion bodies in the cell are surrounded by a membrane, thus indicating an important metabolic relationship between the membrane and PHB."....*(31)*. A similar conclusion was drawn from research by Ellar, *et. al.* in 1968, who utilized electron microcopy and methods of polymer physics to investigate the morphology of PHB in inclusion bodies isolated from *B. megaterium (32)*. They found that...."*In vitro* biochemical studies with native inclusion body suspensions can be performed as long as the membrane is present....The integrity of this surface coat appears essential for biochemical processes involving the native inclusion body.".... They also report a size variation determined by a combination of electron microscopy, dark field light microscopy and light scattering particle size measurement techniques ranging from 0.2 to 0.7µm diameter *(32)*.

The research of Ellar, *et. al.*, focusing on *in vivo* PHB morphology, led them to speculate on a model that incorporated a membrane, the 'fibrillar' PHB and the means by which inclusion bodies may form and function. According to this model, (Figure 4), PHB synthesis was speculated to occur at the surface of the polymerising enzymes. The biosynthetic system then aggregates into a micellar form, with the size and nature of this aggregate possibly reflecting some quaternary structural requirement of the enzyme system. The model proposed a number of chain initiation sites, most likely determined by the number of protein units. As synthesis continues and the size of the inclusion body increases, further

Figure 3: *Electron micrographs of carbon replicas of PHB inclusion bodies isolated from Bacillus megaterium. (a): Illustrating the presence of membrane fragments associated with the inclusions [x40k], (b): Showing a defined core and outer membrane layer [x40k]. Bars = 0.5 μm.(Reproduced from reference 30. Copyright 1964 Society for General Microbiology)*

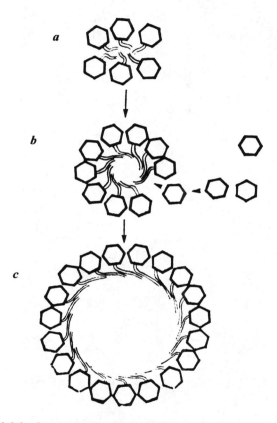

Figure 4: Model for biosynthesis of PHB inclusion bodies based on emulsion polymerisation, proposed by Ellar, et. al. Extended chain fibrils of PHB are formed within a micelle composed of synthetic enzymes. Monomer is transported through the membrane and polymerised at its inner surface. (Reproduced from reference 32. Copyright 1968 Academic Press)

polymerisation sites may be formed by the adsorption of additional PHB synthases *(32)*.

The development of this model was based on a combination of standard emulsion polymerisation reactions and the apparent lack of a bacterial membrane invagination. There are a number of noticeable differences between the postulated model for PHB synthesis and the emulsion polymerisation process, namely; the number of chain initiation sites being determined by a defined number of synthases and the polymerisation proceeding from the surface of the inclusion body. It was originally thought that the origins of inclusion bodies might come from the process of membrane invagination leading to a vesicle, within which the PHB could be synthesised. The authors reported, however, that...."Extensive studies using several organisms, together with the results reported here, have failed to reveal the presence of a typical unit membrane profile surrounding the inclusion body.".... Furthermore, the authors failed to theorise on how the intracellular degradative system fitted into their model *(32)*. A similar model was later proposed by Dunlop & Robards, who utilized freeze-fracture electron microscopy to examine PHB inclusions isolated from *B. cereus* (Figure 5a) *(33)*. Their model suggested an inclusion body where the PHB was composed of an inner core surrounded by an outer layer bounded by a membrane or 'coat'.....''but its precise role in the synthetic and degradative functions of the inclusion bodies is still the subject of investigation.''...(Figure 5b). The observation of a membrane was shown, however, to be dependent upon the sample preparation *(33)*.

By 1974, the concept of PHB inclusion bodies being surrounded by some form of membrane was well established. Shively includes them in his review of prokaryotic inclusion bodies with...."..lipid deposits, like poly-β-hydroxybutyrate, sulfur globules, and gas vacuoles, which are surrounded by a non-unit membrane.''....(*34*). The idea of a non-unit membrane for PHA inclusion bodies was challenged by Fuller, *et. al.* who utilized a combination of Normarski computer enhanced imaging light microscopy, negative staining, freeze-fracturing and electron microscopy to examine isolated PHO inclusion bodies from *Pseudomonas oleovorans* *(35)*. They observed a...."..highly organised pararystalline network located on the surface of the inclusion body.''...(Figure 6). The authors demonstrated the presence of lipids and the proteinaceous nature of this network, and measured the spacing between the repeating units at 135A. The network pattern is similar to that observed for the 'S-layer' of wild-type *P. acidovorans*...circumstantial evidence to the possible origins of PHA inclusion bodies?

One Membrane or Two?

In 1964 Lundgren, *et. al.* reported their investigations regarding the structure of PHB inclusion bodies isolated from *B. megaterium* and *B. cereus* *(30)*. Observations based on electron micrographs of carbon replicas and germanium shadowed 'native' PHB inclusions led them to speculate that...."A highly dense

50

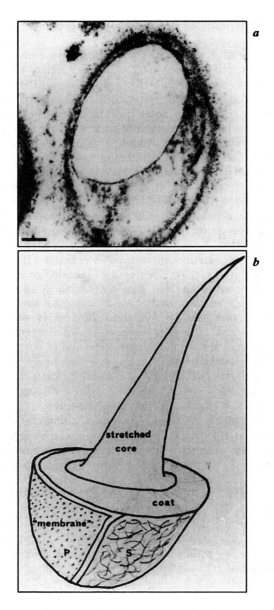

Figure 5: (a): Electron micrograph of thin section of a disrupted Bacillus cereus cell, illustrating a membrane-like structure surrounding a PHB inclusion body, Bar = 0.1 µm, (b): Structural model of PHB inclusion body as isolated from B. cereus and examined using freeze-fracture electron microscopy, proposed by Dunlop & Robards. P = Particulate surface observed in sonicated samples, S = Standed surface shown in samples extracted through hypochlorite. (Reproduced from reference 33. Copyright 1973 American Society for Microbiology)

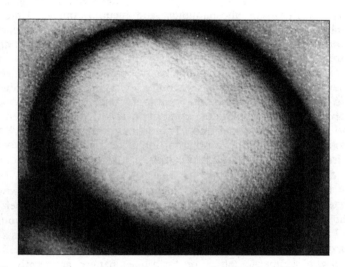

Figure 6: Freeze-fracture electron micrograph of PHO inclusion body isolated from Pseudomonas oleovorans illustrating an organised surface membrane, [x300k]. (Reproduced from reference 35. Copyright 1992 Elsevier Science)

peripheral portion of a membrane which appears to have more than one layer and which extends to the neighbouring inclusion bodies as well."....(Figure 7). In the subsequent confusion regarding the issue of membrane and non-membranous PHB inclusion bodies, the possibility of organised membranes was not considered in any detail until the 1980's.

Prior to the early 1980's PHB had been the focus of PHA research. In 1983, however, Witholt and coworkers examined the refractile intracellular inclusions formed when *P. oleovorans*, a hydrocarbon utilising strain supplied by Exxon and unable to synthesise PHB, was cultivated on octane *(36)*. Their characterisation of the inclusions produced from this fermentation found that the polymer contained was consistent with an aliphatic polyester of the PHA family; 'polyhydroxyoctanoate - PHO'. These PHO inclusions had diameters similar to those reported for PHB inclusions by Ellar, *et. al.*, and were shown to possess approximately 1% (wt/wt) phospholipids and neutral lipids. Thin layer chromatography of these lipids suggested that they were of the same composition as lipids extracted from cell membranes of *P. oleovorans*. The application of freeze-fracturing, carbon replication and electron microscopy to observe isolated PHO inclusion bodies and the behaviour of the biopolyester within, led the authors to tentatively mention the possibility of a bilayer membrane...."A second type of structure, not seen in intact cells, shows convex and concave convoluted fracture faces, indicating a polymer inclusion body with a plied or convoluted surface. The fracture might occur within a bilayer membrane surrounding a polymer inclusion or it might represent a hydrophilic fracture face between a polymer inclusion and the surrounding medium."....(36).

The structural organisation of PHO inclusion bodies from *P. oleovorans* have been extensively investigated by Fuller and coworkers *(35,37)*. Their 1992 report was the first to propose the presence of a more organised membranous coat surrounding PHA inclusions *(35)*. This suggestion was further emphasised by their continued research in which they propose the presence of 2 discrete layers, each possessing a geometric lattice or 'grid-like' pattern apparently having similar repeat spacings between units *(37)*. Curiously, one of the lattice arrays did not appear to completely enclose the polymer, (Figure 8a), these apparently incomplete membranes were considered too regular to constitute random damage, but are speculated to be sites of lattice assembly or disassembly *(37)*.

Elemental analysis confirmed that both the purified PHO inclusion bodies and lipid extracted from them contained phosphorous, suggesting the presence of phospholipids. Consistent with this suggestion was the results of thin layer chromatography which showed two species with distinctly different retention factors. In addition, when these purified inclusions were subjected to a mild methanol wash, (known to solubilise phospholipids), and then examined using negative staining and electron microscopy they showed no apparent distortions of the lattice arrays (Figure 8b). In comparison, when the treatment was followed by gentle washing with water and centrifuging prior to examination, the observations suggested the loss of the 'outer' lattice array (Figure 8c). The authors concluded

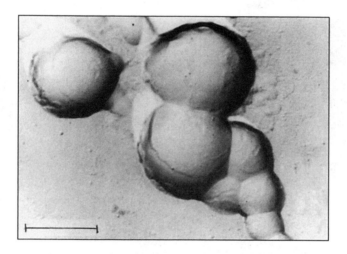

Figure 7: Electron micrograph of germanium shadowed carbon replica of PHB inclusion bodies isolated from Bacillus megaterium illustrating smooth surfaces with a more dense peripheral structure, possibly multilayered [x50k], Bar = 0.5 μm. (Reproduced from reference 30. Copyright 1964 Society for General Microbiology)

54

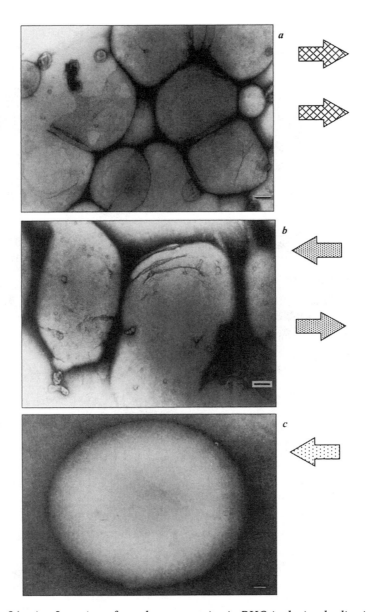

Figure 8(a-c):   Location of membrane proteins in PHO inclusion bodies isolated from Pseudomonas oleovorans, electron micrographs of negatively stained inclusions; (a): Showing the presence of two membrane layers with an incomplete inner layer, (b): After treatment with methanol wash illustrating the presence of the inner membrane only, (c): After subsequent treatment with guanadine-HCl, illustrating the absence of both outer and inner membrane layers, Bars = 0.1 μm. (Reproduced from reference 37. Copyright 1995 National Research Council of Canada)

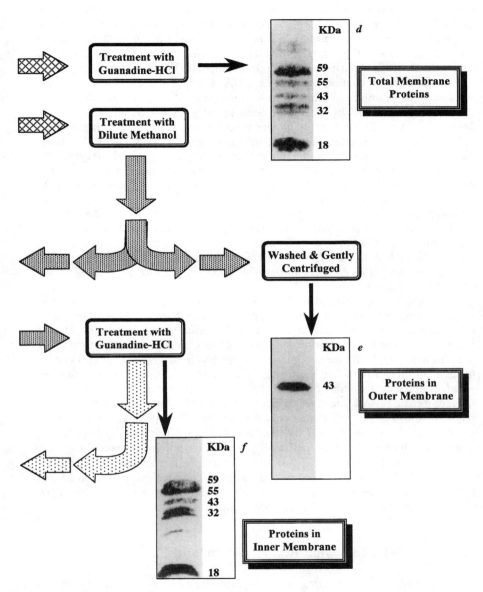

Figure 8(d-f): SDS-PAGE fractionation of PHO inclusion body membrane proteins released after corresponding treatments in (a-c); (d): Total membrane protein content from untreated isolated inclusions - outer and inner layers, (e): Protein content released by methanol wash treatment - outer layer only, (f): Proteins released from inclusions after methanol wash treatment - inner layer only. (Reproduced from reference 49. Copyright 1996 Elsevier Science)

that.....".. methanol-soluble substance; phospholipid, lies between the two layers.".....The remaining lighter array seemed directly associated with the polyester.".... Furthermore that....."This array contained the circular regions evident in the untreated inclusions.".....This inner lattice could be subsequently removed using sodium dodecyl sulphate (Figure 8d) *(37)*.

Ultrastructural studies of isolated PHA inclusions from *P. oleovorans*, *Pseudomonas putida*, *Rhodospirillum eutropha*, *Nocardia corallina* and *Azotobacter vinelandii* have recently been investigated by Stuart, *et. al.* using negative staining, freeze-fracture and electron microscopy *(38)*. The lattice array of PHO inclusions isolated from *P. oleovorans* was essentially identical to those observed from *P. putida,* with both exhibiting highly organised arrays at the polymer-cytosol surface (Figure 9a). In contrast, the membrane surface of PHB inclusions from *R. eutropha* lacked a uniform spacing and appeared to comprise of a single layer rather than the double geometric lattice observed in both the pseudomonads (Figure 9b). Inclusion bodies isolated from *N. corallina* and *A. vinelandii* appeared to possess characteristics from both the PHO and PHB inclusion bodies as observed in the *Pseudomonas* and *Rhodospirillum* species. These PHB inclusions from *N. corallina* (Figure 9c) and *A. vinelandii* (Figure 9d) exhibited organised membranes apparently double in structure but lacking a rigid geometric pattern *(38)*.

The original speculation by Witholt and coworkers of a bilayer for PHO inclusion bodies from *P. oleovorans* appears to have been established by Fuller and coworkers, whose research regarding the macromolecular architecture of PHO inclusion bodies, supplemented with their biochemical studies regarding the proteins associated with these organised membranous coats (see below), have led to their postulated model in Figure 11. In contrast, calculations by Steinbüchel, *et. al.* concerning the size of PHO inclusion bodies, the density of polymer and related measurements found it difficult to correlate the amounts of proteins required for a bilayer with the amount detected by electropherograms of a crude extract from *P. oleovorans (39)*. The original 1983 observations by Witholt and coworkers, leading to their tentative proposal of a bilayer have also been contradicted by Preusting, *et. al.* in 1991, who, in a like manner, used freeze-fracture techniques *(40)*; a pertinent reminder of the conclusion by Dunlop and Robards that the preparation and examination techniques used to examine PHA inclusion bodies greatly influence the observations.

PHA Inclusion Bodies - Form and Function

The hydrophobic nature of PHB combined with a natural cellular requirement to sequester the, sometimes large, volumes of the storage polyester to prevent disruption of cell function; promoted the view that the inclusion bodies obviously served a containment function. Kepes and Peaud-Lenoel, however, predicted the scenario with amazing foresight in 1952 when they stated....."One imagines only with difficulty that a molecule as hydrophobic as poly(D-(-)-hydroxybutyric acid) could exist without being in a permanent state of

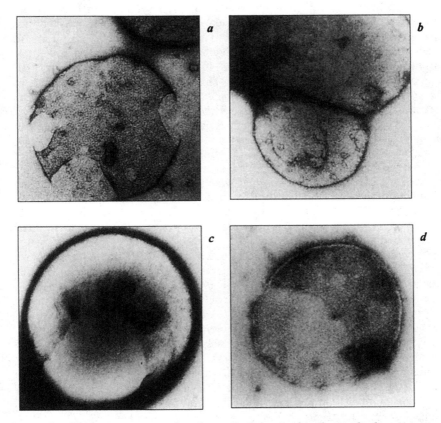

Figure 9: Electron micrographs of negatively stained inclusion bodies; (a): PHO inclusion bodies isolated from Pseudomonas putida [x140k], (b): PHB inclusions from Rhodococcus eutropha [x140k], (c): PHV inclusions from Nocardia corallina [x105k] and (d): PHB inclusion bodies isolated from Azotobacter vinelandii [x86k]. (Reproduced from reference 38. Copyright 1998 Elsevier Science)

complexing with some element of the enzyme system responsible for its synthesis and degradation."......(41). Subsequent research by Merrick & Doudoroff strongly supported this model when they demonstrated that the inclusions may play a more involved role in the PHB metabolic pathways (42,43). In examining the biosynthesis of PHB from D(-)-β-hydroxybutyl-coenzyme A in cell-free extracts of both R. rubrum and B. megaterium, they demonstrated that the enzymes responsible for PHB synthesis in both species were associated with their isolated inclusion bodies. A subsequent paper by the same authors demonstrated that isolated PHB inclusion bodies from B. megaterium, when added to a reaction mixture containing the crude extract of R. rubrum cells which had been cultivated to produce and expend their PHB accumulation, were readily degraded (43). This in vitro 'digestion' of the isolated B. megaterium inclusions by the R. rubrum extract, suggested the presence of an intracellular PHB depolymerase from the R. rubrum cells.

The research of Merrick & Doudoroff was later complemented by Hippe & Schelgel who demonstrated the presence of an intracellular depolymerase associated with PHB inclusion bodies from Alcaligenes eutrophus, which degraded the polymer to its monomeric component (44). This contrasted with the degradative system for B. megaterium which produced a mixture of monomer and dimer units. An explanation for this was facilitated when, in 1966, Merrick & Yu isolated an intracellular dimer hydrolase from R. rubrum inclusions (45). By the early 1970's the biochemical importance of the membrane surrounding PHB inclusion bodies was readily recognised.

In contrast to the research on PHA synthesis, relatively little remains known about the mechanisms and activities of intracellular PHA depolymerases. The research by Merrick and coworkers in the 1960's was continued in the 1990's, when Saito and coworkers showed that a supernatant fraction from centrifuged PHB inclusion bodies isolated from Z. ramigera when added to 'inactivated' PHB inclusions isolated from A. eutrophus increased their degradation (46,47). The degradation of isolated PHB inclusions from A. eutrophus was highly dependent upon pH, showing two peaks in the rate of polymer loss, leading to the speculation that A. eutrophus possesses two depolymerase enzymes (47). In contrast, the activity and location of the intracellular depolymerase associated with inclusion bodies possessing PHAs with relatively long alkyl and functional chemical group substituents are limited to Foster and coworkers (48-51).

The advent of recombinant DNA technology in the early 1970's has been harnessed by a number of researchers to isolate and characterise the genes responsible for PHB synthesis (52). In a series of papers by Witholt and coworkers from 1988 to 1991, the genes responsible for PHO synthesis in P. oleovorans were elucidated, their results undeniably confirmed the association of 2 polymerases with the inclusion body membrane (53-56). In addition, the resultant nucleotide sequences permitted the deduction of the amino acid sequences for the different gene products, one gene coded for a protein species of approximately 32kDa and was similar in its deduced amino acid series to a known lipase. The function of this protein as the depolymerase was later confirmed by Foster and coworkers (49). The association of these proteins with the membranous coat surrounding the polymer can be visualised using SDS-PAGE from isolated inclusion bodies (Figure 8d).

In addition to the polymerase-depolymerase system, 2 other protein species with molecular weights of approximately 43 and 18kDa are also components of the PHO inclusion body coat (Figure 8d). While the relative concentrations of all the inclusion body proteins have been demonstrated to vary during cultivation of *P. oleovorans*, the 43 and 18kDa species constitute between 70 and 85% of the total protein *(49)*. This, together with the observations of an organised lattice array, led Fuller, *et. al.* to speculate that...."The function of the 43 and 18kDa polypeptides is currently unknown, but they may play a structural role in organising the inclusion body.".... *(35,36)*.

Are these proteins part of the PHA membrane, or a result of the isolation and preparative processes? Evidence to suggest that they may be contaminants, a consequence of the cell lysing, comes from Stuart, *et. al.* where the 43 and 18kDa proteins associated with PHO inclusions from *P. oleovorans* were present even in cells where starvation had occurred and no inclusion bodies were visible *(37)*. Research by Sanders and coworkers demonstrated that biomimetic PHB granules could be easily 'coated' by a range of proteins including bovine serum albumin (BSA) *(26)*. Similarly, Foster, *et. al.* demonstrated that the presence of varying concentrations of BSA when added to isolated PHO inclusion bodies, inhibited the intracellular depolymerase activity in a proportional relationship *(50)*. Furthermore, Liebergesell, *et. al.* found that egg white lysozyme, used to lyse the cells prior to inclusion body isolation, was purified in combination with the inclusion bodies *(57)*. The calculations by Steinbüchel, *et. al.* would also appear to support the possibility of protein contamination *(39)*. This scenario is unlikely, however, when one considers the highly organised appearance of the protein coats with regular repeat spacings and the limitation in mass distribution of the unknown species to 43 and 18kDa. A more irregular distribution and appearance characteristic of a non-unit membrane would be expected from random contamination. Further evidence to confirm that these proteins fulfil a structural function in containing the biopolyester has recently been reported by Stuart, *et. al. (38)* and Valentin, *et. al. (58)*.

The research by Fuller and coworkers suggested the presence of two protein layers separated by a phospholipid membrane *(38)*. In the same paper from 1995 where this arrangement was proposed; they also report the use of gold labelled antibodies generated against the 43kDa protein and the polymerases of 59 and 55kDa to investigate further the distribution of these proteins in PHO inclusion body membranes. They found that the binding patterns of the anti-55 and anti-59kDa IgG labels differed significantly from those observed for the anti-43kDa IgG. The latter was primarily bound to the inner surface of the outer lattice layer, particularly where this lattice appeared stressed or torn (Figure 10a). In contrast, antibodies generated against the polymerases were apparently concentrated on the inner lattice layer where the outer lattice had been removed (Figure 10b). This apparent distribution of the 43kDa protein led the authors to conclude that the protein membrane(s)....."..include structural component(s), fundamental to the specific subcellular assembly associated with the polymer inclusion bodies."....Later work by the same group using SDS-PAGE confirmed the association of the polymerases and determined the location of the intracellular

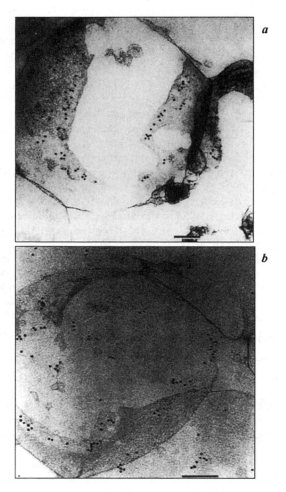

Figure 10: Electron micrographs of negatively stained PHO inclusion bodies isolated from Pseudomonas oleovorans; (a): Inclusions probed with antibodies generated against the 43kDa protein species, showing their accumulation on the inner surface of the outer membrane, (b): inclusions probed with antibodies generated against the 59kDa polymerase protein, illustrating their accumulation on the outer surface of the inner membrane, Bars = 0.1 μm. (Reproduced from reference 37. Copyright 1995 National Research Council of Canada)

depolymerase on the inner lattice layer (Figures 8e & 8f) *(50)*. These results were utilized to propose a model for inclusion body structure as isolated from *P. oleovorans* (Figure 11).

Immunocytochemical analysis was also utilized by Gerngross, *et. al.* who investigated the localisation of the PHB polymerase in *A. eutrophus (59)*. Their results confirmed the 1968 suggestions of Griebel, *et. al.* that the polymerase was localised at the surface, and not within, the PHB inclusion bodies *(60)*. Their 1993 model for PHB inclusion body formation was based on the micellar theory of Griebel, *et. al.* with the addition of hydrophobic or amphipathic metabolites such as phospholipids. The authors were unable to determine if these additional components were a result of contamination ...."It is not clear whether these molecules are essential for inclusion body formation or are simply trapped there by virtue of their amphipathic nature." ....*(59)*.

At the same time that Fuller, *et. al.* were speculating on the function of the unknown proteins comprising part of PHO inclusion body membrane, Liebergesell, *et. al.* utilized SDS-PAGE to investigate PHB inclusion bodies from the anoxygenic phototropic bacterium *Chromatium vinosum (57)*. They reported the presence of a 17kDa protein species associated with the inclusion body membranes also having an unknown function. In 1993 the localisation of a 15.5kDa protein, also of 'unknown function', at the surface of PHB inclusions from *R. ruber* was reported by Pieper-Fürst, *et. al. (61)*. In *A. eutrophus* the predominant protein species associated with PHB inclusions has a molecular mass of approximately 24kDa, this is similar in mass to that reported for *B. megaterium:* 22kDa *(52)*. The exact function/s of these proteins remains unclear. Steinbüchel and coworkers have termed these proteins 'phasins' *(52)*. Their 1993 model for the structure of PHB inclusion bodies expanded further on that originally proposed by Griebel, *et. al. (60)*. and later complemented by Gerngross, *et. al. (59)*. The model for PHB inclusions as isolated from *R.ruber*, suggests an inclusion body membrane consisting of the 15.5kDa 'phasin' protein with the catalytic proteins integrated or bound to the inclusion body surface (Figure 12).

Stuart, *et. al.* have recently utilized immunocytochemical techniques to demonstrate that the 43kDa phasin associated with the PHO inclusion body membranes from *P. oleovorans* is antigenically related to its *P. putida* counterpart possessing a similar mass, but is not related to the phasins of *R. eutropha*, *N. corallina* and *A. vinelandii (38)*. In the same report, and a companion study by Valentin, *et. al. (58)*, the use of transgenic *P. putida* cells have clearly demonstrated that the 43kDa phasin protein...."..is associated with a remarkably more ordered and defined inclusion boundary array which begins to effectively segregate polymer from the cytosol."

Conclusion

It was originally thought that *B. megaterium* and other microbes capable of synthesising the relatively short chain PHAs such as PHB were incapable of

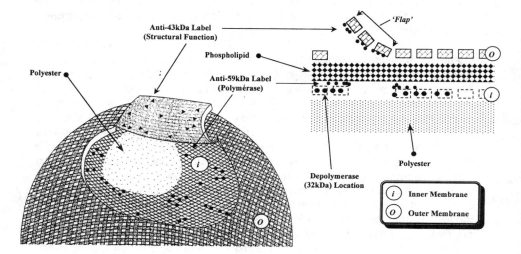

Figure 11:   Structural model of PHO inclusion bodies as isolated from Pseudomonas oleovorans, postulated by Fuller and coworkers with added data from author. The model proposes a complex bilayer with a complete outer membrane consisting of 43kDa proteins and an incomplete inner membrane, also composed of 43kDa proteins but with the polyermases (59 and 55kDa) and depolymerase (32kDa) in close proximity to the biopolyester. The two membranes are separated by a phospholipid layer. (Reproduced in part from reference 37. Copyright 1995 National Research Council of Canada)

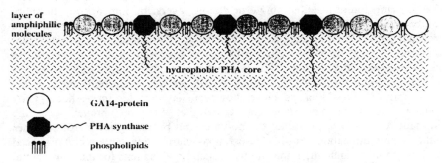

Figure 12: Structural model of PHB inclusion bodies as isolated from Rhodococcus ruber, postulated by Steinbüchel and coworkers. The model proposes a single membrane consisting of 15.5kDa proteins randomly interspersed with the polymerase, and presumably, depolymerase enzymes. Binding for the proteins is facilitated by phopholipids. (Reproduced from reference 61. Copyright 1993 American Society for Microbiology).

producing their more novel counterparts with long alkyl or functional chemical groups in the PHA side chain, and vice versa for organisms such as *P. oleovorans*. This apparent distinction has recently been challenged by research demonstrating the accumulation of the minor component Polyhydroxyhexanoate (PHHex) in both sets of microorganisms and the formation of copolymers of PHB with PHAs possessing 6-12 carbon units *(C. Scholz, Personal Communication 1999,62)*. Nevertheless, this apparent distinction is repeated in the carbon substrate utilising abilities of the microorganisms. *B. megaterium* utilises simple sugars such as glucose and fructose to produce PHB. In contrast *P. oleovorans* requires starting substrates of higher carbon numbers, which it utilises to synthesise PHAs possessing a majority of equinumerable monomeric units, eg: octane for PHO, nonane for PHN etc. It may be that these metabolic differences are reflected by contrasts in the structural organisation of their PHA inclusion bodies, with PHB accumulating bacteria possessing inclusions of relatively simple membranes, as suggested by the model of Steinbüchel and coworkers *(39,61)*, being found in either single or double layers. In comparison, the more complex PHAs, as accumulated by *P. oleovorans* and others, may require a more organised intracellular arrangement. Evidence for these apparently two different structural 'groups' of PHA inclusion bodies is suggested from the research to date as reviewed in this chapter. The recent reports from Fuller and coworkers utilising recombinant DNA technology would appear to strongly support this proposition *(38,58)*. What is clear, however, is that the intracellular PHA inclusion bodies are membrane bound and that these membranes serve the important function of containing the polymer and providing a site for polymer synthesis and utilisation. Further research to clarify this tentatively proposed division and investigate the implications of such organisations for the commercial production of PHAs is required. In so doing we will address the issues raised by Lemoigne nearly 75 years ago.

References

1. Lemoigne, M. *C. R. Acad. Sci.* **1925**, *180*, 5139.
2. Lemoigne, M. *Ann. Inst. Pasteur.* **1925**, *39*, 144-173.
3. Lemoigne, M. *Bull. Soc. Chim. Biol.* **1926**, *8*, 770-782.
4. Lemoigne, M. *Ann. Inst. Pasteur.* **1927**, *41*, 148-165.
5. Foster, L.J.R.; Lenz, R.W. & Fuller, R.C. In: 'Hydrogels and Biodegradable Polymers for Bioapplications', Ottenbrite, R.M.; Huang, S.J. & Park, K. Eds.; ACS Symposium Series, Washington DC, USA.**1996**, *627*, pp.68-92.
6. Steinbüchel, A. In: 'Biotechnology' Roehr, M. Ed. VCH Publishers, New York, USA. **1996**, pp.403-464.
7. Doi, Y. 'Microbial Polyesters' VCH Publishers, New York, USA. **1990**.
8. Baptist, J.N. & Ziegler, J.B. U.S.A. Patent. **1965**, 3-225-766.
9. Howells, E.R. *Chem. Ind.* **1982**, *15*, 508-511.
10. Holmes, P.A. *Phys. Technol.* **1985**, *16*, 32-36.
11. Anon. *Chemistry in Britain.* **1987**, 23(12), 1157.

12. Lee, S.Y.; Choi, J. & Chang, H.N. In: 'Proceedings of the 1996 International Symposium on Bacterial Polyhydroxyalkanoates', Eggink, G.; Steinbüchel, A.; Poirier, Y. & Witholt, B. Eds.; NRC Research Press, Zürich, CH. **1997**, pp.127-136.

13. Weusthuis, R.A.; Huijberts, G.N.M. & Eggink, G. In: 'Proceedings of the 1996 International Symposium on Bacterial Polyhydroxyalkanoates', Eggink, G.; Steinbüchel, A.; Poirier, Y. & Witholt, B. Eds.; NRC Research Press, Zürich, CH. **1997**, pp.102-109.

14. Somerville, C.R.; Nawrath, C. & Poirier, Y. PCT Patent Application. **1995**, WO95/05472.

15. John, M.E. & Keller, G. *Proc. Natl. Acad. Sci. U.S.A.* **1996**, *93*, 12768-12773.

16. Hahn, J.J.; Eschenlauer, A.C.; Narrol, M.H.; Somers, D.A. & Srienc, F. *Biotechnol. Prog.* **1997**, *13*, 347-354.

17. Gilmore, D.F.; Antoun, S.; Lenz, R.W.; Goodwin, S.; Austin, R. & Fuller, R.C. *J. Ind. Microbiol.* **1992**, *10*, 199-206.

18. Mergaert, J.; Webb, A.; Anderson, C.; Wouten, A. & Swings, *J. App. Environ. Microbiol.* **1993**, *59(10)*, 3233-3238.

19. Brandl, H. & Püchner, P. In: 'Novel Biodegradable Microbial Polymers', Dawes, E.A. Ed.; Kluwer Academic Publishers, Dordrecht, NE. **1990**, pp.421-422.

20. Püchner, P. & Müller, W.R. *FEMS Microbiol. Rev.* **1992**, *103*, 469-470.

21. Steinbüchel, A. & Valentin, H.E. *FEMS Microbiol. Lett.* **1995**, *128*, 219-228.

22. Anderson, A.J. & Dawes, E.A. *Microbiol. Rev.* **1990**. *34*, 450.

23. Brandl, H.; gross, R.A.; lenz, R.W. & Fuller, R.C. *Adv. Biochem. Eng. Biotech.* **1990**, *41*, 77.

24. Barnard, G.N. & Sanders, J.K.M. *FEBS Lett.* **1988**, *231*, 16-18.

25. Barnard, G.N. & Sanders, J.K.M. *J. Biolog. Chem.* **1989**, *264(6)*, 3286-3291.

26. Horowitz, D.M. & Sanders, J.K.M. *J. American. Chem. Soc.* **1994**, *116(7)*, 2695-2702.

27. Chapman, G.B. *J. Bact.* **1956**, *71*, 348.

28. Vatter, A.E. & Wolfe, R.S. *J. Bact.* **1958**, *75*, 480.

29. Cohen-Bazire, G. & Kunisawa, R. *J. Cell. Biol.* **1963**, *16*, 401.

30. Lundgren, D.G.; Pfister, R.M. & Merrick, J.M. *J. Gen. Microbiol.* **1964**, *34*, 441-446.

31. Merrick, J.M.; Lundgren, D.G. & Pfister, R.M. *J. Bact.* **1965**, *89(1)*, 234-239.

32. Ellar, D.; Lundgren, D.G.; Okamura, K. & Marchessault, R.H. *J. Mol. Biol.* **1968**, *35*, 489-502.

33. Dunlop, W.F. & Robards, A.W. *J. Bact.* **1973**, *114(3)*, 1271-1280.

34. Shively, J.M. *Annu. Rev. Microbiol.* **1974**, *28*, 167-187.

35. Fuller, R.C.; O'Donnell, J.P.; Saulnier, J.; Redlinger, T.E.; Foster, J. & Lenz, R.W. *FEMS Microbiol. Rev.* **1992**, *103*, 279-288.

36. De Smet, M.J.; Eggink, G.; Witholt, B.; Kingma, J. & Wynberg, H. *J. Bact.* **1983**, *154(2)*, 870-878.

37. Stuart, E.S.; Lenz, R.W. & Fuller, R.C. *Can. J. Microbiol.* **1995**, *41(Suppl. 1)*, 84-93.

66

38. Stuart, E.S.; Tehrani, A.; Valentin, H.E.; Dennis, D.; Lenz, R.W. & Fuller, R.C. *J. Biotech.* **1998**, *64*, 137-144.
39. Steinbüchel, A.; Aerts, K.; Babel, W.; Follner, C.; Liebergesell, M.; Madkour, M.H.; Mayer, F., Pieper-Fürst, U.; Pries, A.; Valentin, H.E. & Wieczorek, R. *Can. J. Microbiol.* **1995**, *41(Suppl. 1)*, 94-105.
40. Preusting, H.; Kingma, J. & Witholt, B. *Enzyme Microbiol. Technol.* **1991**, *13*, 770-780.
41. Kepes, A.O. & Peaud-Lenoel, C. *Bull. Soc. Chim. Biol.* **1952**, *34*, 563.
42. Merrick, J.M. & Doudoroff, M. *Nature.* **1961**, *189*, 890-892.
43. Merrick, J.M. & Doudoroff, M. *J. Bact.* **1964**, *88(1)*, 60-71.
44. Hippe, H. & Schelegel, G. *Arch. Microbiol.* **1967**, *56*, 278-299.
45. Merrick, J.M. & Yu, C.I. *J. Biochem.* **1966**, *5(11)*, 3563-3568.
46. Saito, T.; Saegusa, H.; Miyata, Y. & Fukui, T. *FEMS Microbiol. Lett.* **1992**, *118*, 279-282.
47. Saito, T.; Takizawa, K. & Saegusa, H. *Can. J. Microbiol.* **1995**, *41(Suppl. 1)*, 187-191.
48. Foster, L.J.R.; Lenz, R.W. & Fuller, R.C. *FEMS Microbiol. Lett.* **1994**, *118*, 279-282.
49. Stuart, E.S.; Foster, L.J.R.; Lenz, R.W. & Fuller, R.C. *Int. J. Biolog. Macromol.* **1996**, *19*, 171-176.
50. Foster, L.J.R.; Stuart, E.S.; Tehrani, A.; Lenz, R.W. & Fuller, R.C. *Int. J. Biolog. Macromol.* **1996**, *19*, 177-183.
51. Foster, L.J.R.; Lenz, R.W. & Fuller R.C. *Int. J. Biolog. Macromol.* **1999**, *(in press)*.
52. Steinbüchel, A. & Füchtenbusch, B. *Tibtech.* **1998**, *16*, 41-427.
53. Lageveen, R.G.; Huisman, G.W.; Preusting, H.; Ketelaar, P.; Eggink, G. & Witholt, B. *Appl. Environ. Microbiol.* **1988**, *54*, 2924-2932.
54. Huisman, G.W.; De Leeuw, O.; Eggink G. & Witholt, B. *Appl. Environ. Microbiol.* **1989**, *55*, 1949-1954.
55. Preusting, H.; Nijenhuis, A. & Witholt, B. *Macromolecules.* **1990**, *23*, 4220-4224.
56. Huisman, G.W.; Wonink, E.; Meima, R.; Kazemier, B.; Terpstra, P. & Witholt, B. *J. Biolog. Chem.* **1991**, *4*, 2191-2198.
57. Liebergesell, M.; Schmidt, B. & Steinbüchel, A. *FEMS Microbiol. Lett.* **1992**, *99*, 227-232.
58. Valentin, H.E.; Stuart, E.S.; Fuller, R.C.; Lenz, R.W. & Dennis, D. *J. Biotech.* **1998**, *64*, 145-157.
59. Gerngross, T.U.; Reilly, P.; Stubbe, J.; Sinskey, A.J. & Peoples, O.P. *J. Bact.* **1993**, *175(16)*, 5289-5293.
60. Griebel, R.; Smith, Z. & Merrick, J.M. *J. Biochem.* **1968**, *7*, 3676-3681.
61. Pieper-Fürst, U.; Makour, M.H.; Mayer, F. & Steinbüchel, A. *J. Bact.* **1993**, *176(14)*, 4328-4337.
62. Fukui, T; Kato, M; Matsusaki, H; Iwata, T. & Doi, Y. *FEMS Microbiol. Lett.* **1998**, *164*, 219-225.

Chapter 5

Microbial Synthesis, Physical Properties, and Biodegradability of Ultra-High-Molecular-Weight Poly[(R)-3-hydroxybutyrate]

Tadahisa Iwata, Satoshi Kusaka, and Yoshiharu Doi

Polymer Chemistry Laboratory, The Insitiute of Physical and Chemical Research (RIKEN), 2–1 Hirosawa, Wako-shi, Saitama 351–0198 Japan

Ultra-high-molecular-weight poly[(R)-3-hydroxybutyrate] (P(3HB)) was biosynthesized from glucose by a recombinant *Escherichia coli* XL-1 Blue (pSYL105) harboring *Ralstonia eutropha* PHB biosynthesis genes. Ultra-high-molecular-weight P(3HB) films prepared by slow drying of a solution in chloroform were oriented easily and reproducibly by stretching up to almost 600% of their initial length. Mechanical properties of the stretched P(3HB) films were markedly improved relative to those of the solvent-cast film. When the oriented film was further annealed to increase its crystallinity, the mechanical properties were further improved. The mechanical properties of the stretched-annealed film remained almost unchanged for 6 months at room temperature, suggesting that a high crystallinity of the stretched-annealed film avoids a progress of secondary crystallization. All ultra-high-molecular-weight P(3HB) films were degraded by an extracellular PHB depolymerase from *Alcaligenes faecalis* T1, and it was revealed that the stretched films had the shish-kebab structure.

Poly[(R)-3-hydroxybutyrate] (P(3HB)) is accumulated by a wide variety of microorganisms as intracellular carbon and energy storage material (*1*), and is extensively studied as a biodegradable and biocompatible thermoplastic (*2-4*). However, it is well known that mechanical properties of P(3HB) homopolymer films markedly deteriorate by a process of secondary crystallization (*5,6*). Accordingly, microbial P(3HB) has been regarded as a polymer required to co-polymerize with other monomer components from the viewpoint of industrial applications because of its stiffness and brittleness (*4,7,8*).

Recently, we have succeeded in producing ultra-high-molecular-weight P(3HB) under specific fermentation conditions (9) by using a recombinant *Escherichia coli* XL-1 Blue (pSYL105) (*10*) harboring *Ralstonia eutropha* PHB biosynthesis *phbCAB* genes. The films of ultra-high-molecular-weight P(3HB) are expected to have improved mechanical properties by drawing procedure, as it has been demonstrated for polylactide (*11*) and polyethylene (*12,13*).

We review, herein, the microbial synthesis of ultra-high-molecular-weight P(3HB) from glucose by the recombinant *E. coli* XL-1 Blue (pSYL105), and the physical properties and enzymatic degradability of these P(3HB) films.

Microbial Synthesis

High-molecular-weight P(3HB) samples ($Mw = 1.1 - 11 \times 10^6$) were produced by *E. coli* XL1-Blue (pSYL105) (*10*) containing a stable plasmid harboring the *Ralstonia eutropha* H16 (ATCC 17699) PHB biosynthesis gene operon *phbCAB*. Two-step cultivation of the recombinant *E. coli* was applied for the production of high-molecular-weight P(3HB) (9). The formation of high-molecular-weight P(3HB) was confirmed by analysis of ^{1}H- and ^{13}C-NMR spectra. We reported previously that molecular weights of P(3HB) produced within cells were strongly dependent of the pH value of culture medium containing glucose as a carbon source (9).

Figure 1 shows the time course of cell growth and P(3HB) accumulation during the batch cultivation of *E. coli* XL1-Blue (pSYL105) in Luria-Bertani (LB) media containing 20 g/L glucose at 37°C and different pH values. The pH values of the media remained constant at 6.0 and 7.0, respectively, during the cultivation. Glucose at 20 g/L was consumed within around 12 h, independent of the pH value of the medium. The dry cell weights and amounts of P(3HB) were respectively about 8 g/L and 4 g/L at pH values of 6.0 and 7.0 after cultivation for 12 h. An extremely high-molecular-weight P(3HB) was produced in the medium of pH 6.0. The molecular weight of P(3HB) increased with time to reach a maximum of 14×10^6 after cultivation for 12 h at pH 6.0.

The molecular weight data of P(3HB) samples produced from glucose by a recombinant *E. coli* XL1-Blue (pSYL105) are summarized in Table I. The molecular weights were determined by both gel-permeation chromatography (GPC) and multiangle laser light scattering (MALLS). GPC was used to determine the molecular weight distribution of P(3HB), although the Mw and Mn values are relative to polystyrene standard. The polydispersities (Mw/Mn) of P(3HB) samples produced by a recombinant *E. coli* XL1-Blue (pSYL105) ranged from 1.4 to 3.4. The absolute Mw values of P(3HB) samples were determined by MALLS analysis. Figure 2 shows the relation between Mw(MALLS) and Mw(GPC) values of 1.1×10^6 to 11×10^6 prepared by the recombinant *E. coli*. Furthermore, a P(3HB) sample with Mw(MALLS) of 0.6×10^6 prepared from *Ralstonia eutropha* H16 grown on butyric acid (7) is also plotted in Figure 2. On the basis of the plotting in Figure 2, it was shown that 0.7 is the conversion coefficient between Mw(MALLS) and Mw(GPC) values.

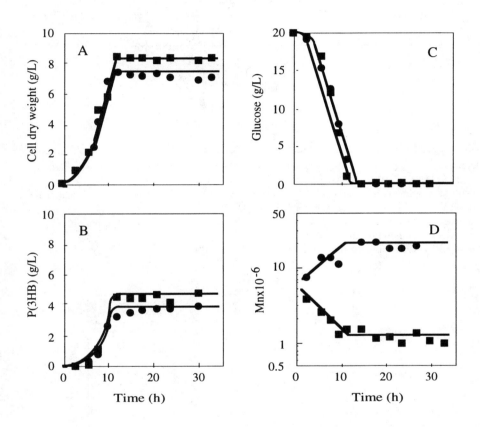

Figure 1. Time-dependent changes in the cell dry weight (A), P(3HB) levels (B), the concentration of glucose (C), and the number-average molecular weight (Mn) of P(3HB) (D) during the batch cultivation of E. coli *XL1-Blue (pSYL105) in LB media containing 20 g/L glucose at 37°C and pH of (●) 6.0 and (■) 7.0.*

Table I. Molecular weights of poly[(R)-3-hydroxybutyrate] produced from glucose by a recombinant *E. coli* XL1-Blue (pSYL105) at 37°C and the reproducibility of hot-drawings for P(3HB) films.

pH of medium	Molecular weights, ×10⁻⁶		Reproducibility[c]
	Mw(GPC)[a]	Mw(MALLS)[b]	
6.5	16	11	O
6.5	13	9.6	O
6.7	7.0	4.9	O
7.0	4.6	3.3	O
7.0	1.3	1.1	×
_[d]	0.76	0.60	×

[a] Determined by gel-permeation chromatography (GPC) relative to polystyrene standard in chloroform at 40°C.

[b] Absolute weight-average molecular weight determined by multiangle laser light scattering (MALLS) in 2,2,2-trifluoroethanol at 25°C.

[c] When all of three P(3HB) films were stretched over 400% against its initial length, reproducibility is represented by "O".

[d] Produced by *Ralstonia eutropha* H16 from butyric acid [7].

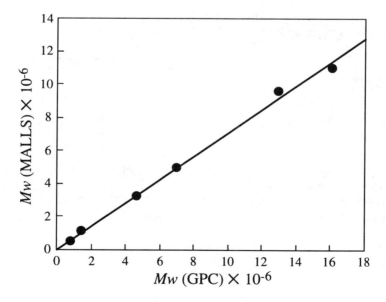

Figure 2. Correlation between Mw(MALLS) and Mw(GPC) values.

Physical Properties

The ultra-high-molecular-weight P(3HB) accumulated within cells of the recombinant *E. coli* grown for 15 h was apparently amorphous and extracted with chloroform over 24 h at room temperature. The solvent-cast films were prepared by slow drying of a solution in chloroform, and they were oriented easily and reproducibly by stretching up to almost 600 % of their initial length in a silicone oil bath at 160°C by loosely hanging weight. It is of important to note that the stability and reproducibility of the hot-drawing process is controlled by molecular weight of P(3HB) refer to Table I. When the hot-drawings were performed against P(3HB) films of various Mw(MALLS) over 0.6×10^6, it was found that the Mw of 3.3×10^6 was a critical value whether hot-drawings over 400 % against its initial length were reproducible or not (see Table I). The stretched films were further annealed at 160°C for 2 h to increase their crystallinity. Typical X-ray diffraction patterns of the solvent-cast and stretched-annealed films of ultra-high-molecular-weight P(3HB) shown in Figure 3, indicate that stretched-annealed film is well oriented and crystallized.

Mechanical properties and X-ray crystallinities of ultra-high-molecular-weight P(3HB) films are summarized in Table II, together with preparation conditions of films. The ultra-high-molecular-weight P(3HB) film was drawn readily and reproducibly to draw ratio 650 % at 160°C, and the stretched film showed the improved mechanical properties as listed in Table II. Elongation to break and tensile strength increased from 7 to 58 % and from 41 to 62 MPa, respectively. In contrast, the Young's modulus decreased from 2.3 to 1.1 GPa. This result demonstrates that the stiff and brittle solvent-cast P(3HB) film is improved to a ductile and flexible material by a hot-drawing process. Furthermore, when an annealing procedure was applied to the stretched film, the mechanical properties of stretched-annealed film were even more improved. Elongation to break and tensile strength of the stretched-annealed film increased approximately 20 % in comparison with those of stretched film.

Table II. Physical properties of three kinds of ultra-high-molecular-weight P(3HB) (Mw(MALLS) = 11×10^6) films.

	Solvent-cast	Stretched	Stretched-Annealed
Draw ratio (%)	0	650	650
Annealing time at 160°C	0	1 sec	2 hrs
Tm (°C)	178	185	188
Long period (nm)	-	7.4	18
X-ray crystallinity (%)	65±5	80±5	>85
Elongation to break (%)	7±3	58±1	67±1
Young's modulus (GPa)	2.3±0.5	1.1±0.1	1.8±0.3
Tensile strength (MPa)	41±4	62±5	77±10

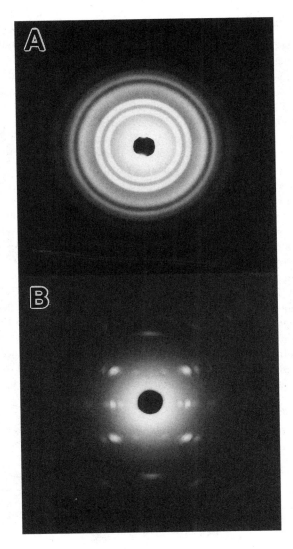

Figure 3. X-ray diffraction patterns of the solvent-cast film (A) and the stretched-annealed film (B) of ultra-high-molecular-weight P(3HB). The stretching direction is vertical and the high oriented film was held perpendicular to the X-ray beam.

The improvement of mechanical properties seems to arise from an increase in the crystallinity and from a decrease in the amount of amorphous polymer chains constrained between lamellar crystals. In dynamic mechanical property, loss tangent (tan δ) peak of β transition shifted from 10°C to 26°C by hot-drawing. Shift of the β relaxation peak of the tan δ to higher temperature would arise from a low mobility of the polymer chains in amorphous region between crystal regions. The melting points of the stretched and stretched-annealed film shifted toward higher temperature over 185°C, suggesting an increase in the thickness of lamellar crystals due to drawing and annealing procedures. This is also confirmed by small-angle X-ray diffraction (SAXD) analysis. Long period calculated from SAXD pattern of the stretched and stretched-annealed films were 7.4 and 18 nm, respectively, in spite of non observation of diffraction peak in solvent-cast film. De Koning and Lemstra (5) reported that the constriction of polymer chains in amorphous phase between lamellar crystals due to the secondary crystallization made P(3HB) films stiff and brittle. In our case, high crystallinity and high orientation of polymer chains in the stretched and stretched-annealed films are likely to avoid the secondary crystallization. The procedures of hot-drawing and annealing against the ultra-high-molecular-weight P(3HB) films are methods which drastically reduce the secondary crystallization.

The stretched film and stretched-annealed film of P(3HB) were stored for about 6 months at room temperature to study the time-dependent change of the mechanical properties, and the stress-strain tests were performed. Mechanical properties, as shown in Table III, of the stretched film remained almost unchanged during 6 months. In addition, the mechanical properties of the stretched-annealed film did not deteriorate during 6 months. It is concluded that a highly oriented and crystallized P(3HB) film keeps superior mechanical properties for long periods.

Table III. Mechanical properties of two kinds of ultra-high-molecular-weight P(3HB) (Mw(MALLS) = 11×10^6) films after stored for 6 months.

	Stretched	Stretched-Annealed
Draw ratio (%)	650	650
Annealing time at 160°C	1 sec	2 hrs
X-ray crystallinity (%)	75±5	>85
Elongation to break (%)	30±1	67±2
Young's modulus (GPa)	2.5±0.2	2.5±0.2
Tensile strength (MPa)	88±8	100±10

Enzymatic Degradation

The enzymatic degradation of three kinds of ultra-high-molecular-weight P(3HB) films with Mw(MALLS) of 5.2×10^6, solvent-cast, 600% stretched, and 600%

stretched-annealed were performed in 0.1M phosphate buffer (pH 7.4) using an extracellular PHB depolymerase from *Alcaligenes faecalis* T1 at 37°C. Figure 4 shows the rate of erosion profiles of the ultra-high-molecular-weight P(3HB) films as a function of time.

The amount of film erosion increased proportionally with time for all the samples. The rate of erosion of solvent-cast film was 95 ± 10 µg/h/cm^2, and this value is almost the same as that of conventional P(3HB) film with Mw(MALLS) of 0.6×10^6 reported by Koyama and Doi (*14*). This result indicates that the molecular weight does not affect the rate of enzymatic degradation. On the other hand, the rates of erosion of the stretched and stretched-annealed films were 25 ± 5 and 3 ± 1 µg/h/cm^2, respectively, suggesting the effect of the crystallinity and long period on the rate of erosion. Some researchers reported the effect of crystallinity (*14,15*) and solid state structure (*14,16*) on enzymatic erosion of P(3HB). In the case of ultra-high-molecular-weight P(3HB), the enzymatic erosion rate seems to be strongly affected by the crystallinity and long period.

Figure 5 shows a scanning electron micrograph of stretched film of P(3HB) with Mw(MALLS) of 5.2×10^6 after enzymatic degradation using an extracellular PHB depolymerase from *Alcaligenes faecalis* T1 at 37°C for 3 h. It is well known that the amorphous region is etched faster than the crystal one (*15,16*). Accordingly, this micrograph expresses the unetched core along the draw direction and lamellar crystals perpendicular to the core. The PHB at the surface of the film seems to have a shish-kebab morphology similar as found polyethylene crystallized in agitated solution (*17*), extruded high modulus fiber of polyethylene (*18-21*), extruded fiber of nylon 6,6 (*22*), and crystallized natural rubber under strain (*23*). The same morphology was revealed in the surface of enzymatically degraded stretched-annealed film. The high tensile strength of both films might be due to a stretched chain core in the shish-kebab morphology.

Conclusion

We reported the microbial synthesis of ultra-high-molecular-weight P(3HB) produced from glucose using a recombinant *E. coli* and the physical properties and enzymatic degradation of its film. It was revealed that the molecular weights of P(3HB) produced within cells were strongly dependent on the pH of the culture medium containing glucose as a carbon source. When cultured in the medium of pH 6.0, the molecular weight of P(3HB) reached to Mw(MALLS) of 14×10^6. Solvent-cast film prepared from a solution of chloroform was stretched easily and reproducibly by hot-drawing in a silicone oil bath at 160°C for 1 second. The stretched film had a high crystallinity and acceptable mechanical properties, and these properties were further improved by an annealing treatment. In addition, the properties remain unchanged during the aging for 6 months. It has been suggested that the embrittlement of P(3HB) film due to secondary crystallization is avoided by its high crystallinity and high degree of orientation. All ultra-high-molecular-weight P(3HB) films were

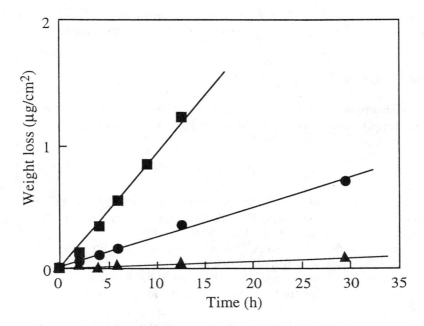

Figure 4. Enzymatic degradation of ultra-high-molecular-weight P(3HB) films in an aqueous solution of PHB depolymerase from Alcaligenes faecalis *T1 at 37°C; (■) solvent-cast film, (●) stretched film, and (▲) stretched-annealed film.*

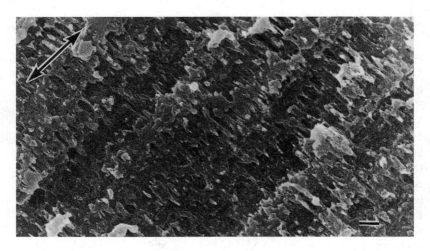

Figure 5. Scanning electron micrograph of the stretched film of ultra-high-molecular-weight P(3HB) after enzymatic degradation by an extracellular PHB depolymerase from Alcaligenes faecalis *T1. The scale bar represents 1 μm and the arrow indicates the draw direction.*

degraded by an extracellular PHB depolymerase purified from *Alcaligenes faecalis* T1, and it was revealed that the stretched films had the shish-kebab structure.

Acknowledgment

This work has been supported by NEDO (New Energy and Industrial Technology Development Organization) International Joint Research Grant of Japan.

References

1.   Dawes, E. A.; Senior, P. J. *Adv. Microbiol. Physiol.* **1973**, *10*, 135.
2.   Doi, Y. *Microbial Polyesters*; VHC Publishers, Inc.: New York, 1990.
3.   Anderson, A. J.; Dawes, E. A. *Microbiol. Rev.* **1990**, *54*, 450.
4.   Holmes, P. A. *Developments in Crystalline Polymers - 2*; Elsevier Applied Science: London and New York, 1988, pp. 1.
5.   De Koning, G. J. M.; Lemstra, P. J. *Polymer* **1993**, *34*, 4089.
6.   Scandola, M.; Ceccorulli, G.; Pizzoli, M. *Makromol. Chem. Rapid Commun.* **1989**, *10*, 47.
7.   Nakamura, S.; Doi, Y.; Scandola, M. *Macromolecules* **1992**, *25*, 4237.
8.   Pizzoli, M.; Scandora, M.; Ceccorulli, G. *Macromolecules* **1994**, *27*, 4755.
9.   Kusaka, S.; Abe, H.; Lee, S. Y.; Doi, Y. *Appl. Microbiol. Biotechnol.* **1997**, *47*, 140.
10. Lee, S. Y.; Lee, K. M.; Chang, H. N.; Steinbüchel, A. *Biotechnol. Bioeng.* **1994**, *44*, 1337.
11. Postema, A. R.; Luiten, A. H.; Pennings, A. J. *J. Appl. Polym. Sci.* **1990**, *39*, 1265.
12. Pennings, A. J. *J. Polym. Sci., Polym. Symp.* **1977**, *59*, 55.
13. Smith, P.; Lemstra, P. J. *J. Mater. Sci.* **1980**, *15*, 505.
14. Koyama, N.: Doi, Y. *Macromolecules* **1997**, *30*, 826-832.
15. Kumagai, Y.; Kanesawa, Y.; Doi, Y. *Macromol. Chem.* **1992**, *193*, 53.
16. Tomasi, G.; Scandola, M.; Briese, B. H.; Jendrossek, D. *Macromolecules* **1996**, *29*, 507.
17. Pennings, A. L.; Kiel, A. M. *Kolloid-Z* **1965**, 205, 160.
18. Odell, J. A.; Grubb, D. T.; Keller, A. J. *Polymer* **1978**, *19*, 617.
19. Bashir, Z.; Odell, J. A.; Keller, A. *J. Mater. Sci.* **1984**, *19*, 3713.
20. Bashir, Z.; Hill, M. J.; Keller, A. *J. Mater. Sci. Lett.* **1986**, *5*, 876.
21. Bashir, Z.; Odell, J. A.; Keller, A. *J. Mater. Sci.* **1986**, *21*, 3993.
22. Keller, A. J. *J. Polym. Sci.* **1956**, *21*, 363.
23. Andrews, E. H. *Proc. Roy. Soc.* **1964**, A*277*, 562.

Chapter 6

Polyhydroxyalkanoate Production by Recombinant *Escherichia coli:* New Genes and New Strains

Sang Yup Lee[1] and Jong-il Choi

Department of Chemical Engineering and BioProcess Engineering Research Center, Korea Advanced Institute of Science and Technology, 373–1 Kusong-dong, Yusong-gu, Taejon 305–701, Korea

Recombinant *Escherichia coli* has been considered as a strong candidate for the production of polyhydroxyalkanoates (PHAs) because it possesses several advantages over other bacteria. However, PHA productivity obtainable by recombinant *E. coli* harboring the *Ralstonia eutropha* PHA biosynthesis genes was much lower than that can be obtained by employing *Alcaligenes latus*. To endow the superior PHA biosynthetic machinery of *A. latus* to *E. coli*, the PHA biosynthesis genes of *A. latus* were cloned and several recombinant *E. coli* strains harboring the *A. latus* PHA biosynthesis genes were constructed. In flask culture, recombinant *E. coli* harboring the *A. latus* PHA biosynthesis genes produced more poly(3-hydroxybutyrate) [P(3HB)] than recombinant *E. coli* harboring the *R. eutropha* PHA biosynthesis genes. By the pH-stat fed-batch cultures of several recombinant *E. coli* strains harboring the *A. latus* PHA biosynthesis genes, much higher cell and P(3HB) concentrations, and productivity could be obtained. When this P(3HB) production employing recombinant *E. coli* harboring the *A.*

[1]Corresponding author: E-mail: leesy@sorak.kaist.ac.kr

latus PHA biosynthesis genes is coupled with recovery method of simple alkali digestion, the production cost of P(3HB) could be $ 3.32/kg P(3HB).

Recently, the problems concerning the global environment have created much interest in the development of biodegradable polymers. Polyhydroxyalkanoates (PHAs) are the polyesters of hydroxyalkanoates that are synthesized and intracellularly accumulated as an energy and/or carbon storage material by numerous microorganisms (*1-4*). PHAs are considered to be good candidates for biodegradable plastics and elastomers since they possess material properties similar to those of synthetic polymers currently in use, and are completely biodegradable when disposed (*3,5*). However, commercialization of PHAs has been hampered by their high production cost (*6,7*). Much effort has been devoted to lower the production cost by developing better bacterial strains and more efficient fermentation/recovery processes (*3,8*).

Recombinant *Escherichia coli* harboring the *Ralstonia eutropha* (formerly known as *Alcaligenes eutrophus*) PHA biosynthesis genes can accumulate a large amount of PHA (*9,10*). Production of PHA by recombinant *E. coli* has been investigated by several groups (*10,11*). There appeared a number of papers that described the development of host-plasmid systems and the strategies for the production of PHA to a high concentration and content (*9,12,13,14*). Poly(3-hydroxybutyrate) [P(3HB)], the best studied member of PHAs, could be produced to a level as high as 157 g/L by the pH-stat fed-batch culture in a defined medium (*15*). Recombinant *E. coli* has been considered to be a strong candidate as PHA producer due to several advantages over wide type PHA producers such as a wide range of utilizable carbon sources (*16,17*), P(3HB) accumulation to a high content (up to 90% of cell dry weight), fragility of cells allowing easy recovery of P(3HB) (*18,19*), production of ultrahigh molecular weight P(3HB) (*20*) and no degradation of P(3HB) due to lack of PHA depolymerase. However, the highest productivity of P(3HB) by recombinant *E. coli* was 3.2 g/L-h (*15*), which is relatively high but much lower than that can be obtained with *Alcaligenes latus* (*21*). From the economical analysis of the process for the production of P(3HB) by recombinant *E. coli*, the lower P(3HB) productivity is one reason for the higher production cost of P(3HB) compared with that by *A. latus* (*7,22*). It was assumed that *A. latus* possesses more efficient PHA biosynthetic enzymes. Therefore, it was thought that recombinant *E. coli* harboring the *A. latus* PHA biosynthesis genes might produce PHA more efficiently, while retaining several advantages mentioned above.

Recently, we have cloned the *A. latus* PHA biosynthesis genes (*23*). In this paper, we report development of several recombinant *E. coli* strains harboring the *A. latus* PHA biosynthesis genes and characteristics of these strains. The results from the fed-batch cultures of these recombinant strains and process analysis designed based on these actual fermentation performances are also reported.

Material and Method

Bacterial strains, plasmids and growth conditions

Plasmid pSYL104 (*9*) harboring the *R. eutropha* PHA biosynthesis genes and the *hok/sok* gene, and plasmid pSYL107 (*24*) harboring the *R. eutropha* PHA biosynthesis genes, the *hok/sok* gene and the *E. coli ftsZ* gene have been described previously. *E. coli* XL1-Blue was used as a host strain. Plasmid pJC1 (Figure 1(a)) contains the *A. latus* PHA biosynthesis genes and ORF4 in pUC19. To stabilize the recombinant plasmid of pJC1, a plasmid pJC2 was constructed by cloning the 6.4-kb *Eco*RI-*Hind*III fragment containing the *A. latus* PHA biosynthesis genes and ORF4 into pSYL104 digested with the same restriction enzymes, thus replacing the *R. eutropha* PHA biosynthesis genes. Plasmid pJC3 was constructed by removing 1076-bp (*Bst*EII site) unnecessary DNA fragment (containing the ORF4) in front of the promoter region of the *A. latus* PHA biosynthesis genes in pJC1. Stable plasmid pJC4 (Figure 1(b)) was constructed by cloning the 5.4-kb *Eco*RI-*Hind*III fragment containing the *A. latus* PHA biosynthesis genes in pJC3 into pSYL104 digested with the same restriction enzymes.

Culture condition for the production of P(3HB)

Chemically defined R (*24*) or MR (*15*) medium was used for flask and fed-batch cultures of recombinant *E. coli*. Fed-batch cultures were carried out at 30 °C in a 6.6 L jar fermentor (Bioflo 3000, New Brunswick Scientific Co., Edison, NJ) initially containing 1.2 L of MR medium (*15*). Culture pH was controlled by automatic addition of 28% (v/v) NH_4OH solution. The feeding solution used for the fed-bath culture contained per liter: glucose, 700g; $MgSO_4 \cdot 7H_2O$, 15g; and thiamine, 250 mg. The pH-stat feeding strategy was employed in fed-batch culture.

Analytical procedures

Cell concentration, defined as cell dry weight per liter of culture broth, was determined by weighing dry cells as described previously (*15*). P(3HB) concentration was determined by gas-chromatography (HP 5890, Hewlett-Packard, Wilmington, DE) with benzoic acid as an internal standard (*25*). P(3HB) content was defined as the percentage of the ratio of P(3HB) to dry cell concentration. P(3HB) productivity is defined as the amount of P(3HB) produced per unit volume per unit time. The results of flask cultures are reported as the average values of three repeated experiments

(a)

(b)

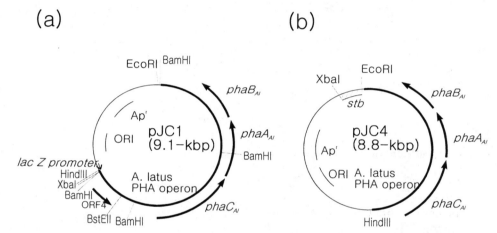

Figure 1. Plasmids pJC1 (a) and pJC4 (b). The abbreviations are: stb, the parB locus of plasmid R1; Ap^r, ampicillin resistance; ORI, origin of replication. The bold line represents the DNA fragment from A. latus (Redrawn with permission from reference 23).

Process analysis and economic evaluation

Process analysis and economic evaluation for P(3HB) production and recovery were carried out as previously described (*7,18,22*).

Results and Discussion

Cell growth and P(3HB) synthesis in various recombinant *E. coli*

Recombinant *E. coli* strains harboring 4 different plasmids (pJC1, pJC2, pJC3 and pJC4) containing the *A. latus* PHA biosynthesis genes and recombinant *E. coli* harboring pSYL104 and pSYL107 containing the *R. eutropha* PHA biosynthesis genes were cultivated in a defined R medium. The cell and P(3HB) concentrations, and P(3HB) content of these recombinant *E. coli* strains were shown in Table I. *E. coli* XL1-Blue(pJC1) and *E. coli* XL1-Blue(pJC2) grew to lower cell concentrations and accumulated less P(3HB) than *E. coli* XL1-Blue(pSYL107). However, *E. coli* XL1-Blue(pJC3) and *E. coli* XL1-Blue(pJC4) showed higher cell and P(3HB) concentrations than *E. coli* XL1-Blue(pSYL107). Especially, *E. coli* XL1-Blue(pJC4) showed the highest cell and P(3HB) concentrations among the recombinant *E. coli* strains examined.

The superior ability of recombinant *E. coli* strains harboring the *A. latus* PHA biosynthesis genes for the production of P(3HB) can be easily seen when *E. coli* XL1-Blue(pJC4) and *E. coli* XL1-Blue(pSYL104) are compared. The only difference between pJC4 and pSYL104 is the source of the PHA biosynthesis genes. pJC4 contains the PHA biosynthesis genes of *A. latus*, while pSYL104 contains those of *R. eutropha*. *E. coli* XL1-Blue(pJC4) grew to a higher cell concentration and accumulated more P(3HB) than *E. coli* XL1-Blue(pSYL104). From these results, it was thought that recombinant *E. coli* harboring the *A. latus* PHA biosynthesis genes can be used for more efficient production of PHA. To confirm this superiority, fed-batch cultures of recombinant *E. coli* strains were carried out for the production of high concentration of P(3HB).

Fed-batch cultures of recombinant *E. coli* strains harboring the *A. latus* PHA biosynthesis genes

The pH-stat fed-batch cultures of recombinant *E. coli* strains harboring 4 different plasmids (pJC1, pJC2, pJC3 and pJC4) containing the *A. latus* PHA biosynthesis genes were carried out. The results are summarized in Table II. By the fed-batch culture of *E. coli* XL1-Blue(pJC1), the final cell and P(3HB) concentrations of 177.9 g/L and 76.2 g/L, respectively, were obtained with P(3HB) content of 42.8%. The final cell and P(3HB) concentrations, and P(3HB) content obtained with *E. coli* XL1-Blue(pJC2) were 186.0 g/L, 103.8 g/L, and 55.8%, respectively. Although *E. coli* XL1-Blue(pJC1) and *E. coli* XL1-Blue(pJC2) could be grown to high cell

Table I. Cell growth, P(3HB) concentration and P(3HB) content obtained with different recombinant *E. coli* strains cultured at 30 °C for 66 h in a defined R medium containing 20 g/L glucose and 20 mg/L thiamine (data taken from reference 23).

	pJC1	pJC2	pJC3	pJC4	pSYL104	pSYL107
Cell dry weight concentration (g/L)	5.31	3.88	6.83	7.45	5.78	6.48
P(3HB) concentration (g/L)	2.84	2.10	4.83	5.30	3.47	4.33
P(3HB) content (%)	53.5	54.1	70.7	71.1	60.0	66.8

Table II. Summary of fed-batch cultures of recombinant *E. coli* harboring the *A. latus* PHA biosynthesis genes.

	pJC1	pJC2	pJC3	pJC4
Fermentation time (h)	34.58	45.47	57.43	30.60
Cell dry weight concentration (g/L)	177.9	186.0	202.6	194.1
P(3HB) concentration (g/L)	76.2	103.8	148.4	141.6
P(3HB) content (%)	42.8	55.8	73.3	73
Productivity (g P(3HB)/L-h)	2.20	2.28	2.58	4.63

concentrations, P(3HB) content was much lower than that could be obtained with recombinant *E. coli* harboring the *R. eutropha* PHA biosynthesis genes. Much enhancement was achieved when recombinant *E. coli* strains harboring pJC3 or pJC4 were employed. With *E. coli* XL1-Blue(pJC3), the final cell and P(3HB) concentrations obtained were 202.6 and 148.4 g/L, respectively (Figure 2). The final P(3HB) content was 73.3%. When fed-batch culture of *E. coli* XL1-Blue(pJC4) was carried out, cell concentration of 194.1 g/L, P(3HB) concentration of 141.6 g/L and P(3HB) content of 73% were obtained with P(3HB) productivity as high as 4.63 g/L-h (Figure 3). This P(3HB) productivity is much higher than the highest productivity obtained with recombinant *E. coli* harboring the *R. eutropha* PHA biosynthesis genes (3.2 g/L-h) (*15*).

From these results, recombinant *E. coli* harboring the *A. latus* PHA biosynthesis genes can be considered as a good candidate for the production of P(3HB) with high productivity. The beneficial effect of this enhanced P(3HB) production is analyzed below.

Economic evaluation for the process employing recombinant *E. coli* harboring the *A. latus* PHA biosynthesis genes

The processes for the production of P(3HB) by fed-batch cultures of two different recombinant *E. coli* strains from glucose coupled with the recovery method of simple alkali digestion were designed (Figure 4) and analyzed. The actual fermentation results obtained with recombinant *E. coli* harboring the *R. eutropha* PHA biosynthesis genes (*15*) and with recombinant *E. coli* harboring the *A. latus* PHA biosynthesis genes were used in the simulations. The targeted amount of purified P(3HB) was 100,000 tonnes per year, and the total operating time was assumed to be 7,920 h (330 days) per year. The detailed procedures for process design and simulations have been described previously (*7,18,22*). Fermentation performance and the operating costs for the production of 100,000 tonnes of P(3HB) are summarized in Table III. In comparison of the fermentation performance obtained with two different recombinant *E. coli* strains, P(3HB) productivity obtained with recombinant *E. coli* harboring the *A. latus* PHA biosynthesis genes is higher than that with recombinant *E. coli* harboring the *R. eutropha* PHA biosynthesis genes. The calculated production cost when recombinant *E. coli* harboring the *A. latus* PHA biosynthesis genes was employed was $ 3.32/kg P(3HB), which is $ 0.34/kg lower than that obtained with recombinant *E. coli* harboring the *R. eutropha* PHA biosynthesis genes. In operating costs, direct-fixed capital dependent items and utilities costs of the process employing recombinant *E. coli* harboring the *A. latus* PHA biosynthesis genes are lower than that of recombinant *E. coli* harboring the *R. eutropha* PHA biosynthesis genes. This is because for the production of the same amount of PHA per year, the process with higher productivity requires smaller equipments.

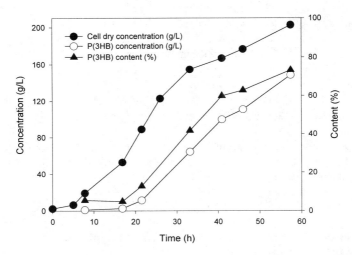

Figure 2. Time profiles of cell concentration, P(3HB) concentration and P(3HB) content during the fed-batch culture of E. coli XL1-Blue(pJC3) in a chemically defined medium.

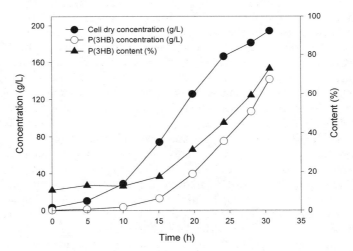

Figure 3. Time profiles of cell concentration, P(3HB) concentration and P(3HB) content during the fed-batch culture of E. coli XL1-Blue(pJC4) in a chemically defined medium (Redrawn with permission from reference 23).

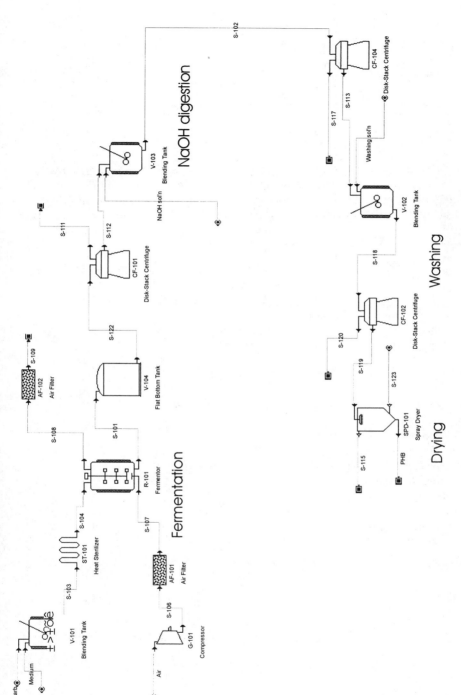

Figure 4. Process flowsheet for the production of P(3HB) with recovery method of simple alkali digestion.

Table III. Comparison of fermentation performance and annual operating costs for the production of P(3HB) (100,000 tonnes per year) by recombinant *E. coli* harboring the *R. eutropha* PHA biosynthesis genes and recombinant *E. coli* harboring the *A. latus* PHA biosynthesis genes. P(3HB) was recovered by simple alkali digestion.

	Recombinant *E. coli* harboring the *R. eutropha* PHA biosynthesis genes	Recombinant *E. coli* harboring the *A. latus* PHA biosynthesis genes
Fermentation performance		
Culture time (h)	49	30.6
Cell concentration (g/L)	204.3	194.1
P(3HB) concentration (g/L)	157.1	141.6
P(3HB) content (%)	77	73
P(3HB) productivity (g/L-h)	3.2	4.63
P(3HB) yield (g P(3HB)/g carbon)	0.27	0.28
Economic evaluation		
Directed-fixed capital dependent cost	0.95	0.66
Labor-dependent cost	0.16	0.15
Administration and overhead	0.08	0.08
Raw material cost	2.10	2.10
Utilities	0.34	0.30
Waste treatment/disposal	0.03	0.03
Total production cost (US $ /kg P(3HB))	**3.66**	**3.32**

Conclusion

Biodegradable polymer PHA is drawing much attention as a substitute for conventional petrochemical based polymer. However, its production cost is much higher than that of conventional polymer, thus preventing the application of this biodegradable polymer. From the economical analysis of PHA production by bacterial fermentation, PHA productivity is an important factor affecting the overall production cost.

Recombinant *E. coli* is considered as a strong candidate for PHA producer because of several advantages over other bacteria. However, the P(3HB) productivity obtained with recombinant *E. coli* was much lower than that obtainable with *A. latus*. The high PHA productivity obtained with *A. latus* was assumed to be due to that *A. latus* had superior PHA biosynthetic enzymes. The PHA biosynthesis genes were cloned from *A. latus,* and several recombinant *E. coli* strains harboring the *A. latus* PHA biosynthesis genes were constructed. Recombinant *E. coli* harboring the *A. latus* PHA biosynthesis genes accumulated a large amount of P(3HB) with higher P(3HB) productivity than recombinant *E. coli* harboring the *R. eutropha* PHA biosynthesis genes. Process design and analysis showed that the production cost of P(3HB) could be reduced by the increase of the productivity.

Acknowledgements

This work was supported by the Ministry of Science and Technology and by the LG Chemicals, Ltd.

References

1.   Anderson, A. J.; Dawes, E. A. *Microbiol. Rev.* **1990**, *54*, 450-472.
2.   Doi, Y. *Microbial polyesters*. VCH: New York, 1990.
3.   Lee, S. Y. *Biotechnol. Bioeng.* **1996**, *49*, 1-14.
4.   Steinbuchel, A. *Biomaterials: novel materials from biological sources*; Byrom, D. Ed.; Stockton: New York, 1991, pp. 124-213.
5.   Holmes, P. A.. *Developments in crystalline polymers*; Bassett, D.C., Ed; Elsevier: London, 1988, Vol. 2, pp. 1-65.
6.   Byrom, D;. *Trends Biotechnol.* **1987**, *5*, 246-250.
7.   Choi. J.; Lee, S. Y. *Bioprocess. Eng.* **1997**, *17*, 335-342.
8.   Lee, S. Y. *Trends Biotechnol.* **1996**, *14*, 431- 438.
9.   Lee, S. Y.; Yim, K. S.; Chang, H. N.; Chang Y. K. *J. Biotechnol.* **1994**, *32*, 203-211.
10.   Fidler, S.; Dennis D. *FEMS Microbiol. Rev.* **1992**, *103*, 231-236.
11.   Lee, S. Y. *Nature Biotechnol.* **1997**, *15*, 17-18.
12.   Lee, S. Y.; Chang, H. N. *Can. J. Microbiol.* **1995**, *41(Suppl. 1)*, 207-215.

88

13. Lee, S. Y.; Lee, K. M.; Chang, H. N.; Steinbuchel, A. *Biotechnol. Bioeng.* **1994**, *44*, 1337-1347.

14. Kidwell, J; Valentin, H. E.; Dennis, D. *Appl. Environ. Microbiol.* **1995**, *61*, 1391-1398.

15. Wang, F.; Lee, S. Y. *Appl. Environ. Microbiol.* **1997**, *63*, 4765-4769.

16. Lee, S. Y.; Chang, H. N. *Biotechnol. Lett.* **1993**, *15*, 971-974.

17. Lee, S. Y.; Middelberg, A. P. J.; Lee, Y. K. *Biotechnol. Lett.* **1997**, *19*, 1033-1035.

18. Choi, J.; Lee, S. Y. *Biotechnol. Bioeng.* **1999**, *62*, 546-553.

19. Lubitz, W. Deutsches Patentamt DE 40 03 827 A1, 1991.

20. Kusaka, S.; Abe, H.; Lee, S. Y.; Doi, Y. *Appl. Microbiol. Biotechnol.* **1997**, *47*, 140-143.

21. Wang, F.; Lee, S. Y. *Appl. Environ. Microbiol.* **1997**, *63*, 3703-3706.

22. Lee, S. Y.; Choi, J. *Polymer Degrad. Stabil.* **1998**, *59*, 387-393.

23. Choi, J.; Lee, S. Y.; Han, K. *Appl. Environ. Microbiol.* **1998**, *64*, 4897-4903.

24. Lee, S. Y. *Biotechnol. Lett.* **1994**, *16*, 1247-1252.

25. Braunegg, G.; Sonnleitner, B.; Lafferty, R. M. *Eur. J. Appl. Microbiol. Biotechnol.* **1978**, *6*, 29-37.

Biodegradable Polyesters:
Synthesis and Characterization

Enzymatic Synthesis

Chapter 7

Introduction to Enzyme Technology in Polymer Science

David L. Kaplan[1] and Richard A. Gross[2]

[1]Center for Enzymes in Polymer Science, Department of Chemical Engineering and Biotechnology Center, Tufts University, Medford, MA 02155
[2]Department of Chemistry, Chemical Engineering and Materials Science, Polytechnic University, Brooklyn, NY 11201

Enzyme technology has significantly expanded in scope and impact over the past ten years. Historically, the specificity, high substrate turnover rates, and the ability to function under mild conditions have provided the impetus for enzyme utilization in a wide range of reactions. The most common examples of uses for enzymes in industry have been in food, pharmaceutical, detergent and textile processes. While these traditional uses of enzymes have sustained a vibrant scientific and technology community, recent advances in a number of areas suggest that an improved fundamental understanding of enzyme mechanisms, as well as new opportunities for applications for enzymes, are on the horizon. Illustrative examples of important recent advances in enzyme technology include: advancements in rapid screening and molecular evolution methods, new avenues for 'biomining' to obtain novel enzymes, genetic material from diverse and extreme environments, observations on reaction equilibria in nontraditional environments, and the push toward 'green chemistries'.

There are many reviews in the literature that can be accessed for additional details on pharmaceutical or small molecule modifications using enzymes [for examples, see *1-3*]. While these studies represent the majority of information on enzyme technology, there is a growing literature that has emphasized key issues of interest to many polymer scientists within the context of enzymology [*4*]. Hence, this book contribution focuses on the latter, and refers readers to other reviews (see above) for additional information on enzyme-catalyzed organic transformations of small molecules.

90

Within the context of polymer science and engineering, there are some key attributes of enzymes that can be highlighted:

- Polymerization Kinetics – High monomer turnover rates with reasonable enzyme stability in polymer synthesis under ambient conditions.

- Control of Polymer Structure – Stereoselectivity, regioselectivity, molecular weight, polydispersity, block size and architectural control are all feasible attributes in polymer synthesis with enzyme technology. In most cases, this control of structure can be achieved in much simpler ways than with traditional synthetic catalysts.

- Versatility of Enzymes in Polymer Synthesis – Enzymes can be engineered with a suite of powerful methods to carry out new reactions or to improve inefficient reactions. Similarly, the reaction environment can be engineered to facilitate these goals.

- Polymer Life-Cycles - Polymers synthesized using enzymes should be biodegradable, thus geochemical cycles can be maintained in balance without undue accumulation of nondegradable materials; often a concern with many synthetic polymers.

Some selective observations to expand on this listing are provided in the subsequent pages. In addition, speculation is offered on future developments in the field.

Polymerization Kinetics

Concepts related to enzyme specificity and rapid turnover rates are well documented in most biochemistry textbooks. Billions of years of evolution have provided the selective pressure for these features. Probably the most compelling advances in recent years with respect to enzyme reaction kinetics have been in the area of enzyme stability. This has been a long-standing issue often raised as a concern or limitation of enzymes in certain applications. Advances are being made on at least two fronts:

(1) An improved fundamental understanding of the mechanisms involved in enhancing enzyme stability over a wide range of environmental conditions, such as high temperature, high salt concentrations and pH extremes. In general, insights being gained through the study of thermophilic enzymes have led the way in this area. The recent availability of data from mesophiles and thermophiles to provide molecular-level differences in primary sequences at genetic and protein levels has been very valuable. Insights are also being gained through studies of enzyme molecular evolution, aided by rapid screening [5-7]. These studies have already expanded the limited understanding of the importance of amino acid residues not present in the

active site in influencing stability and activity. Examples of recent work that has provided enhanced enzyme stability includes altering hydrophobic interactions, creating salt-bridges, hydrogen bonding [8].

(2) Chemical and physical methods to enhance stability – In recent years there has been a resurgence of activity aimed at improving enzyme stability through the use of physical and chemical methods. As examples, glutaraldehyde cross-linked enzyme crystals [9-12], enzymes immobilized in polymer foams or sponges [13,14], and 'plastic' enzymes [15] are some examples of these recent advancements. In all cases, significant improvements in enzyme stability are obtained and in many cases these processes can be carried out even with crude enzyme preparations. Chemical modification of enzymes to enhance organic solvent compatibility also has been used to enhance reactions, such as through surfactant ion pairing [16-20]. Enzymes have also been used to form cross-links to modify surface features and enhance stability of protein films [21].

Anticipated future advancements in this area may include the identification and use of smaller catalytic domains, stabilized by some of the mechanisms outlined above. This option would be particularly attractive where a stable environment, such as a specific organic solvent or supercritical fluid, is used for the reaction. Closer coupling between materials design and its potential enzymatic degradation will also become more important as material life-cycle and disposal concerns continue to increase. It would be useful to build in enzyme-recognition sites that would be accessed by enzymes common in a specific environment to initiate the degradation process. This design approach may be more suitable for biomedical materials in the near term, such as for controlled release and controlled degradation. However, future needs may inspire the development of such materials for specialty or even commodity polymers. Many current biomaterial designs are based on controlled changes in structure, driven by shifts in physiological conditions *in vivo*, such as pH.

Control of Polymer Structure

Molecular recognition in polymerizations is an important advantage derived from the specificity of enzyme reactions. This feature is critical to create desired biological interactions as well as to control macromolecular assembly. Biopolymer interactions to attain higher-order and more complex structure and function are driven by molecular recognition that originates at the level of enzyme specificity (e.g., linkage sites, regioselectivity, stereoselectivity).

Linkage sites - New insights have been gained into enzyme selectivity based on the studies with specially designed monomers. For example, monomer design dictated the selectivity in horseradish peroxidase-catalyzed free radical polymerizations as well as for oligo- and polysaccharide synthesis using cellulases and other glycosidases [22-25].

Similarly, phenolics modified with aliphatic side chains were preorganized at air-water interfaces to promote horseradish peroxidase free radical polymerization through ortho-ortho coupling [26].

Regioselectivity – Early studies with carbohydrates clearly illustrated regioselective acylation of sugars by proteases in organic solvents [27]. These studies have been extended to oligomer synthesis for the formation of complex carbohydrates [2], the acylation of the primary hydroxyl groups of cellulose, amylose and many other polysaccharides by proteases reacting with vinyl esters in organic solvents [20], and the ethyl glycoside initiated ring-opening polymerization of lactones by lipases at the primary hydroxyl groups on the sugar [28]. Redox polymers such as poly(hydroquinone) have been synthesized in aqueous environments by taking advantage of the regioselective formation of hydroquinone glycoside followed by peroxidase polymerization to form the polymer [29].

Stereoselectivity – Single step enzymatic polymerization reactions can be much simpler to carry out instead of the multistep protection-deprotection schemes that are often required in polymerization reactions in order to control stereo- or regiospecificity.. For example, stereoselective control of synthetic polymer synthesis often requires multistep processes or heavy metal catalysts, usually ending in low yields if a stereochemically pure product is the goal. Imprinting to enhance the stereoselectivity was reported [30], and stereo-elective ring-opening polymerization of lactones using lipases was demonstrated [31].

Future advances are anticipated that will lead to even greater levels of selectivity over a broader spectrum of reaction substrates and conditions.. Enzyme engineering to optimize or enhance selectivity will be a very useful option to gain additional insight into binding sites and protein-structure related to catalytic activity. The modulation of enzyme-activity, such as with optical or electrical signals, are examples of interesting options with which to control over enzyme-based processes. For example, the regulation of solvent-polymer hybrids with photoirradiation was shown to control subtilisin activity in organic solvents [32]. Recent studies with bacterial epimerases involved in the conversion of mannuronic acid to guluronic acid in alginate structures suggest a surprising level of control of the primary sequence of these two sugars in the polysaccharide [33,34]. Furthermore, these structural features have direct impact on the functional properties of the alginates that form, leading to different uses in the biological system in which they are synthesized as well as in potential biotechnological applications. The ability to engineer binding domains to take advantage of advances in modular construction of enzymes, such as cellulose binding and catalytic domains, suggests important directions in control of stability, selectivity in substrate reactions, and enzyme-incorporated materials as future directions. These domains could be modified with current genetic techniques to enhance substrate selectivities or perhaps broaden these preferences, depending on the specific applications desired.

Versatility of Enzymes in Polymer Synthesis

The range of reaction environments in which enzymes can function and carry out polymerization reactions has expanded substantially since the original discoveries by Klibanov and his coworkers [35-39]. Organic solvents, biphasic organic solvent-aqueous systems, reversed micellar systems, solventless systems, and supercritical fluids are some examples of the breadth of conditions studied to date.

The exploration of novel sources of enzymes or enzyme-encoding genetic material obtained from novel environments has been instrumental in many of the recent developments in the field. The classic case that prompted major thrusts in 'biomining' has been the identification of thermophilic DNA polymerases, a polymer (DNA)-forming enzyme that functions at high temperature and helped to significantly expand options in molecular biology through cost effective Polymerase Chain Reaction (PCR) procedures. Similarly, enzymes from extremeophiles that are active over a wider range of temperatures, pressures, organic solvents, salt and pH are being exploited. Environmental mining for new genetic material, even in cases where the microorganisms may be unculturable, is an option that has become feasible based on rapid screening methods.

Molecular evolution methods, including error prone PCR and DNA shuffling, have also expanded options for the rapid engineering of enzymes for a specific target activity in a specific environment [5-7,40]. A major advantage of these approaches is that there is no information required on sequence, the crystalline structure or the active site of the enzyme in order to perform these techniques [41]. Currently, much of the focus in enzyme structure-function studies is aimed at understanding the fundamentals of these relationships and exploring optimization of activity for specific reaction conditions. As we continue to expand the fundamental understanding between protein sequence and enzyme structure/activity, significant movement can be anticipated into the rational design of catalytic systems that carry out reactions that do not exist today [7]. Combinatorial methods using enzymes to carry out complex organic syntheses are also expanding the field of options [42].

New production environments, such as plant systems, are also being explored through metabolic engineering of enzyme pathways. One of the more vigorous efforts in recent years has focused on enzymes involved in the biosynthesis and degradation of (polyhydroxyalkanoates) in bacteria [43]. These efforts have led to transgenic plants for production of these polymers through the cloning of genes encoding these enzymes from bacteria [44].

Reaction environment engineering is also an active area of study aimed at changing the nature of the polymer products generated by enzymes. This approach was pioneered by Klibanov and coworkers about ten years ago when they reported on the extensive array of enzyme reactions in which equilibria could be reversed by carrying out the reactions in organic solvents. Early studies on the use of enzymes such as chymotrypsin and horseradish peroxidase in solvents date back thirty years. However, the detailed study of enzymatic catalysis in monophasic organic solvents is a relatively recent development [38,45]. Altering solvent composition provides a unique window into control of enzyme selectivity [46]. Predictive models to explain some of the differences between physiochemical properties of the solvents and the change in enzyme structure in a specific solvent have been described. However, for polymers, these types of studies have not been explored in as much detail. Nonetheless, it is clear that solvent composition has a profound influence on molecular weight and polydispersity in the case of peroxidase-based free-radical coupling reactions [see 39,47]. Recent studies in supercritical fluids and in reversed micelles further illustrate the versatility of these reactions [48]. In addition, specific control of molecular weight and polydispersity was achieved with lipase-catalyzed polyester synthesis by modulation of the supercritical reaction environment, such as control of pressure [49].

These types of studies have been extended to the synthesis of new polymers. For example, polyesters have been formed by lactone ring-opening polymerizations catalyzed by lipases in organic solvents or in solventless (bulk) reactions [50-53]. Other examples include the synthesis of polycarbonates by lipase-catalysis in similar environments [54] and the preparation of polyacrylamides catalyzed by peroxidases [55,56].

With the research tools that are know available, continued study of enzymes for new reactions in challenging environments will surely lead to important discoveries. Major benefits from these studies will include better insights into structure-function relationships related to improved enzyme design for particular reactions and reaction environments. With this understanding, major advances in enzyme-mimetics can be expected as well, such as the synthesis of enzyme analogs that would be simple to make, highly stable, and could carry out a robust set of reactions [e.g., see57]. It is also conceivable that, based on the understanding of the mechanisms involved in enzyme activity, as well as the loss of this activity, enzymes or their mimics may be generated that are significantly more stable than current versions. For example, improving selectivity may allow the design of active sites that can more efficiently avoid access by inhibitors. This may also reduce reaction rates, and this balance will have to be assessed to determine the benefits. Finally, hybrid enzymes [58], and multistep reactions from enzyme-hybrids or complex assemblies of enzymes would be a logical next step, so entire complex biosynthesis pathways can be recapitulated from a single enzyme complex [e.g., see 59]. This would be analogous to the modular or megaenzyme approach to polyketide antibiotic synthesis [60]. Combinatorial approaches using enzymes as catalysts to identify new lead compounds and polymers with specific structural or functional features is also a new direction in the field [61]. Enzyme designs capable of switching between different types of reactions, level of activity, or perhaps capable of providing options to regenerate enzyme activity enzyme (e.g., unfolding and refolding to 'clean' the active site) may be important future directions in the field. Similar strategies to facilitate efficient recovery of enzymes from reactions to provide options for recycling or reuse would add options in terms of economic benefits to these approaches.

Polymer Life-Cycles

The use of environmentally compatible strategies during the synthesis and processing of polymers is key to meeting current and future environmental regulations. Enzymes are an important option in addressing these needs due to their efficient function in aqueous systems or in low solvent or solventless systems. Other environmental benefits to enzyme-catalysis include operation at ambient conditions of temperature and pressure and the avoidance of heavy metal catalysts (other than the metals bound in the protein structure of the enzyme). Furthermore, the products of enzyme reactions, whether polyesters, certain polyphenols, proteins or other polymers, are often biodegradable. Thus, under the appropriate environmental conditions, the elemental components of the polymers can be returned to natural geochemical cycles without the problems associated with nondegradable synthetic polymers that accumulate in landfills or in the environment as litter.

It is worth speculating that any polymer synthesized through an enzymatic reaction will in turn be degradable in the environment. It can be argued that this would be the case, otherwise there would be an accumulation of metabolically unavailable carbon over time in the biosphere that would result in a disproportionate alteration in geochemical balances. The biosphere has evolved a careful balance with respect to this issue illustrated with the degradation of wood. Certain polymeric components in wood, such as the cellulose and hemicellulose, are rapidly degraded by fungi and bacteria in many environments to provide readily available nutrients for microbial and invertebrate communities. However, the lignin component of wood, a complex heteropolymer, degrades or turns over only very slowly, thus providing an important soil stabilizer to maintain organic content in soils, to hold moisture and metals, and to provide slow conversion to humics and related substances in soils. This balance generates a suitable mix leading to the recycling of nutrients on time scales that promote an appropriate ecological balance. While advances in enzymology described earlier suggest that novel polymers will be forthcoming, it will remain to be shown that these materials are also degradable. A recent study of the biodegradation of a limited set of peroxidase-catalyzed polyaromatics, as mimics of lignins and humic acids, support the notion that even when complex extensively cross-linked polyaromatic products are formed via free-radical polymerization reactions, they remain biodegradable under appropriate conditions [62].

Future directions in this area may include complex waste-regeneration schemes based on the use of enzymes optimized for stability and activity against specific organic waste components, the construction of multiplex enzymes to handle complex mixtures of waste materials, and the formation of new enzyme activities against normally recalcitrant components in waste streams. Goals for these systems may include the generation of monomers, gases such as methane, or heat for energy recovery.

References

1. *Enzymes in Synthetic Organic Chemistry;* Wong, C. H.; G. M. Whitesides; Pergamon, Oxford, England, 1994.
2. Wang, P. G.; W. Fitz; C-H. Wong. *Chemtech.* **1995**, *4*, 22.
3. *Enzymatic Reactions in Organic Media*; Koskinen, A. M. P.; A. M. Klibanov; Chapman and Hall, Glasgow, Scotland, 1996.
4. *Enzymes in Polymer Synthesis*; Gross, R. A.; D. L. Kaplan; G. Swift. ACS Symp. Series 684, 1998.
5. Stemmer, W. P. C. *Nature* **1994**, *370*, 389.
6. Stemmer, W. P. C. *Proc. Natl. Acad. Sci.* **1994**, *91*, 10747.
7. Shao, Z.; F. H. Arnold. *Curr. Opinion Structural Biol.* **1996**, *6*, 513.
8. Vielle, C.; J. G. Zeikus. *Trends Biotech.* **1996**, *14*, 183.
9. Khalaf, N.; C. P. Govardhan; J. J. Lalonde; R. A. Perscihettii; Y.-F. Wang; A. L. Margolin. *J. Am. Chem. Soc.*, **1994**, *118*, 5494.
10. Persichetti, R. A.; N. L. St. Clair; J. P. Griffith; M. A. Navia; A. L. Margolin. *J. Am. Chem. Soc.* **1995**, *117*, 2732.
11. Margolin, A. L. *Trends Biotech.* **1996**, *14*, 223.
12. Vilenchik, L. Z.; J. P. Griffith; N. S. Clair; M. A. Navia; A. L. Margolin. *J. Am. Chem. Soc.* **1998**, *120*, 4290.
13. Havens, P. L.; H. F. Rase. *Ind. Eng. Chem. Res.* **1993**, *232*, 2254.
14. Yang, Z.; A. J. Mesiano; S. Venkatasubramanian; S. H. Gross; J. M. Harris; A. J. Russell. *J. Am. Chem. Soc.* **1995**, *117*, 4843.
15. Wang, P.; M. V. Sergeeva; L. Lim; J. S. Dordick.. *Nat. Biotechnol.* **1997**, *15*, 789.
16. Okahata, Y.; K. Ijiro. *J. Chem. Soc. Chem. Commun. Japan* **1988**, 1392.
17. Paradkar, V. M.; J. S. Dordick. *Biotech. Bioeng.* **1994**, *43*, 529.
18. Paradkar, V. M.; J. S. Dordick. *J. Am. Chem. Soc.* **1994**, *116*, 5009.
19. Ito, Y.; H. Fujii; Y. Imanishi. *Biotechnol. Prog.* **1994**, *10*, 398.
20. Bruno, F. F.; J. A. Akkara; M. Ayyagari; D. L. Kaplan; R. Gross; G. Swift; J. S. Dordick. *Macromolecules* **1995**, *28*, 8881.
21. Hansen, D. C.; S. G. Corcoran; J. H. Waite. **1998**. *Langmuir* 1998, *14*, 1139.
22. Kobayashi, S.; K. Kashiwa; T. Kawasaki; S.-I. Shoda. *J. Am. Chem. Soc.* **1991**, *113*, 3079.
23. Kobayashi, S.; X. Wen; S.-I., Shoda. *Macromolecules* **1996**, *29*, 2698.
24. Shoda, S.-I.; S. Kobayashi. *Trends in Polym. Sci.* **1997**, *5*, 109.
25. Shoda, S.-I.; K. Obata; O. Karthaus; S. Kobayashi. *J. Chem. Soc. Chem. Commun.* **1993**, *18*, 1402.
26. Bruno, F. F.; J. A. Akkara; L. A. Samuleson; D. L. Kaplan; B. K. Mandal; K. A. Marx; J. Kumar; S. K. Tripathy. *Langmuir* **1995**, *11*, 889.
27. Riva, S.; J. Chopineau; A. p. G. Kieboom; A. M. Klibanov. *J. Am. Chem. Soc.* **1988**, *110*, 584.

28. Bisht, K. S.; F. Deng; R. A. Gross; D. L. Kaplan; G. Swift. *J. Am. Chem. Soc.* **1998**, *120*, 1363.
29. Wang, P.; B. D. Martin; S. Parida; D. G. Rethwisch; J. S. Dordick. *J. Am. Chem. Soc.* **1995**, *117*, 12885.
30. Stahl, M.; U. Jeppsson-Wistrand; M. Mansson; K. Mosbach. *J. Am. Chem. Soc.* **1991**, *113*, 9366.
31. Svirkin, Y. Y.; J. Xu; R. A. Gross; D. L. Kaplan; G. Swift. *Macromolecules* **1996**, *29*, 4591.
32. Ito, Y.; N. Sugimura; O. H. Kwon; Y. Imanishi. *Nature Biotech.* **1999**, *17*, 73.
33. Rehm, B. H.; H. Ertesvag; S. Valla. *J. Bacteriol.* **1996**, *178*, 5884.
34. Ertesvag; H., H. K. Hoidal; G. Skjak-Braek; S. Valla. *J. Biol. Chem.* **1998**, *273*, 30927
35. Zaks, A.; A. M. Klibanov. *J. Biol. Chem.* **1988**, *263*, 8017.
36. Zaks, A.; A. J. Russell. *J. Biotechnol.* **1988**, *8*, 259.
37. Zaks, A.; A. M. Klibanov. *Proc. National. Acad. Sci. USA* **1985**, *82*, 3192.
38. Dordick, J. S.; M. A. Marletta; A. M. Klibanov. *Biotechnol. Bioeng.* **1987**, *30*, 31.
39. Dordick, J. S. *Enzyme Microb. Technol.* **1989**, *11*, 194.
40. Moore, J. C.; F. H. Arnold. *Nature Biotechnol.* **1996**, *14*, 456.
41. Arnold, F. H. 1998. *Nature Biotechnol.* **1998**, *16*, 617.
42. Wright, C. W.; G. F. Joyce. *Science* **1997**, *276*, 614.
43. Steinbuchel, A.; H. E. Valentin. *FEMS Microbial. Lett.* **1995**, *128*, 219.
44. Nawrath, C.; Y. Poirier; C. Somerville.. *Proc. Natl. Acad. Sci.* **1994**, *91*, 12760.
45. Akkara, J. A.; K. J. Senecal; D. L. Kaplan. *J. Polym. Sci.* **1991**, *29*, 1561.
46. Carrea, G.; G. Ottolina; S. Riva. *Trends Biotech.* **1995**, *13*, 63.
47. Ayyagari, M. S.; K. A. Marx; S. K. Tripathy; J. A. Akkara; D. L. Kaplan. *Macromolecules* **1995**, *28*, 5192.
48. Hammond, D. A.; M. Karel, A. M. Klibanov; V. J. Krukonis. *Appl. Biochem. Biotechnol.* **1985**, *11*, 393.
49. Chaudhary, A. K.; E. J. Beckman; A. J. Russell. *J. Am. Chem. Soc.* **1995**, *117*, 3728.
50. Macdonald, R. T.; S. K. Pulapura; Y. Y. Svirkin; R. A. Gross. *Macromol.* **1995**, *28*, 73.
51. Uyama, H.; K. Takeya; N. Hoshi; S. Kobayashi. *Macromol.* **1995**, *28*, 7046.
52. Nobes, G. A. R.; R. J. Kazlauskas; R. H. Marchessault. *Macromol.* **1996**, *29*, 4829.
53. Henderson, L.A.; Y. Y. Svirkin; R. A. Gross; D. L. Kaplan; G. Swift. *Macromolecules,* **1996**, *29*, 7759.
54. Bisht, K. S.; Y. Y. Svirkin; L. A. Henderson; R. A. Gross; D. L. Kaplan; G. Swift. *Macromolecules* **1997**, *30*, 7735.
55. Emery, O.; T. Lalot, M. Brigodiot; E. Marechal. *J. Polym. Sci.* **1997**, *35*, 3333.
56. Teixeira, D.; T. lalot; M. Brigodiot; E. Marechal. *Macromol.* **1999**, *32*, 70.
57. Breslow, R. *Anal. N. Y. Academy Sci., New York,* **1986**, *471*, 60.
58. Nixon, A. E.; M. Ostermeier; S. J. Benkovic. *Trends in Biotech.* **1998**, *16*, 258.

59.   Campbell, R. K; E. R. Bergert; Y. Wang; J. C. Morris; W. R. Moyle. *Nature Biotech.* **1997**, *15*, 439.
60.   Rawls, R. L. *Chem. Eng. News*, **1998**, *March*, 29.
61.   Michels, P. C.; Y. L. Khmelnitsky; J. S. Dordick; D. S. Clark. *Trends Biotech.* **1998**, *16*, 210.
62.   Farrell, R.; M. Ayyagari; J. Akkara; D. L. Kaplan. *J. Environ. Polym. Degradation* **1998**, *6*, 115.

Chapter 8

Lipase-Catalyzed Ring-Opening Polymerization of ω-Pentadecalactone: Kinetic and Mechanistic Investigations

Lori A. Henderson[1] and Richard A. Gross[2]

[1]Novo Nordisk BioChem North America Inc., 77 Perry Chapel Church Road, Raleigh, NC 27525
[2]Herman F. Mark Professor, Polytechnic University, Six Metrotech Center, Brooklyn, NY 11201 .

Abstract

The bulk polymerization of ω-pentadecalactone (PDL) at 50°C catalyzed by an immobilized lipase from a *Pseudomonas sp.* (I-PS-30) deviated from an *ideal* living system.   Analysis of the kinetic data indicated that the polymerizing system consisted of propagating chains that were slowly increasing in number and that M_n is not a simple function of the $[monomer]_0$ to $[initiator]_0$ ratio.   It was also shown that the M_w/M_n (MWD) becomes more narrow with increasing conversion and that only a fraction of the water available in the system was used to initiate the formation of chains.   Based on these results, comparative studies between *slow-initiation* and *slow exchanges* involving dormant and active chains were made.   The current model used to describe these observations is that which governs the slow dynamics of exchange during propagation.   Interestingly, the rate at which monomer was consumed was found to vary from first to second order when examining the kinetics at different stages of the reaction (i.e., 3-13% and 13-40% conversion, respectively).   The concentration of polymer chains ($[R{\sim}OH]$) increasing with monomer conversion may explain the increased sensitivity of the reaction rate to monomer concentration as the reaction progresses.   In summary, the kinetic analyses along with the analysis of products at low conversion indicates that the PDL polymerization using I-PS-30 proceeds by a chain-reaction mechanism with a slower rate of initiation relative to the rate of propagation.   An enzyme-activated monomer mechanism was defined for PDL based on the following: *(i)* no termination, *(ii)* a R_p that is dependent on $[R{\sim}OH]_0$, $[Catalyst]_0$ and [Monomer] and *(iii)* the formation of 15-hydroxypentadecanoic acid prior to chain growth.

Introduction

Recent research efforts exploiting enzymes as catalyst in polymer synthesis has lead to significant technological advances[1]. Several studies on designing architectural polymers without tedious protective/deprotective chemistry, well-controlled block copolymers, high optically pure polymers and macromolecules with controlled stereochemistry have been successfully demonstrated in our laboratory. As the advantages of enzymes in polymer synthesis become more evident, it is clear that the rates of chemical catalysis greatly exceed those for similar transformations using enzymes in organic media. Thus, the goals and objectives in our research laboratory also include developing an understanding of mechanistic pathways to enhance reaction rates. Early studies indicate that the reaction rates were strongly dependent on the enzyme-substrate specificity. It is now known that lipases (of similar origins) will ring-open a range of lactone monomers (4, 5, 6, 7, 12, 13 and 16 membered rings)[2,3,4]. Pioneering work carried out by MacDonald on the ring-opening polymerization of ε-caprolactone (ε-CL) using porcine pancreatic lipase as the catalyst demonstrated that product molecular weight can be controlled as well as the chain end structure by proper selection of the initiator[2]. Kinetic studies were later pursued by Henderson et. al.,[5] to develop an understanding of factors that govern the reaction rates and M_n for the lipase catalyzed polymerization of ε-CL. This latter investigation also demonstrated that by studying the propagation kinetics while utilizing tools developed from classical polymer syntheses, mechanisms for enzyme-catalyzed polymerizations can be proposed. This protocol was subsequently applied to evaluate the ring-opening of ε-CL and trimethylene carbonate using Novzym 435 as the lipase catalyst[6]. It was found that by varying the monomer to catalyst ratio via the concentration of the enzyme, the rate of propagation varied significantly for both polymerization processes. The above protocol defined by Henderson et. al.[5] was based on the kinetic models for *living, controlled* and *immortal* polymerizations (see definitions below). The fitting of experimental data to these kinetic models, developed on a hypothesized acyl-enzyme mechanism for transesterification in organic media, was also the premise of this work involving the ring-opening of PDL.

Over the past few decades, research in polymer synthesis has been directed towards achieving living conditions to obtain well-defined architectural polymers. It is one of the most versatile methodology used to engineer macromolecules with well-defined topographies. In addition to controlling and producing uniform size polymers, it provides the simplest and most convenient method for the preparation of architectural polymers. Living polymerizations techniques give the synthetic chemist two particularly powerful tools for polymer chain design: *1)* the synthesis of architectural polymers by the sequential additions of monomers and *2)* the synthesis (in-situ) of functional polymers by selective termination of living ends with appropriate reagents. Examples of the type of polymers that can be prepared include graft and multiblock copolymers, comb-like, star, and ladder polymers. In general, the well-behaved "living systems" need only an initiator and monomer.

The only criteria for living is that the chain polymerizations proceed without irreversible chain breaking reactions; that is, termination and chain transfer. This includes any reactions proceeding by slow initiation, undergoing exchanges between species of various reactivities (reversible), reversible deactivation such as the equilibrium between dormant and active species and reversible transfer reactions. The effect of these reactions on molecular properties have been studied by Matyjaszewski[7] for several cationic polymerizations (illustrated in Table 1).

Table I. Models that deviate from an ideal living system and their effect on kinetics, molecular weight and molecular weight distributions

	Rate	**MW**	**MWD**
Slow initiation	slower than ILS[a] (increasing slope)	initially higher but approaches ILS	narrows with conversion and is less than 1.35
Termination	slower than ILS (with deceleration)	no effect (limited conversion)	broader than ILS
Transfer	no effect	lower than ILS	usually broader than ILS
Slow exchange	no effect	may be equal or higher than ILS	may be very broad even bimodal or narrow and unimodal[b]

[a]ILS is "ideal living system".
[b]MWD decreases with conversion.
Chart derived from the material written by Krzystof Matyjaszewski[7].

In a truly living system where initiation is also complete and fast relative to propagation, molecular weights can be controlled by the $[monomer]_0/[initiator]_0$ ratio and the resulting MWD are indeed narrow ($M_w/M_n \sim 1.0$). However, living systems do not necessarily produce well-defined polymers with control over the molecular weight and narrow MWD. To achieve this, the rates of initiation and corresponding rates of exchange must be greater or equal to the rates of propagation. When chain transfer does exist, and is faster than propagation, the propagating reactions are thus described as "controlled" polymerizations.

It is important to note that the terms; *living, control* and *immortal* polymerizations have recently been redefined due to controversy and confusion in the literature. As of April 1997, the Nomenclature Committee of the ACS Division of Polymer Chemistry has adopted a description of these polymerizations proposed by Matyjaszewski *et. al.* Some features relevant to this investigation are: living is as previously described above, control represents a synthetic method to prepare polymers where transfer and termination may occur but at a low level such that control of the molecular properties are not altered (narrow MWD expected) and it is suggested that immortal be named as "living with reversible transfer" where narrow distributions may result if these reactions are fast compared to propagation.

In this investigation, the propagation kinetics were ascertained from the direct initiation of ω-pentadecalactone using the lipase from a *Pseudomonas sp.* immobilized on Celite 521. Theoretical and experimental R_p expressions were derived and used to *(i)* generate analytical solutions for assessing living characteristics and *(ii)* confirm the proposed lipase- catalyzed mechanisms for initiation and propagation via an acyl-enzyme intermediate. Kinetic studies were carried out to less than 40% monomer conversion based on our preliminary work on PDL ring-opening[4] where the propagation kinetics appeared to be diffusion-controlled at higher levels. Since these processes are also heterogeneous (respect to the catalyst only), many additional factors can govern the rate of propagation at some point during the polymerization. Therefore, the kinetic data was evaluated for livingness initially from 3-40% conversion and then at two different stages of the reaction; 3-13% and 13-40%. Comparisons were subsequently made to models that deviated from a true living polymerization (i.e., *slow-initiation* and *slow exchanges* during propagation). Overall, this chapter represents a minor segment of the dissertation research carried out by the author to develop an understanding of enzyme-catalyzed polymerization mechanisms. Additional scientific adventures like: *(i)* mechanisms underlying lipase action during polymerization, *(ii)* an activated monomer approach to synthesis, and *(iii)* rate constants for initiation, propagation as well as the rate of polymerization, have been postulated for both ε-CL and PDL polymerizations catalyzed by lipases[8].

Experimental

Materials

The monomer, ω-pentadecalactone (98%, Aldrich), was used without further purification. The lipase PS-30 was a gift from Amano enzymes (USA). This enzyme is from a *Pseudomonas sp.* and has a specified activity of 30,000 μ/g at pH = 7.0. The adsorption of this lipase onto Celite 521 (referred to as I-PS-30) was as previously described[4]. The resulting lipase preparation was also assayed according to a procedure given elsewhere[4].

Bulk Polymerization of PDL

All reactions were carried out in the bulk at 50°C. The ratio of monomer to I-PS-30 catalyst was 2/1 w/w. The polymerization procedure using water from the reaction mixture as the initiator is as follows. The monomer (400 mg) and lipase (200 mg I-PS-30) were transferred into separate 6 mL reaction vials and then dried over P_2O_5 in a vacuum desiccator using a diffusion pump apparatus (0.06 mm Hg, RT, 38 h). The contents of the vials were mixed under dry argon atmosphere, securely capped, and immediately stored in dry ice until all transfers were completed. The reaction vials were placed in a constant oil bath at 50°C and removed at predetermined reaction times. The reactions were terminated by removal of the enzyme. This involved dissolution of the residual monomer and polymer in chloroform-d and vacuum filtration. The insoluble enzyme was washed and the filtrates combined. Proton (^{1}H) NMR was then used to analyze the chloroform-d soluble fraction.

Solution Polymerization of PDL

The solution polymerization of PDL was conducted utilizing the same transfer and drying techniques described above. To a reaction vial containing PDL (400 mg) and I-PS-30 (200 mg), 1 mL of chloroform-d (anhydrous) was added via a syringe under argon. The vial was placed in an oil bath at 50°C and removed after 1 h. The sample was then processed as described above and resulted in <3.0% conversion. Solubility studies of a commercial sample of 15-hydropentadecanoic acid, HPA, (99%, Aldrich) were also carried out for identifying reaction products at low conversion.

Instrumental Methods

^{1}H NMR spectra were recorded in chloroform-d for polymerization samples and in methanol-d for HPA analysis using a Bruker ARX-500 spectrometer at 250 MHz. The method used for the determination of %-monomer (PDL) conversion and M_n was analogous to a previous study published by us[4]. The details on the method used to calculate total chain number (N_p) was similar to another previous study by our laboratory[5]. The M_n and M_w/M_n (MWD; molecular weight distributions) were determined by gel permeation chromatography (GPC) based on a conventional calibration technique (relative to polystyrene standards).

Initiator Concentrations

The amount of water (wt.-%) in the reaction mixture was analyzed by titration using a Mettler DL 18 Karl Fischer titrator. Individual solutions/suspensions of lipase I-PS-30 (200 mg in 5.0 mL DMSO) and PDL (400 mg in 4.0 mL DMSO) were prepared and analyzed relative to a DMSO control[8].

Results and Discussions

Many experiments were carried out to determine the factors that influence the conversion of PDL with time. Studies included varying the concentration of water (initiator), catalyst and the monomer to catalyst ratio in the bulk polymerization of PDL. The water content in the reaction media, for example, was varied by implementing different drying techniques for removing water from the enzyme[4,8]. It was found that water greatly affects the rate as well as the M_n of the polymer. It is believed that modulating water content can alter not only the total number of chains being formed (N_p) during the process but also the catalytic activity of the enzyme. Additional exploratory work was also done to try and understand how water plays a role in the overall polymerization process. The investigative reports of such work has been documented[8] and will not be presented here. This also applies to studies evaluating the effect of catalyst concentration upon the rate of reaction. It was concluded, however, that the rates at which PDL is converted to products is greatly enhanced in the presence of I-PS-30. In summary, the rate of polymerization changes with changes in the [water], [Catalyst] and [Monomer]$_0$ to [Catalyst]$_0$ ratio and thus considered in the derivation of an overall rate equation.

The propagation kinetics (conversion-time data) obtained from the ring-opening polymerization of PDL were analyzed using the diagnostic tools for living polymerizations. Interestingly, the results deviate significantly from linearity as shown in Figures 1 and 2. Analysis of these graphs shows that the ring-opening of PDL deviates from an *ideal* living system with: *(i)* a slow increase in propagating chains (see slope in Figure 1 and assuming k_p to be constant), *(ii)* M_n values that are higher than the theoretical M_n, *(iii)* MWD decreasing with increasing conversions and *(iv)* a low initiator efficiency (comparing M_n values). In addition, the N_p also increased with conversion from 0.6×10^{-5} to 3.1×10^{-5} moles based on end group analysis (see experimental).

Previous studies[4,8] in our laboratory also suggested that the total number of chains being produced can effect the rates. For example, polymerization processes with high [water] formed many chains with fewer repeat units compared to much lower levels. The slope of these curves for monomer-time data decreased and increased respectively[4,8]. Interestingly, the M_w/M_n values in this investigation became more narrow with time as opposed to any other enzyme-catalyzed polymerization studied in the literature thus far. Hence, it is believed that irreversible chain breaking reactions are not taking place or they occur so infrequently that the molecular properties of the polymer are not adversely affected. Unfortunately, the kinetic data

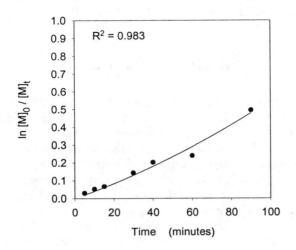

Figure 1. *A plot of the ln [M]$_0$/[M]$_t$ versus time for I-PS-30 catalyzed PDL polymerizations at 50 °C. Low conversion data were obtained in triplicate.*

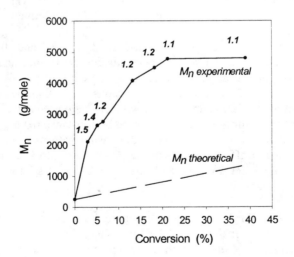

Figure 2. *A plot of M$_n$ versus conversion for PDL polymerizations conducted at 50 °C. The values above the data points are the MWD. The theoretical M$_n$ values were calculated from the conversion data and [M]$_0$/[I]$_0$ ratio after correcting for the amount of water bound to the immobilized lipase using Shimizu's equation[9].*

does not describe an *ideal* living system, that is, a *living with control* polymerization.

There are unique characteristics encompassing the kinetic treatment presented above that led us to consider comparing this polymerization to other models that deviate from an ideal living system (Table 1). The two living models describing *slow-initiation* and *slow exchanges* between dormant and active species were selected based on the aforementioned analysis of Figures 1 and 2 (see items i - iv). Since a maximum value of 1.5 for the MWD was obtained in our results and exceeds that theoretically possible (1.35), we can not conclude that this is a slow-initiating, living polymerization. We also know that the MWD values are not exact due to the fact that a conventional calibration technique was used in the GPC analysis. A more reliable approach would be to correct for the hydrodynamic volume (poly(PDL) versus polystyrene) and thus developing a universal calibration in which to obtain M_n and MWD. Work is in progress exploring this and alternate techniques to gather such data and re-evaluate this hypothesis. Meanwhile, the model that currently describes the above mentioned observations the best (MWD decreasing with increasing chain lengths), is that which considers slow exchanges between dormant and active species during propagation. This discussion is extremely relevant to the mechanisms proposed thus far in ring-opening polymerizations using enzymes for catalysis[2,5,10]. Here, the dynamic exchange between polymer chain ends at the catalytic site of the enzyme are crucial for chain growth. We hypothesize that those chains bound at the active site of the enzyme will propagate (active) while the others, that are not properly associated with the enzyme, are considered dormant. Since the experimental M_n values are higher than the theoretical values and the MWD decreases with conversion (Poisson distribution), the rate of exchange between the dormant and active chains is believed to be slow relative to propagation (Table 1). One plausible explanation for these results is that the initiator efficiency (low) as well as the number of monomer units that are added during the activation period governs the MWD. This is consistent with concepts in conventional polymer reactions involving Ziegler-Natta chemistry and cationic ring-opening where chain ends of equal reactivity persist[11].

Polymerization Mechanism

Experiments (bulk and solution) were also carried out to low conversion to analyze the products formed during the reaction and to obtain initial rate data. From this and the kinetic analysis describe herein, the subsequent mechanism for initiation and propagation was proposed. In the direct initiation of PDL, the enzyme reacts with the monomer to give an enzyme-activated monomer (serine-acylated species) which then reacts with water to form the monoadduct, 15-hydroxypentadecanoic acid (HPA). Interestingly, the conceptual formation of a covalent serine-acylated substrate is now widely accepted within the modern literature for organic transformations including lipase-catalyzed polymerizations[5,8].

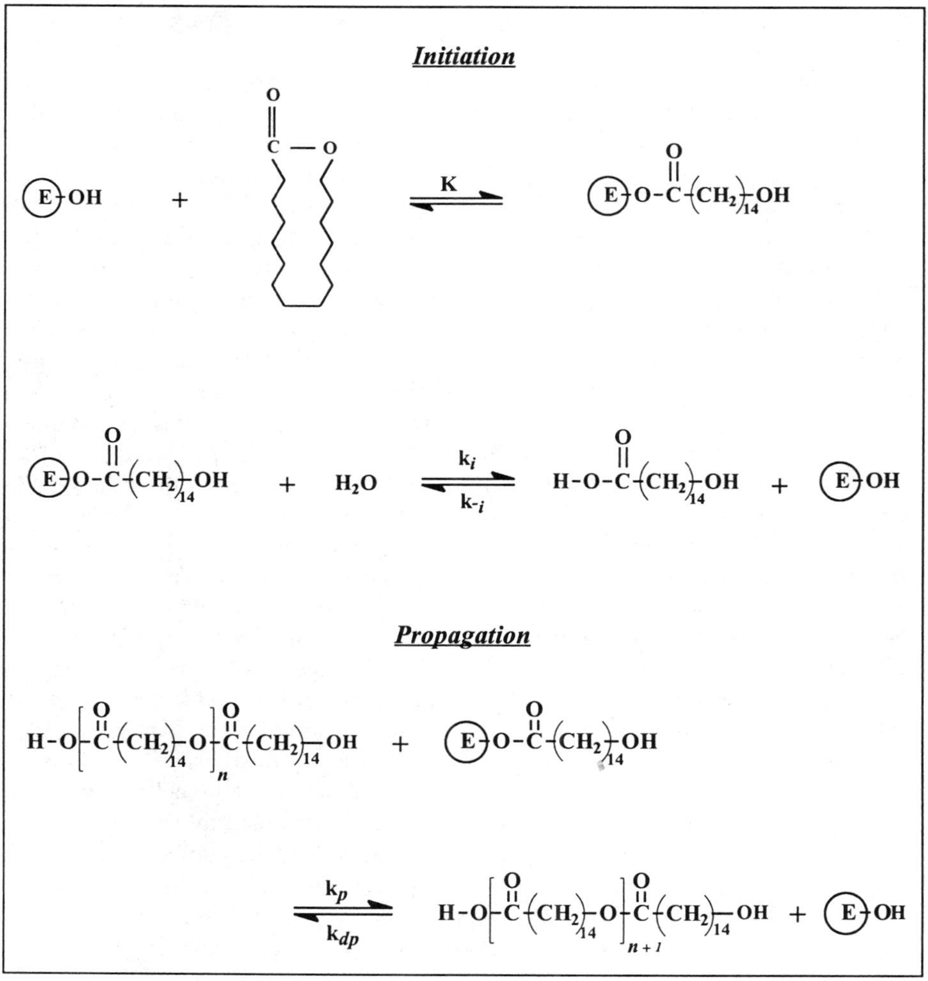

Initiation

$$\text{(E)}-\text{OH} \;+\; \begin{matrix}\text{O}\\ \| \\ \text{C}-\text{O}\end{matrix} \quad \overset{K}{\rightleftharpoons} \quad \text{(E)}-\text{O}-\overset{\text{O}}{\overset{\|}{\text{C}}}-(\text{CH}_2)_{14}-\text{OH}$$

$$\text{(E)}-\text{O}-\overset{\text{O}}{\overset{\|}{\text{C}}}-(\text{CH}_2)_{14}-\text{OH} \;+\; \text{H}_2\text{O} \quad \underset{k_{-i}}{\overset{k_i}{\rightleftharpoons}} \quad \text{H}-\text{O}-\overset{\text{O}}{\overset{\|}{\text{C}}}-(\text{CH}_2)_{14}-\text{OH} \;+\; \text{(E)}-\text{OH}$$

Propagation

$$\text{H}-\text{O}-\left[\overset{\text{O}}{\overset{\|}{\text{C}}}-(\text{CH}_2)_{14}-\text{O}\right]_n \overset{\text{O}}{\overset{\|}{\text{C}}}-(\text{CH}_2)_{14}-\text{OH} \;+\; \text{(E)}-\text{O}-\overset{\text{O}}{\overset{\|}{\text{C}}}-(\text{CH}_2)_{14}-\text{OH}$$

$$\underset{k_{dp}}{\overset{k_p}{\rightleftharpoons}} \quad \text{H}-\text{O}-\left[\overset{\text{O}}{\overset{\|}{\text{C}}}-(\text{CH}_2)_{14}-\text{O}\right]_{n+1} \overset{\text{O}}{\overset{\|}{\text{C}}}-(\text{CH}_2)_{14}-\text{OH} \;+\; \text{(E)}-\text{OH}$$

Propagation then proceeds by the reaction between propagating chain ends and that of the activated monomer. The proposed enzyme-activated mechanism (EAM) is also consistent with the rate equation derived experimentally. This kinetic work also suggest that the rate of initiation is slow relative to propagation. Additional work was carried out studying the effect of other reactive nucleophiles (primary amines, alcohols and thiols) as the initiator upon the rates of reaction. Significant variations in PDL conversion was observed depending not only on the type of initiator used but also its' concentration[8]. From this and Michaelis-Menten kinetic studies[8], the rate-determining step for chain initiation is the second step of the initiation mechanism. This step is also defined as an equilibrium reaction because

Shimizu *et. al.*[9] were able to form PDL starting with HPA (the reaction product, X_n = 1) using *Psuedomonas fluorescens* as the catalyst at 40°C.

Kinetics at different stages of the reaction

Given that the I-PS-30 catalyzed polymerizations of PDL are heterogeneous (insoluble) with respect to the catalyst and utilizes an immobilized form of lipase PS-30, we know that particle size, enzyme aggregation as well as its' distribution onto the Celite support are crucial factors that can govern the rate of propagation at any time during polymerization. Therefore, the kinetics were assessed at different stages of the reaction using the analytical solutions derived from the following rate expressions:

Experimental R_p. An experimental rate equation was derived from the conversion-time data for experiments varying different reactants in the bulk synthesis of PDL:

$$-d[M]/dt = k_p[Cat]^x[M]^y[R{\sim}OH]^z \qquad (1a)$$

where k_p, [Cat] and [R~OH] are the observed rate constant, the concentration of the lipase preparation and total chains, respectively. Rearranging and integrating the above with time knowing that N_p changes with conversion gives:

$$\ln [M]_0/[M]_t = k_p[Cat]^x \int_0^t ([I_f]_0 - [I]_t)dt \qquad (1b)$$

where $[I_f]_0$ and $[I]_t$ represent the amount of water available for initiation at time zero and t, respectively. Note that a linear relationship should still exist as the [I] varies with time.

Theoretical R_p. A theoretical rate equation was also developed from the proposed EAM mechanism to give $-d[M]/dt = k_p[AM][R{\sim}OH]$ where [AM] is the concentration of the activated monomer. Substituting into this equation the equilibrium concentrations for [AM] and [R~OH] (scheme above) yields:

$$-d[M]/dt = k_p(K_1)^2K_2[Cat]^x[I]^z[M]^{y'} \qquad (2a)$$

where K_1 and K_2 are the equilibrium constants defined by steps 1 and 2 in initiation (shown above), respectively, and [I] is the initiator concentration at time t. Assuming that the equilibrium [AM] depends on $[M]^1$ resulting in a y' of 2, equation 2a can be rearranged and integrated to give an analytical solution which is *second order* with respect to monomer consumption:

$$(1/[M]_t - 1/[M]_0) = k_p(K_1)^2K_2[Cat]^{x'} \int_0^t [I]^z dt \qquad (2b)$$

Assuming that the [Cat] is constant during the course of the reaction, a linear relationship by plotting the left side of Equation 2b versus $\int_0^t [I]^z dt$ would indicate that the R_p was second order with respect to monomer consumption. Moreover, a value of 1 for z was used in the calculations based on experiments varying the [I].

The kinetic data was re-evaluated from 3-13% and 13-40% PDL conversion by constructing plots of $\ln[M]_0/[M]_t$ versus $\int_0^t \{[I_f]_0 - [I]_t\}$ and $(1/[M]_t - 1/[M]_0)$ versus $\int_0^t [I]^z dt$ respectively. The first order plot (equation 1b) shown in Figure 3 was indeed linear with a correlation coefficient of 0.995. Interestingly, the plot was also linear at higher conversions when using the second order monomer equation 2b (Figure 4).

Thus, it is concluded that as the number of propagating chains increase, R_p varies from first to second order with respect to monomer consumption. The [Cat] also represents a constant which is indicative of the role of a catalyst. Moreover, the linear relationship illustrated in Figure 4 suggests that reverse equilibrium reactions should be considered as the water content decreases in the reaction mixture with time. Although the rational for the differential in monomer consumption behavior is not clearly defined, it is consistent with the proposed *slow exchange* model where the rate is not affected by the exchange process. Interestingly, the polymerization at early stages of the reaction appears to exhibit living characteristics much like some of the chemical preparative routes in ring-opening polymerizations via an activated monomer mechanism (AMM). For example, the cationic ring-opening of ethylene oxide or epichlorohydrin using an alcohol for initiation can propagate by the AMM. These systems should be living since the growing chains ends which are usually responsible for transfer/termination are neutral. Moreover, the growth of the material chains are not directly linked to the propagation of kinetic chains. Indeed, the living nature of these reactions were also observed but only during the early stages of polymerization. As the concentration of chains increased, the formed macromolecules would abstract the protons from the monomer and thus limiting polymer growth[12].

In summary, there are two possible polymerization routes that describe the observations present in this study; slow-initiation and slow-exchanges. The dynamic exchange between chain ends at the active site is conceptually very interesting since this ensures the growth of polymer chains. The experimental rate equation was consistent with the proposed EAM for synthesis of poly(PDL). The data also supports a derived theoretical equation identifying equilibrium reactions taking place. The challenge for the future is to develop a process which results in high M_n products and assess its' kinetics in order to test our postulates/mechanisms for lipase-catalyzed ring-opening polymerizations further.

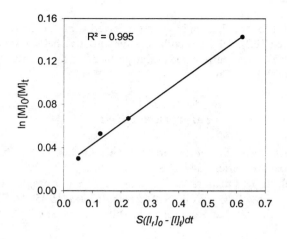

Figure 3. The ln[M]₀/[M]ₜ versus $\int_0^t \{[I_f]_0-[I]_t\}dt$ plot derived from the experimental R_p.

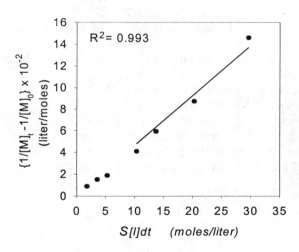

Figure 4. The second order plot; $(1/[M]_t - 1/[M]_o)$ versus $\int_0^t [I]dt$, derived from the theoretical R_p.

112

Acknowledgment

We are grateful for the financial support received from the Center for Enzymes in Polymer Science (EPS). EPS is a collaborative initiative between the Polytechnic University and Tufts. The current industrial partners of this Center include Novo Nordisk and Rohm & Haas Co.

Literature Cited

(1.)   *Enzymes in Polymer Synthesis*; Gross, R. A.; Kaplan, D. L.; Swift, G., Eds.; ACS symposium series 684, ACS: Washington, D.C., 1998.

(2.)   MacDonald, R. T.; Pulapura, S. K.; Svirkin, Y. Y.; Gross, R. A.; Kaplan, D. L. Akkara, A.; Swift, G.; Wolk, S. *Macromolecules*, **1995**, 28, 73-78.

(3.)   Uyama, H.; Takuya, K.; Kobayashi, S. Bull. Chem. Soc. Jpn. **1995**, 68, 56-61.

(4.)   Bisht, K. S.; Henderson, L. A.; Gross, R. A.; Kaplan, D.L.; Swift, G. *Macromolecules*, **1997**, 30, 2705-2711.

(5.)   Henderson, L. A.; Svirkin, Y. Y.; Gross, R. A. Kaplan, D.L.; Swift, G. *Macromolecules*, **1996,** 29, 7759-7766.

(6.)   Deng, F.; Henderson, L. A.; Gross, R. A., *Polymer Preprint*, **1998**, 39 (2), 144-145.

(7.)   Matyjaszewski, K.; Sawamoto, M. In: *Cationic Polymerizations, Mechanisms, Synthesis and Applications;* Matyjaszewski, K., Ed.; Marcel Dekker, Inc.: New York, NY, 1996, pp. 265-270, 279-285.

(8.)   Henderson, L. A. Ph. D. thesis, University of Massachusetts Lowell, Lowell, MA, 1998.

(9.)   Yamane, T.; Ichiryu, T.; Nagata, M.; Ueno, A.; Shimizu, S. *Biotechnology and Bioengineering,* **1990**, 36, 1063-1069.

(10.)   Bisht, K. S.; Henderson, L. A.; Gross, R. A.; Kaplan, D.L.; Swift, G. *Macromolecules*, **1997**, 30, 7735.

(11.)   Matyjaszewski, K.; Sawamoto, M. In: *Cationic Polymerizations, Mechanisms, Synthesis and Applications;* Matyjaszewski, K., Ed.; Marcel Dekker, Inc.: New York, NY, 1996, pp. 277-285, 217-219.

(12.)   Penczek, S.; Kubisa,P. *Polymer Preprint,* April 1997. pp.316-317.

Chapter 9

Enzymatic Polymerization of Natural Phenol Derivatives and Enzymatic Synthesis of Polyesters from Vinyl Esters

Hiroshi Uyama[1], Ryohei Ikeda[2], Shigeru Yaguchi[3], and Shiro Kobayashi[1]

[1]Department of Materials Chemistry, Graduate School of Engineering, Kyoto University, Kyoto 606–8501, Japan
[2]Joint Research Center for Precision Polymerization, Japan Chemical Innovation Institute, NIMC, Tsukuba, Ibaraki 305–8565, Japan
[3]Department of Materials Chemistry, Graduate School of Engineering, Tohoku University, Sendai 980–8579, Japan

In the first part of this chapter, enzymatic oxidative polymerizations of natural phenolic compounds, syringic acid and cardanol, have been described. Laccase and peroxidase induced the polymerization of syringic acid to give a poly(1,4-phenylene oxide) bearing a carboxylic acid at one end and a phenolic hydroxyl group at the other. The polymerization is a new type of oxidative polymerization involving elimination of not only hydrogen but also carbon dioxide from the monomer. The effects of solvent composition have been systematically investigated. The cardanol polymerization produced the polymer having an unsaturated alkyl group in the side chain, which was subjected to hardening reactions by cobalt naphthenate catalyst. The second part deals with lipase-catalyzed polymerization using vinyl esters as monomer to polyesters. The polycondensation of dicarboxylic acid divinyl esters and glycols proceeded in the presence of lipase catalyst under mild reaction conditions. The polymerization of divinyl adipate and 1,4-butanediol using *Pseudomonas cepacia* lipase as catalyst produced the polyester with molecular weight of more than 2×10^4. Lipase catalysis also induced the polymerization of oxyacid vinyl esters.

Enzymes have several remarkable catalytic properties such as high catalytic power and high selectivities under mild reaction conditions, as compared with those of other chemical catalysts. In the field of organic synthesis, enzymes have been employed as catalyst for many years. By utilizing such characteristic catalysis of enzymes, highly selective organic reactions have been developed, in certain cases producing functional materials (*1-3*).

Production of all naturally occurring polymers is in vivo catalyzed by enzymes. Recently, reports on in vitro synthesis of not only biopolymers but also non-natural synthetic polymers through enzymatic catalysis have appeared (4-8). These enzyme-catalyzed polymerizations receive much attention as new methodology of polymer syntheses, since in recent years structural variation of synthetic targets on polymers has begun to develop highly selective polymerizations for the increasing demands in the production of various functional polymers in material science. In the first part of this chapter, enzymatic polymerizations of natural phenolic compounds are described. The second part deals with enzymatic synthesis of polyesters from a combination of dicarboxylic acid vinyl esters and glycols and from vinyl hydroxyalkanoates.

Enzymatic Oxidative Polymerization of Natural Phenolic Compounds

Enzymatic Synthesis of Poly(phenylene oxide)

Poly(2,6-dimethyl-1,4-oxyphenylene) (poly(phenylene oxide), PPO) is widely used as high-performance engineering plastics, since the polymer has excellent chemical and physical properties, e.g., a high glass transition temperature (ca. 210 °C) and mechanically tough property. PPO was first prepared from 2,6-dimethylphenol monomer using a copper/amine catalyst system (9). 2,6-Dimethylphenol was also polymerized by horseradish peroxidase (HRP) as catalyst (10).

Recently, enzymatic synthesis of polyaromatics has been extensively developed. Phenol and alkylphenols were oxidatively polymerized by peroxidases in aqueous organic solvents to produce novel polymeric materials consisting of a mixture of phenylene and oxypheneylene units (11-16). The resulting polymers exhibited relatively high thermal stability. This process is expected to be an alternative for production of conventional phenol resins (novolak and resol resins), which involves use of toxic formaldehyde.

Syringaldehyde is abundantly present in plants as its glycosidic derivative. The enzymatic polymerization of syringic acid, an acidic form of syringaldehyde, has been examined by using peroxidase catalyst (Figure 1) (17,18). HRP and soybean peroxidase (SBP) showing the high catalytic activity toward the oxidative polymerization of various phenol derivatives were used as catalyst. The polymerization was started by the addition of hydrogen peroxide (oxidizing agent). In the polymerization using HRP catalyst in a mixture of acetone/acetate buffer (pH 5) (40:60 vol%), powdery materials were formed during the polymerization. The yield of the methanol-insoluble part was 79 %. The isolated polymer was soluble in common polar organic solvents and its molecular weight was determined by size exclusion chromatography (SEC) to be 1.3×10^4. It is to be noted that a polymer was not obtained from non-substituted 4-hydroxybenzoic acid under the similar reaction conditions. SBP catalyst also afforded the polymer with higher molecular weight than HRP.

The effects of the solvent composition have been systematically investigated by using HRP catalyst (Table I). No polymer formation was observed in a mixture of acetone and buffer of pH 3 (40:60 vol %). This is because HRP becomes inactive in this buffer. When the buffer pH was 5 or 7, the polymerization took place in the aqueous acetone of 60 % buffer to give the polymer having more than 1×10^4 molecular weight. The polymer yield and molecular weight decreased in buffer of pH 9. This is probably due to the low catalytic activity of HRP in alkaline region. Interestingly, the yield and molecular weight of the polymer obtained in buffer of pH 11 were larger than

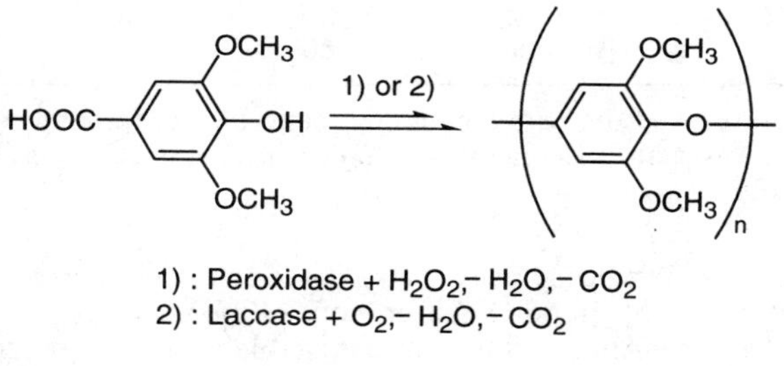
OCH3
HOOC
OH
OCH3
1) or 2)
OCH3
O
OCH3
n
1) : Peroxidase + H2O2,– H2O,– CO2
2) : Laccase + O2,– H2O,– CO2

Figure 1. Enzymatic oxidative polymerization of syringic acid.

those in pH 9. This may be because the oxidation-reduction potential of the monomer is smaller in an alkaline solution.

Table I. Laccase-Catalyzed Polymerization of Syringic Acid[a]

		Solvent		Polymer	
Enzyme	Buffer pH	Organic Solvent	Buffer Content (%)	Yield[b] (%)	Mol. Wt.[c]
HRP	3	Acetone	60	0	
HRP	5	Acetone	20	47	2500
HRP	5	Acetone	40	72	3800
HRP	5	Acetone	60	79	13200
HRP	5	Acetone	80	71	2100
HRP	5	Acetonitrile	51	51	4100
HRP	5	1,4-Dioxane	68	68	10200
HRP	5	Methanol	60	25	2000
HRP	7	Acetone	60	65	15100
HRP	9	Acetone	60	12	9200
HRP	11	Acetone	60	60	12900
SBP	5	Acetone	60	72	14700

[a] Polymerization was carried out using hydrogen peroxide as oxidizing agent at room temperature for 24 h under air. [b] Methanol-insoluble part. [c] Molecular weight of the peak-top determined by SEC.

The polymerization behavior was also dependent on the mixed ratio between acetone and the buffer (pH 5). In 60 % buffer, both of the polymer yield and molecular weight were the largest. Among the water-miscible organic solvents examined, acetone afforded the polymer of the highest molecular weight in the highest yield. The polymerization in the aqueous 1,4-dioxane produced the polymer with molecular weight of more than 1×10^4. Acetonitrile and methanol were not suitable as cosolvent for the polymerization

Polymer structure was analyzed by ^{1}H and ^{13}C NMR and IR spectroscopies as well as MALDI-TOF MS. In the ^{1}H NMR spectrum of the polymer, two large peaks at δ 6.2 and 3.7 ascribable to protons of 2,6-dimethoxy-1,4-oxyphenylene unit were observed. Figure 2 shows expanded ^{13}C NMR spectrum of the polymer, in which there are four main peaks ascribed to aromatic carbons of 2,6-dimethoxy-1,4-oxyphenylene unit. A small characteristic peak (peak A) due to carbonyl carbon of the terminal benzoic acid moiety was observed at δ 169.8. Assignment of other small eight peaks which are due to the terminal units is shown in Figure 2. In the IR spectrum, there is a characteristic peak at 1724 cm^{-1} due to the C=O stretching vibration of the terminal carboxylic acid. These data indicate that the polymer consisted of exclusively 1,4-oxyphenylene unit and had phenolic hydroxyl group at one terminal end and benzoic acid group at the other.

Recently, MALDI-TOF MS has been used for characterization of synthetic polymers, especially determination of the terminal structure. Figure 3 shows the mass spectrum of the polymer obtained in a mixture of acetone/acetate buffer (pH 5) (60:40 vol%). The terminal structures of phenolic and benzoic acid groups were supported since the expected molecular weight is given by $152n + 46$, where n is the degree of polymerization. The mass of the peak-top (1568) agreed with the calculated molecular weight value ($n=10$, mol.wt.=1567) of cations (M+H)$^+$. The peak-to-peak distance was

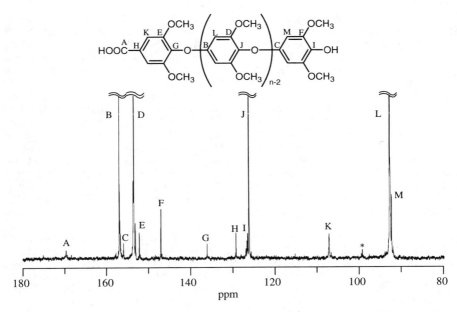

Figure 2. Expanded ^{13}C NMR spectrum of the polymer from syringic acid in deuterium chloroform.

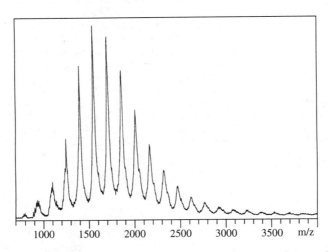

Figure 3. Positive MALDI-TOF spectrum of the polymer using dithranol matrix. The sample was obtained from syringic acid in a mixture of acetone/acetate buffer (pH 5) (60:40 vol%).

152 which is the molecular weight of the polymer repeating unit. These data support the polymer structure as described above.

Thermal properties were evaluated by using differential scanning calorimetry (DSC) and thermogravimetry (TG). In the DSC measurement under nitrogen, glass transition temperature (Tg) was observed at 169 °C and the polymer did not show clear melting point below 300 °C. TG analysis under nitrogen showed that temperatures at 5 and 10 weight % loss were 358 and 376 °C, respectively, and the residual ratio at 1000 °C was 19 %.

Laccases derived from *Pycnoponus coccineus* (PCL) and *Myceliophthore* (MYL) were also active for the polymerization of syringic acid. In the polymerization in a mixture of acetone and acetate buffer (pH 5) (40:60 vol%), both enzymes afforded the polymer in more than 80% yield and the molecular weight was in the range of several thousands. The effect of the buffer pH was examined in the equivolume of acetone and buffers by using PCL catalyst (Figure 4). The polymer yield and molecular weight were the largest in the buffer of pH 5, which is close to the pH region which PCL shows the highest catalytic activity in an aqueous solution. Minimum points of the yield and molecular weight were observed at pH of 10. Interestingly, both values increased as increasing the pH value in the range from 10 to 12.

Figure 5 shows polymerization time versus the monomer conversion and molecular weight of the polymer in the PCL-catalyzed polymerization in a mixture of acetone and the acetate buffer (pH 5) (40:60 vol%). The conversion gradually increased as a function of the polymerization time. The molecular weight rapidly increased in the initial stage of the polymerization, and its increase was small after the monomer conversion was beyond 60%. This polymerization behavior is different from that using a copper/amine as catalyst system; the molecular weight rapidly increased when reaching to the end of the reaction (condensation-type polymerization). This is probably because the polymeric precipitates were formed during the polymerization.

Based on the polymerization behavior and terminal structure of the polymer, the present polymerization mechanism is considered as follows (Figure 6). The monomer is first subjected to the one-electron oxidation by the oxidoreductase to produce three radically resonating intermediates (**Ia-c**). Among the intermediates, **Ia** and **Ib** are reacted to give dimer **II**. The dimer is converted by elimination of carbon dioxide to the phenylene oxide dimer **III**. In the propagation stage, the radical coupling between the radical species of monomer and polymer will take place, leading to the polymer of one-more-elongated unit.

As a possible application, the polymer from syringic acid was converted to a new PPO derivative, poly(2,6-dihydroxy-1,4-oxyphenylene) by demethylation with an excess of boron tribromide in methylene chloride (Figure 7) (*19*). The resulting polymer was soluble in N,N-dimethylformamide, dimethyl sulfoxide, and acetone, however, insoluble in chloroform, methanol, and benzene. NMR and IR analyses showed that the extent of the demethylation was 93 % and the polymer was composed of 2,6-dihydroxy-1,4-oxyphenylene unit. The polymer was stable below 300 °C under nitrogen. The resulting polymer is useful as a starting material for synthesis of functional polymers since it has a reactive phenolic group and a PPO backbone having high thermal stability and chemical resistance.

By utilizing terminal heterogeneous two functional groups of the polymer, a new functional polymeric material containing PPO unit was synthesized. Polycondensation of bisphenol-A, isophthalic acid, and the polymer in the presence of triphenylphosphine / hexachloroethane (coupling agent) afforded PPO - aromatic polyester multiblock copolymers (Figure 8) (*20*). From TG analysis, the multiblock copolymer was found to show relatively high thermal stability.

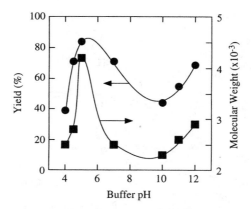

Figure 4. Effects of buffer pH on the yield (●) and molecular weight (■) of the polymer in the PCL-catalyzed polymerization of syringic acid using a mixture of acetone and buffer (50:50 vol%) as solvent.

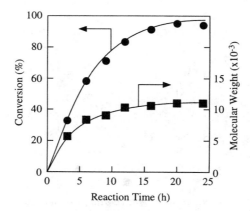

Figure 5. Polymerization time versus the monomer conversion (●) and molecular weight (■) of the polymer in the PCL-catalyzed oxidative polymerization of syringic acid in a mixture of acetone and acetate buffer (pH 5.0) (40:60 vol%).

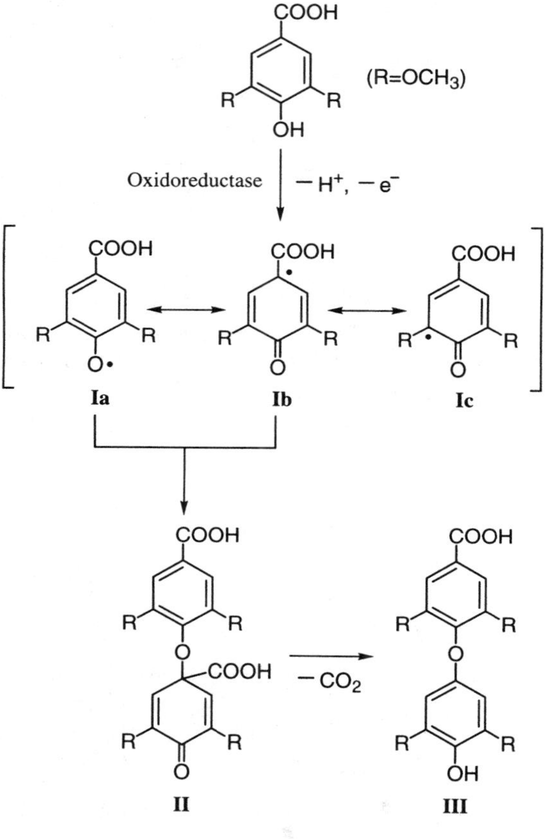

Figure 6. Postulated mechanism of the enzymatic polymerization of syringic acid.

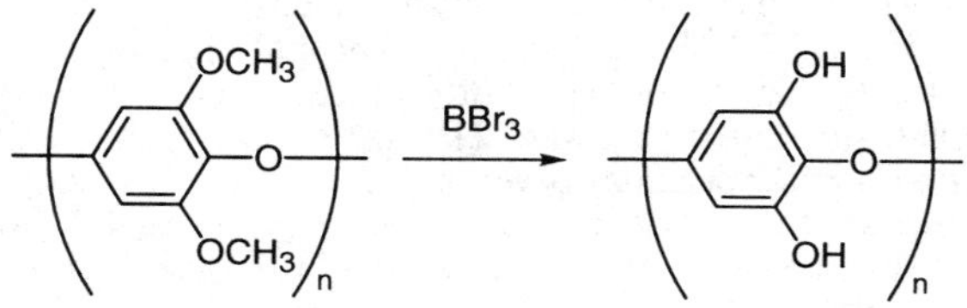

Figure 7. Synthesis of poly(2,6-dihydroxy-1,4-oxyphenylene).

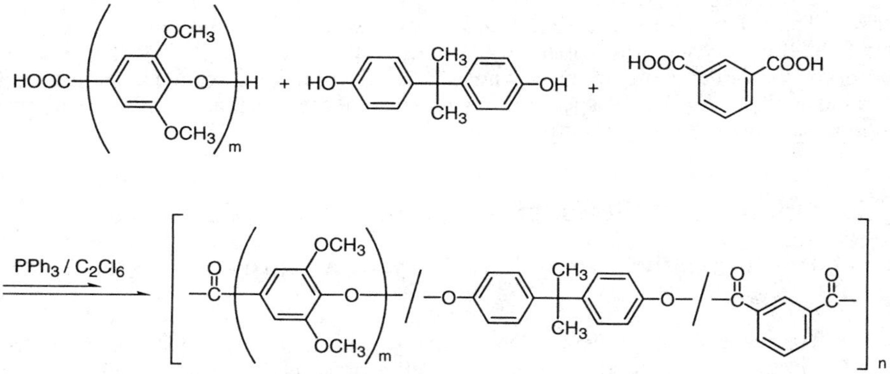

Figure 8. Synthesis of PPO - aromatic polyester multiblock copolymer.

Enzymatic Synthesis of Crosslinkable Polyphenol

Very recently, we have found that a phenol derivative having a methacryloyl group was chemoselectively polymerized by HRP catalyst to give the polymer having the methacryloyl group in the side chain (21). The polymer was subjected to thermal and photochemical hardening. The polymer enzymatically obtained from bisphenol-A was also thermally crosslinked to give the insoluble polymeric materials (22).

Cardanol is obtained by thermal treatment of cashew nut shell liquid. It has the meta substituent of a C15 unsaturated hydrocarbon chain with 1-3 double bonds. Cardanol and its hydrogenated derivative were reported to be polymerized through HRP catalysis, however, the resulting polymers were not characterized in detail (23). We have synthesized a new crosslinkable polymer by the SBP-catalyzed polymerization of cardanol (Figure 9). The ratios of mono-, di-, and tri-unsubstitution in the monomer side chain were 60, 30, and 10 %, respectively. The polymerization in a mixture of acetone/buffer of pH 7 (75:25 %) produced oily products and the yield and molecular weight of the methanol-insoluble part were 21 % and 4800, respectively. The polymer was soluble in common polar organic solvents.

The structure of the polymer was analyzed by using FT-IR spectroscopy. Appearance of new peaks at 1241 and 1190 cm^{-1} ascribed to the asymmetric vibrations of the C-O-C linkage and to the C-OH vibration suggests the formation of the polymer consisting of a mixture of phenylene and oxyphenylene units. Peak intensity at 3010 cm^{-1} due to CH=CH was almost unchanged in comparison with that of the monomer, indicating that the carbon-carbon unsaturated group in the side chain was not reacted during the polymerization.

The curing of poly(cardanol) was examined by using cobalt naphthenate as catalyst. The mixture of the polymer and catalyst spread on a glass slide under air. After 2 h, the polymer was cured to give the crosslinked film. Under the similar conditions, the cardanol monomer was not subjected to the hardening. The monitoring of the curing by using FT-IR spectroscopy shows that the mechanism of the hardening reaction was similar to that of autooxidation of oils (24).

Enzymatic Synthesis of Polyesters from Vinyl Esters

Enzymatic Polymerization of Dicarboxylic Acid Divinyl Esters with Glycols

Transesterifications using lipase catalyst are often very slow owing to the reversible nature of the reactions. An irreversible procedure for the lipase-catalyzed acylation using vinyl esters as acylating agent has been developed, where a leaving group of vinyl alcohol tautomerizes to acetaldehyde. In these cases, the reaction with the vinyl ester proceeds much faster to produce the desired compound in higher yields, in comparison with the alkyl esters. We briefly reported that the enzymatic polymerization of divinyl adipate (1a) with α,ω-glycols proceeded under mild reaction conditions (25).

The enzymatic polymerization of dicarboxylic acid divinyl esters with α,ω-glycols has been systematically investigated (Figure 10). Table II summarizes results of the polymerization of 1a and α,ω-glycols (2a-d). Lipases of different origin were used as catalyst: lipases derived from *Candida antarctica* (lipase CA), from *Candida cylindracea* (lipase CC), *Mucor miehei* (lipase MM), *Pseudomonas cepacia* (lipase PC), *Pseudomonas fluorescens* (lipase PF), and porcine pancreas (PPL). These enzymes showed high catalytic activity for the ring-opening polymerization of lactones (26-28).

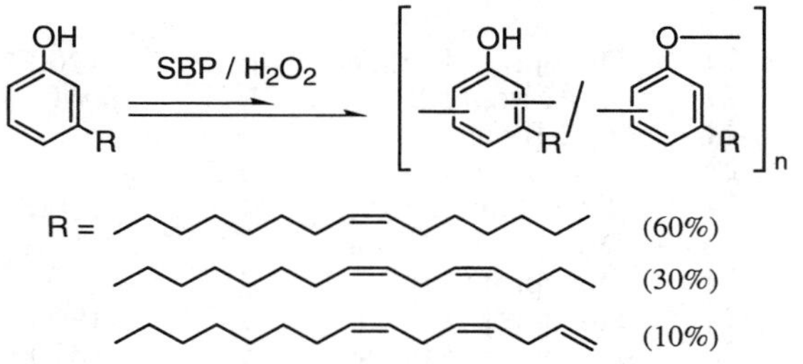
OH
SBP / H2O2
R
OH
O
R
R
n
R =
(60%)
(30%)
(10%)

Figure 9. Enzymatic oxidative polymerization of cardanol.

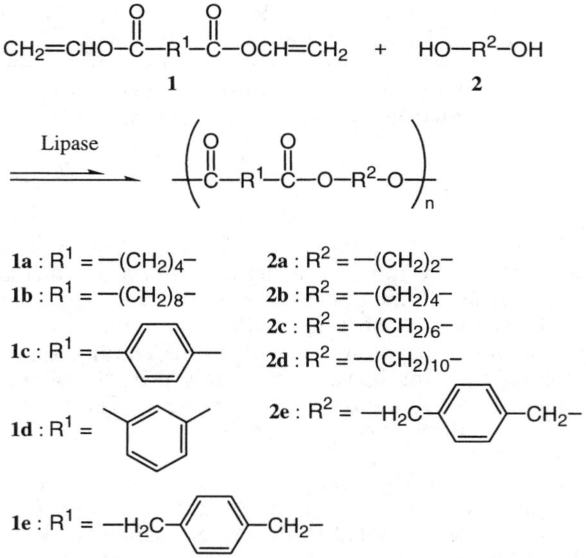
CH2=CHO-C-R1-C-OCH=CH2 + HO-R2-OH
1
2
Lipase
C-R1-C-O-R2-O
n
1a : R1 = -(CH2)4-
1b : R1 = -(CH2)8-
1c : R1 =
1d : R1 =
1e : R1 = -H2C- -CH2-
2a : R2 = -(CH2)2-
2b : R2 = -(CH2)4-
2c : R2 = -(CH2)6-
2d : R2 = -(CH2)10-
2e : R2 = -H2C- -CH2-

Figure 10. Lipase-catalyzed polycondensation of dicarboxylic acid divinyl esters and glycols.

Table II. **Lipase-Catalyzed Polymerization of Divinyl Adipate (1a) and Glycols[a]**

Enzyme	Glycol	Solvent	Yield[b] (%)	Mn[c]	Mw/Mn[c]
Lipase CA	2b	Diisopropyl Ether	41	13000	1.5
Lipase CC	2b	Diisopropyl Ether	0		
Lipase MM	2b	Diisopropyl Ether	40	12000	1.4
Lipase PC	2b	Acetonitrile	0		
Lipase PC	2b	2-Butanone	0		
Lipase PC	2b	Chloroform	0		
Lipase PC	2b	Cyclohexane	34	12000	1.4
Lipase PC	2a	Diisopropyl Ether	0		
Lipase PC	2b	Diisopropyl Ether	43	21000	1.7
Lipase PC	2c	Diisopropyl Ether	35	19000	1.8
Lipase PC	2d	Diisopropyl Ether	55	4700	1.6
Lipase PC	2b	Heptane	24	6400	1.2
Lipase PC	2b	Isooctane	52	15000	1.4
Lipase PC	2b	Tetrahydrofuran	0		
Lipase PC	2b	Toluene	0		
Lipase PF	2b	Diisopropyl Ether	40	17000	1.6
PPL	2b	Diisopropyl Ether	0		

[a] Polymerization of divinyl adipate and glycol (each 2 mmol) using 50 mg of lipase as catalyst at 45 °C for 48 h. [b] Methanol-insoluble part. [c] Determined by SEC.

The polymerization was carried out in organic solvents at 45 °C. The polymer was isolated by reprecipitation using chloroform and methanol as good and poor solvents, respectively. In the polymerization of 1a with 1,4-butanediol (2b) in diisopropyl ether, lipases CA, MM, PC, and PF were highly active; the polymer with molecular weight of more than 1×10^4 was obtained. The highest molecular weight (2.1×10^4) was achieved by using lipase PC as catalyst. Both monomers were recovered unchanged in the absence of lipase (control experiment). These data indicate that the polymerization proceeded through enzyme catalysis.

The effects of the organic solvent was examined in the lipase PC-catalyzed polymerization of 1a and 2b. Hydrophobic solvents, cyclohexane, diisopropyl ether, heptane, and isooctane, were suitable for the production of the polyester. From 1,6-hexanediol (2c) and 1,10-decanediol (2d), the corresponding polyesters were formed in moderate yields, whereas the polymer formation was not observed in the polymerization of 1a and ethylene glycol (2a).

In the lipase PC-catalyzed polymerization of 1a and 2b, 90 % of the monomers were consumed for 3 h to give the oligomer. The molecular weight gradually increased till 36 h, and afterwards the molecular weight was almost constant. These data suggest that the mechanism of the enzymatic polymerization of divinyl ester and glycol is similar to that of conventional condensation-type polymerization.

Lipase also catalyzed the polycondensation of divinyl sebacate (1b) and glycols (Table III). The polymerization behaviors were similar to those using 1a. The yield of the polymer from 1b was higher than that from 1a under the similar reaction conditions, on the other hand, the molecular weight from **1b** was smaller. Interestingly, lipase CA produced the polymer from **2a**; no polymer formation was observed in the lipase PC-catalyzed polymerization of divinyl esters and **2a**.

Table III. Lipase-Catalyzed Polymerization of Divinyl Sebacate (1b) and Glycols[a]

Enzyme	Glycol	Solvent	Yield[b]	Mn^c (%)	Mw/Mn^c
Lipase CA	2a	Diisopropyl Ether	52	2300	1.3
Lipase CA	2b	Diisopropyl Ether	60	14000	1.7
Lipase CA	2c	Diisopropyl Ether	36	4000	2.0
Lipase CA	2d	Diisopropyl Ether	63	3600	2.1
Lipase MM	2b	Diisopropyl Ether	63	9700	1.6
Lipase PC	2b	Cyclohexane	54	7600	1.4
Lipase PC	2a	Diisopropyl Ether	0		
Lipase PC	2b	Diisopropyl Ether	60	13000	1.8
Lipase PC	2c	Diisopropyl Ether	46	11000	2.4
Lipase PC	2d	Diisopropyl Ether	55	7300	1.5
Lipase PC	2b	Isooctane	69	6500	1.3
Lipase PF	2b	Diisopropyl Ether	60	10000	1.4

[a] Polymerization of divinyl sebacate and glycol (each 2 mmol) using 50 mg of lipase as catalyst at 45 °C for 48 h. [b] Methanol-insoluble part. [c] Determined by SEC.

Polyesters containing aromatic moiety in the backbone were enzymatically synthesized. Lipase induced the polymerization of **1b** with p-xylylene glycol (**2e**). In the polymerization using lipase PC catalyst in heptane at 75 °C for 48 h, the corresponding polyester with molecular weight of 4×10^3 was formed in 74 % yield. Enzymatic polymerization of divinyl esters of aromatic diacids (**1c-1e**) and **2c** was examined. Lipase CA showed the high catalytic activity to give the polymers with molecular weight of several thousands.

Enzymatic Polymerization of Vinyl Hydroxyalkanoate

Vinyl ester of 12-hydroxydodecanoic acid was used as a new monomer of the enzymatic polymerization (Table IV). The monomer was almost consumed for 3 h. Lipase MM afforded the polymer in the highest yield (74 %) and the molecular weight of the polymer obtained by using lipase CA in isooctane was the largest. The terminal groups of the linear polymer were observed in the NMR spectra. The formation of the cyclic oligomers was not examined. 12-Hydroxystearic acid vinyl ester was also polymerized by lipase PC catalyst.

Terminal Functionalization of Polyesters by Vinyl Esters

Structural control of polymer terminal has been extensively studied since terminal-functionalized polymers, typically macromonomers and telechelics, are often used as prepolymers for synthesis of functional polymers. Various methodologies for synthesis of these polymers have been developed, however, most of them required

Table IV. Lipase-Catalyzed Polymerization of Vinyl 12-Hydroxydodecanoate[a]

Enzyme	Solvent	Yield[b]	Mn[c] (%)	Mw/Mn[c]
Lipase CA	Cyclohexane	57	4000	1.5
Lipase CA	Isooctane	66	7100	1.5
Lipase CA	Toluene	59	3300	1.5
Lipase MM	Isooctane	74	4500	1.6
Lipase PC	Isooctane	56	4600	1.7

[a] Polymerization of vinyl 12-hydroxydodecanoate (1 mmol) using 100 mg of lipase as catalyst at 60 °C for 48 h. [b] Methanol-insoluble part. [c] Determined by SEC.

elaborate and time-consuming procedures. Recently, we have achieved a single-step, facile production of methacryl-type polyester macromonomers by using vinyl methacrylate as terminator (29,30).

We have systematically investigated enzymatic ring-opening polymerization of lactones and found that the polymerization behavior strongly depended on the ring-size of the lactones (27,28); macrolides showing much lower reactivity and polymerizability than ε-caprolactone by traditional catalysts were enzymatically polymerized much faster than ε-CL.

The polymerization of 12-dodecanolide (13-membered lactone, DDL) was carried out by using lipase PF as catalyst in the presence of vinyl methacrylate. In using vinyl methacrylate (12.5 or 15 mol% based on DDL), the quantitative introduction of methacryloyl group at the polymer terminal was achieved to give the methacryl-type macromonomer. This methodology was expanded to the synthesis of the telechelic polyester having a carboxylic acid group at both ends by using divinyl sebacate as terminator.

Conclusion

Natural phenolic compounds were subjected to enzymatic oxidative polymerizations. A new type of oxidative polymerization involving elimination of hydrogen as well as carbon oxide from the monomer was developed in using syringic acid as monomer. The resulting polymer was converted to a new reactive polymer, poly(2,6-dihydroxy-1,4-oxyphenylene) and PPO -aromatic polyester multiblock copolymer showing high thermal stability. A novel crosslinkable polyphenol was obtained by the SBP-catalyzed polymerization of cardanol. The polycondensation of dicarboxylic acid divinyl esters and glycols in the presence of lipase catalyst proceeded under mild reaction conditions to give aliphatic and aromatic polyesters. The polymerization behavior depended on the lipase origin, solvent, and monomer structure. Vinyl hydroxyalkanoates were also enzymatically polymerized.

Acknowledgements

This work was supported by a Grant-in-Aid for Specially Promoted Research (No. 08102002) from the Ministry of Education, Science, and Culture, Japan, and Culture, Japan and from NEDO for the project on Technology for Novel High-Functional Materials in Industrial Science and Technology Frontier Program, AIST.

Literatures

1. Jones, J. B. *Tetrahedron* **1986**, *42*, 3351.
2. Klibanov, A. M. *Acc. Chem. Res.* **1990**, *23*, 114.
3. Santaniello, E.; Ferraboschi, P.; Grisenti, P.; Manzocchi, A. *Chem. Rev.* **1992**, *92*, 1071.
4. Kobayashi, S.; Shoda, S.; Uyama, H. In *Catalysis in Precision Polymerization*; Kobayashi, S., Ed.; John Wiley & Sons: Chichester, Chap. 8 (1997).
5. Gross, R. A.; Kaplan, D. L.; Swift, G., Ed.; *ACS Symp. Ser.* **1998**, *684*.
6. Kobayashi, S.; Shoda, S.; Uyama, H. *Adv. Polym. Sci.* **1995**, *121*, 1.
7. Ritter, H. *Trends Polym. Sci.* **1993**, *1*, 171.
8. Ritter, H. In *Desk Reference of Functional Polymers, Syntheses and Applications*, Arshady, R., Ed.; American Chemical Society: Washington, pp 103-113 (1997).
9. Hay, A. S. *J. Polym. Sci., Polym. Chem. Ed.* **1998**, *36*, 505.
10. Ikeda, R.; Sugihara, J.; Uyama, H.; Kobayashi, S. *Macromolecules* **1996**, *29*, 8702.
11. Dordick, J. S.; Marletta, M. A.; Klibanov, A. M. *Biotechnol. Bioeng.* **1987**, *30*, 31.
12. Akkara, J. A.; Senecal, K. J.; Kaplan, D. L. *J. Polym. Sci., Polym. Chem. Ed.* **1991**, *29*, 1561.
13. Uyama, H.; Kurioka, H.; Kaneko, I.; Kobayashi, S. *Chem. Lett.* **1994**, 423.
14. Uyama, H.; Kurioka, H.; Sugihara, J.; Kobayashi, S. *Bull. Chem. Soc. Jpn.*, **1996**, *69*, 189.
15. Kobayashi, S.; Kurioka, H.; Uyama, H. *Macromol. Rapid Commun.* **1996**, *17*, 503.
16. Uyama, H.; Kurioka, H.; Sugihara, J.; Komatsu, I.; Kobayashi, S. *J. Polym. Sci., Polym. Chem. Ed.* **1997**, *35*, 1453.
17. Ikeda, R.; Uyama, H.; Kobayashi, S. *Macromolecules* **1996**, *29*, 3053.
18. Ikeda, R.; Sugihara, J.; Uyama, H.; Kobayashi, S. *Polym. International* **1998**, *47*, 295.
19. Ikeda, R.; Uyama, H.; Kobayashi, S. *Polym. Bull.* **1997**, *38*, 273.
20. Ikeda, R.; Sugihara, J.; Uyama, H.; Kobayashi, S. *Polym. Bull.* **1998**, *40*, 367.
21. Uyama, H.; Lohavisavapanich, C.; Ikeda, R.; Kobayashi, S. *Macromolecules* **1998**, *31*, 554.
22. Kobayashi, S.; Uyama, H.; Ushiwata, T.; Uchiyama, T.; Sugihara, J.; Kurioka, H. *Macromol. Chem. Phys.* **1998**, *199*, 777.
23. Alva, K. S.; Nayak, P. L.; Kumar, J.; Tripathy, S. K. *J. Macromol. Sci.-Pure Appl. Chem.* **1997**, *A34*, 665.
24. Wicks, Z. W. Jr. In *Encyclopedia of Polymer Science and Engineering*, 2nd ed.; John Wiley & Sons: New York, vol. 5, pp 203-214 (1986).
25. Uyama, H.; Kobayashi, S. *Chem. Lett.* **1994**, 1687.
26. Uyama, H.; Kobayashi, S. *Chem. Lett.* **1993**, 1149.
27. Uyama, H.; Takeya, K.; Kobayashi, S. *Bull. Chem. Soc. Jpn.* **1995**, *68*, 56.
28. Uyama, H.; Takeya, K.; Hoshi, N.; Kobayashi, S. *Macromolecules* **1995**, *28*, 7046.
29. Uyama, H.; Kikuchi, H.; Kobayashi, S. *Chem. Lett.* **1995**, 1047.
30. Uyama, H.; Kikuchi, H.; Kobayashi, S. *Bull. Chem. Soc. Jpn.* **1997**, *70*, 1691.

Chemical Synthesis

Chapter 10

High-Molecular-Weight [L]-Polylactides Containing Pendant Functional Groups

Xianhai Chen[1,4], Youqing Shen[2], and Richard A. Gross[3]

[1]Polymer Group, Aspen Systems, Inc., 184 Cedar Hill Street, Marlborough, MA 01752
[2]Department of Chemical Engineering, McMaster University, 1280 Main Street West, Hamilton, Ontario L8S 4L7, Canada
[3]Polymer Research Insitiute, Polytechnic University, 6 Metrotech Center, Brooklyn, NY 11201

This paper reviews our previous work on the functional carbonate monomers design, synthesis and their applications for introducing functional groups into polylactides. A number of cyclic carbonate monomers with functional groups have been synthesized and used for the copolymerization with [L]-lactide ([L]-LA). High molecular weight copolymers have been prepared by using Sn(Oct)$_2$ as catalyst. Structural investigations by NMR revealed that [L]-LA/carbonate copolymers had short average carbonate repeat unit segment lengths. This work has resulted in a new family of [L]-PLA-based copolymers that contain C=C, epoxy, ketal, and vicinal diol functional groups. These new copolymers will be of great value for the development of bioresorbable medical materials.

Functional Carbonate Monomers and Homopolymers

We have prepared a series of 6-membered cyclic carbonate monomers, i.e., 2, 2-[2-pentene-1, 5-diyl] trimethylene carbonate (^{c}HTC), 1, 2-o-isopropylidene-[D]-xylofuranose-3, 5-cyclic carbonate (IPXTC), 9, 9-dimethyl-2, 4, 8, 10-tetraoxaspiro [5, 5] undecane-3-one (DMTOS), and 2, 4, 8, 10-tetraoxaspiro [5, 5] undecane-3-one (DOXTC) (Scheme 1). The monomers ^{c}HTC, IPXTC, DMTOS and DOXTC were synthesized in 81%, 41%, 70% and 80% yields respectively by a one-pot reaction in THF at 0°C starting from the corresponding diols (Scheme 2). It should be mentioned that the diol for DMTOS was synthesized from pentaerythritol by i) total ketal protection of OH group by reacting with (CH$_3$)$_2$C(OCH$_3$)$_2$ in the presence of p-toluenesulfonic acid in > 95% yield, and ii) partial deprotection catalyzed by

---

[4]Current address: 3 River Place, A–1106, Lowell, MA 01852.

CF_3COOH in about 20 % yield. 1, 2-O-Isopropylidene-[D]-xylofuranose can also be synthesized by one step reaction from xylose in high yield (~ 80%) under appropriate [H^+] concentration without deprotection procedure.

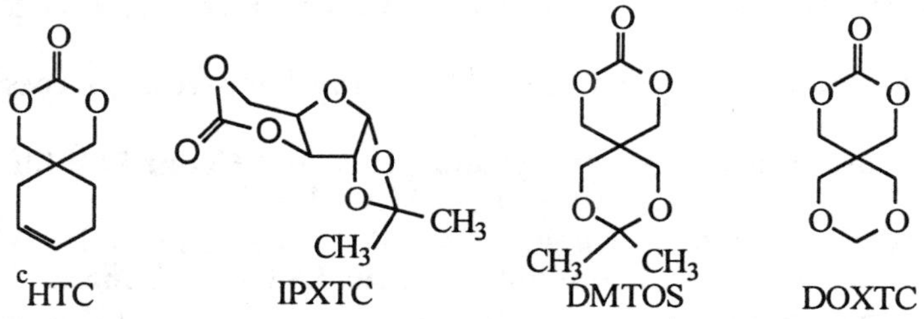

Scheme 1. Structure of functional cyclic carbonate monomers.

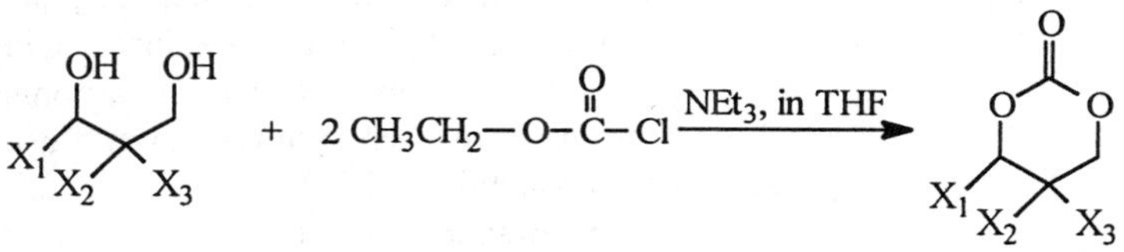

Scheme 2. Synthesis of cyclic carbonate monomers. See Scheme 1 for corresponding X_1, X_2 and X_3 structures.

The structure and purity of these monomers were confirmed by FTIR, ^{1}H-NMR, ^{13}C-NMR and melting point analysis. The melting points of ^{c}HTC, IPXTC, DMTOS and DOXTC were 79.5~80.5°C, 143~144°C, 154~156°C and 126.5~127.5°C respectively. Compared to the yields of ^{c}HTC (1), DMTOS or 2,4,8,10-tetraoxaspiro [5,5] undecane-3-one (DOXTC) (2), the yield of IPXTC was relatively lower. This may be explained by the followings: i) the equilibrium of 5- and 6-membered ring monosaccharides where the latter does not form the desired cyclic carbonate monomer and ii) while ^{c}HTC, DMTOS and DOXTC are formed by the reaction of two primary hydroxy groups, IPXTC synthesis involves the reaction of a relatively lower reactivity secondary 1,2-O-isopropylidene-D-xylofuranose hydroxyl functionality (3).

Polymerizations of ^{c}HTC using organometallic catalysts including aluminoxanes, $ZnEt_2$ and in-situ $ZnEt_2$-H_2O resulted in the corresponding functional homopolymer in high molecular weight (M_n from 80,000 to 260,000). The pendant vinyl groups of P(^{c}HTC) offer a wide range of opportunities for further modification and/or functionalization. For example, P(^{c}HTC) vinyl groups were converted to epoxides with little molecular weight decrease (1). The C=C and epoxy would provide useful handles for further functionalization or crosslinking. The polymerization of IPXTC gave a homopolymer with M_n up to 13,200 by using yttrium isopropoxide as catalyst.

Three different repeat unit linkages (i.e., head-head, head-tail, and tail-tail) have been observed by ^{13}C-NMR spectroscopy. The prepared PIPXTC has a melting transition temperature at 228°C and a glass transition temperature at 128°C. Subsequent deprotection of ketal in PIPXTC was successfully carried out by using CF_3COOH/H_2O (4). The vicinal diols may impart unique properties to the prepared materials that facilitate a variety of potential applications. For example, linear and crosslinked hydroxyl containing polyesters can permit the formation of strong hydrogen bonding and non-bonding interactions with organic and inorganic species (5). Also, hydroxyl containing aliphatic polycarbonates have been reported to be bioerodible and of importance in medical and pharmaceutical applications (6). DMTOS or DOXTC can also be polymerized by using various organometallic catalysts such as aluminoxanes, *in-situ* AlR_3-H_2O or ZnR_2-H_2O, Tin(IV) catalysts, and $Sn(Oct)_2$. However, these two homopolymers were not soluble in common organic solvents.

Preparation of [L]-Polylactides with Pendant Functional Groups

Table I. Copolymerization of [L]-Lactide with Cyclic Carbonate Monomers[a]

Monomer	f_{LA}/f_C[b]	t(h)	yield(%)	M_n[c]	M_w/M_n	F_{LA}/F_C[d]
LA	100/0	6	99	86,000	2.1	100/0
LA-[c]HTC	82/18	6	82	99,700	2.0	93/ 7
	67/33	6	68	82,200	2.1	86/14
	49/51	6	47	49,800	2.2	72/28
	30/70	6	43	35,300	1.9	48/52
	0/100	6	91	55,700	2.0	0/100
LA-IPXTC	83/17	6	82	78,400	1.9	93/ 7
	71/29	6	71	64,500	1.7	91/ 9
	66/34	6	70	44,500	2.0	85/15
	52/48	6	54	27,900	1.6	79/21
	36/64	24	48	13,900	1.7	61/39
LA-DMTOS	93/7	22	95	60,000	3.2	96/ 4
	85/15	22	84	48,800	3.8	92/ 8
	77/23	22	76	28,500	3.6	88/12
	66/34	22	67	21,200	4.0	78/22
	53/47	22	54	17,100	2.8	n.d.

[a]120°C, $Sn(Oct)_2$ as catalyst, M/C = 200. [b]Monomer feed ratio in mol/mol. [c]GPC results, polystyrene as standard. [d]Copolymer composition in mol/mol determined by ^{1}H-NMR.

It is well known that $Sn(Oct)_2$ is a preferred catalyst for the formation of high molecular weight polylactide via ring opening polymerization. We have screened a variety of organometallic catalysts including (methyl and isobutyl) aluminoxanes, *in-situ* AlR$_3$-H$_2$O, *in-situ* ZnEt$_2$-H$_2$O, Sn(IV)-compounds, and Sn(Oct)$_2$ for the copolymerization of [L]-lactide with the prepared functional carbonate monomers except DOXTC. As expected, Sn(Oct)$_2$ gave higher molecular weight copolymers all copolymerization systems especially at high [L]-LA monomer feed ratios.

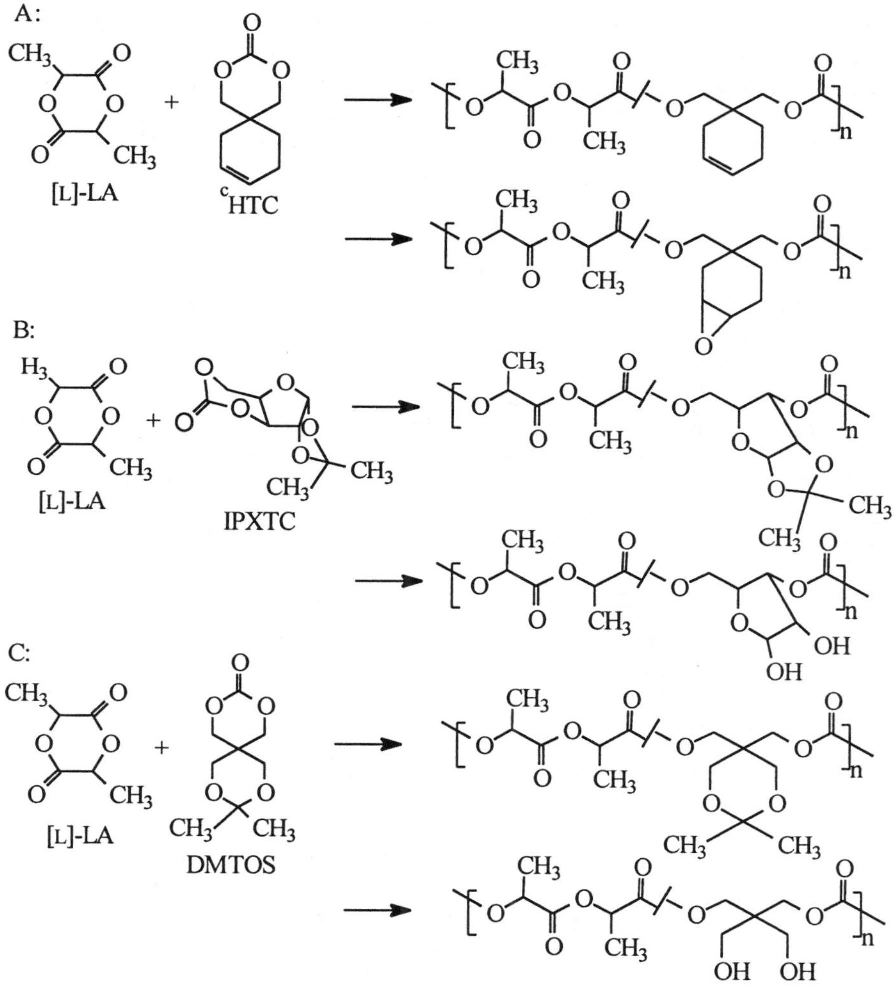

Scheme 3. Preparation of [L]-polylactides with various pendant functional groups.

The monomer reactivity ratio of [L]-lactide is much higher than those of cyclic carbonate monomers. For example, the reactivity ratios were 4.15/0.255 for [L]-LA/IPXTC system and 8.8/0.52 for [L]-LA/^{c}HTC system by using Sn(Oct)$_2$ as catalyst at 120°C. Table 1 summarizes the [L]-LA-cyclic carbonates copolymerization results with Sn(Oct)$_2$ as catalyst at variable monomer feed ratios. The copolymer yields and molecular weights were inversely proportional to the cyclic carbonate monomer feed ratios. However, since our objective was to prepare [L]-PLA based polymers containing controlled low level of functional groups, high molecular weight copolymers could be readily prepared with desired structure.

The prepared copolymers can be easily modified into variable functionality, as illustrated in Scheme 3. The C=C group in [L]-LA-co-^{c}HTC can be converted into epoxy by using 3-(chloroperoxy)benzoic acid with only little molecular weight decrease (7). It is anticipated that epoxy group can be further functionalized into diols, -CH(OH)CH(NHR)-, and –CH(OH)CH(OR)- under mild reaction conditions. Furthermore, the epoxy containing [L]-PLA can be used as macro-initiator to prepare graft copolymers. The ketal structures of [L]-LA-co-IPXTC or [L]-LA-co-DMTOS copolymers can be removed by using CF$_3$COOH with slight molecular weight decrease. The vicinal-diol groups in PLAs provide a variety of opportunities for their use as functional biomedical materials (3,8).

Structure and Properties of Functional [L]-Polylactides

Even though the monomer reactivity ratios of cyclic carbonate monomers were much lower than that of [L]-lactide, ^{1}H-NMR and ^{13}C-NMR results showed that the prepared copolymers had randomness coefficient R values in the range of 0.65 to 0.87 where R is 0 for a diblock copolymer and 1 for a statistically random copolymer (9). Such copolymer structures may be explained by i) cyclic carbonate polymerization occurs by intrachain insertion reaction or intermolecular exchange reactions, and ii) chain transfer reactions occur at high conversion. For [L]-LA-co-^{c}HTC and [L]-LA-co-IPXTC copolymers, the average lengths of ^{c}HTC and IPXTC repeat units were 1.0~1.7 and 1.04~1.13 respectively (3,7). The random distribution of functional groups on [L]-PLA backbone provides desired structure for their applications in biomedical areas.

The thermal properties of the prepared copolymers have been extensively characterized by differential scanning calorimetry (DSC). The presence of carbonate units in copolymer disrupted ordering of the [L]-PLA crystalline phase. Decreased melting transition temperature (T$_m$) and enthalpy (ΔH$_f$) were found for the copolymers relative to pure [L]-PLA. For the copolymers containing high percentage carbonate units, totally amorphous phases were observed. Thus, one outcome of introducing cyclic carbonate units into [L]-PLA is a method to regulate the melting temperature and crystallinity of [L]-PLA based materials and to improve their processability. Depending on the structure of carbonate, the glass transition temperature (T$_g$) can be varied in a wide range. For example, the T$_g$'s of [L]-LA-co-^{c}HTC and [L]-LA-co-IPXTC were in the range of 33~60°C (7) and 60~128°C (3,4,10) respectively. The deprotected [L]-LA-co-IPXTC copolymers showed very similar thermal property compared to their corresponding original copolymers.

Potential Applications

This work has explored new routes to prepare high molecular weight functional PLAs. These pendant functional groups will facilitate covalent prodrug attachment and a variety of other potential applications. For example, the presence of hydroxyl groups in vicinal positions of the deprotected repeating units imparts unique properties to the prepared materials that allows one to establish strong hydrogen bonding and non-bonding interactions with organic and inorganic species (5). The epoxy structure in the polymer can be readily converted into other functionality, such as di-hydroxyl, alkoxyl-hydroxyl, amine-hydroxyl, etc. This will expand the availability of novel [L]-PLA with pendant functional groups. Therefore, these high molecular weight functional [L]-PLAs will have many valuable applications in biomedical field. The examples include i) biocompatible matrix materials which provide available sites for modifying their surfaces with biological active moieties for tissue engineering; ii) implantable biomaterials; and iii) drug delivery systems. In addition, the hydroxyl- or epoxy- containing PLAs will provide handles to form macro-initiator for grafting copolymerization. One application of these graft copolymers will be for the design of improved interfacial agents for biodegradable blends.

References

1. Chen, X.; McCarthy, S. P.; Gross, R. A. *Macromolecules* **1997**, *30*, 3470.
2. Chen, X.; McCarthy, S. P.; Gross, R. A. *J. Appl. Polym. Sci.* **1998**, *67*, 547.
3. Chen, X.; Gross, R. A. *Macromolecules* **1999**, *32*, 308.
4. Shen, Y.; Chen, X.; Gross, R. A. *Macromolecules* **1999**, *32*, in press.
5. Chiellini, E.; Bemporad, L; Solaro, R. *J. Bioactive and Biocompatible Polymers* **1994**, *9*, 152.
6. Acemoglu, M.; Bantle, S.; Mindt, T.; Nimmerfall, F. *Macromolecules* **1995**, *28*, 3030.
7. Chen, X.; McCarthy, S. P.; Gross, R. A. *Macromolecules* **1998**, *31*, 662.
8. Shen, Y.; Chen, X.; Gross, R. A. *Manuscript in preparation*.
9. Kasperczyk, J.; Bero, M. *Makromol. Chem.* **1993**, *194*, 913.
10. Shen, Y.; Chen, X.; Gross, R. A. submitted to *Macromolecules*.

Chapter 11

Macrocyclic Polymerizations of Lactones: A New Approach to Molecular Engineering

Hans R. Kricheldorf, Soo-Ran Lee, Sven Eggerstedt, Dennis Langanke, and Karsten Hauser

Institut für Technische und Makromoleculare Chemie, Bendesstrasse 45, D–20146 Hamburg, Germany

Cyclic tin alkoxides like 2,2-dibutyl-2-stanna-1,3-dioxepane were used as initiator for the ring-opening polymerization of various lactones. In this way series of macrocyclic polyesters were obtained without any competitive formation of linear polymers. Under optimized conditions these macrocyclic polymerizations obey the pattern of "living polymerizations". Therefore, the number average molecular weight (M_n) can be controlled by the monomer/initiator ratio, and macrocyclic block copolyesters can be prepared by sequential copolymerization of two lactones. Macrocyclic oligomers and polymers having various types of backbones were prepared by "ring-closing polycondensation" of $Bu_2Sn(OMe)_2$ with various oligomeric or polymeric diols, such as poly(ethylene glycol)s. These tin-containing macrocycles were, inturn, used as "macroinitiators" for polymerizations of lactones. Ring-opening by acylation with functional acid chlorides or anhydrides yielded telechelic polylactones. Regardless of size and structure all tin-containing macrocycles can be used as bifunctional monomers for polycondensations, for instance, with dicarboxyl acid dichlorides. These "ring-opening polycondensations" are useful for syntheses of biodegradable multiblock copolymers, particularly for thermoplastic elastomers.

Polyesters prepared by ring-opening polymerization of lactones and aliphatic cyclocarbonates play a great role among the biodegradable materials, particularly polyesters based on L- or D,L-lactic acid. Such polyesters comprise homo-polyesters, random copolyesters, triblockcopolyesters, multiblock copolyesters and even star-shaped polyesters. The ring-opening polymerization of the monomers may proceed by three mechanisms: anionic, cationic and coordination-insertion mechanisms. From the preparative point of view the coordination-insertion mechanisms are particularly important, because they may yield high molecular weights and high yields of all lactones or cyclocarbonates, and because they proceed without racemization of L-lactide or L- and D-β-butyrolactone. The most widely used initiators are tin compounds, such as $SnCl_2 \bullet$ n H_2O, Sn (2-ethyl hexanoate)$_2$, Bu_2SnO, $Bu_2Sn(OMe)_2$ or Bu_3SnOMe. Our ongoing research and the present review deal with cyclic tin alkoxides, such as 1 - 7, part of which (1 - 3) has been described in the literature (1 - 5), but never used as initiator before. When cyclic tin alkoxides are used as initiators (6 - 16) they yield macrocyclic oligoesters and polyesters without any linear byproducts. Hence, these "macrocyclic polymerizations" offer an easy access to macrocycles with a broad variation of structure and molecular weight. Furthermore, the high reactivity of the Sn-O bond also allows a broad variation of post-reactions with elimination of the R_2Sn group.

Macrocyclic Polymerizations

Five- and sixmembered cyclic tin alkoxides have the tendency to dimerize (eq. 1) to alleviate the ring strain (2 - 5). The dimers (e.g. 1b, 2b, 3b or 4b) may possess high melting points (> 150°C) (2 - 5), and are poorly soluble in lactones. When a rapid homogenization of initiator and monomer cannot be achieved, the chain growth is faster than the entire initiation process, and average degrees of polymerization (DPs) are higher than the monomer/initiator (M/I) ratios (6 - 8). Apparently for thermodynamic reasons the sevenmembered ring of 3a is a mainly monomeric liquid which is stable on storage at ≤ 20°C and which is miscible with most monomers. Therefore, this heterocycle, 2,2-dibutyl-2-stanna-1,3-dioxepane. (DSDOP), was used as initiator for most preparative purposes (10 - 15) (eq.2). When DSDOP-initiated polymerizations of ε-caprolactone (ε-CL) or δ-valerolactone (δ-VL) are conducted at temperatures ≤ 80°C with an optimized reaction time, the DPs parallel the M/I ratios, and number average molecular weights (M_ns) up to 120 000 were obtained (10). The polydispersities (1.3 -1.6 depending on the temperature) are higher than those of a classical "living polymerization". Reasons for the broader polydispersities are a rather slow initiation ,on the one hand, and transesterification including "back biting degradation", on the other. These side reactions are difficult to avoid completely, because tin compounds belong to the best transesterification catalysts. The polydispersities not only depend on the reaction conditions, but also on the lactone. Polydispersities as low as 1.25 were found for DSDOP-initiated polymerizations of β-

D,L-butyrolactone (*17*) and further studies will probably show that even lower polydistersities can be obtained.

A particular advantage of tin alkoxide-initiated polymerizations is the rather high chemical and thermal stability of the Sn-O bonds. They are sensitive to hydrolysis or alcoholysis, but they can be stored in closed flask for months and years. Therefore, the tin containing macrocyclic polyesters can initiate a second lactone, so that macrocyclic block copolyesters (9) can be prepared (eq. 3)[1,18]. In this case time and temperature need again to be optimized to minimize transesterification. However, when mixtures of two lactones or lactones and trimethylene carbonate are copolymerized by initiation with DSDOP also random copolymers can be obtained[1,8]. When the 2-stanna-thiolane (4) is used as initiator at moderate temperature ($\leq$ 120°C), the insertion of lactones exclusively occurs at the Sn-O bond and macrocyclic polyesters of the structure 10 will result (*7*).

The formation of Sn-S bond is not only kinetically favored, it is also thermodynamically more stable than a Sn-O bond, and thus, its reactivity towards lactones is significantly lower. Furthermore, it should be mentioned that initiation with the unsaturated heterocycles 5 and 6 yields polylactones having a reactive cis-double bond somewhere in the middle of the chain (*19*). Initiation with the spirocompound produces polylactones in the form of "nanopretzels" (11).

Cyclization of polymeric diols

The rapid equilibration of the initiators 1a - 4a/1b - 4b and the NMR spectra of Bu$_3$SnOMe (*18*) demonstrate that the exchange of RO groups between different Sn-OR compounds may be fast above room temperature, and thus, the reaction products are thermodynamically controlled. This characteristic property of Sn-OR compounds has several interesting consequences. First, when a tin-containing macrocycles are large enough so that no ring strain exists anymore, they are thermodynamically stable. Therefore, the macrocyclic oligo and polyesters 8, 9 and 10 will not oligomerize in such a sense that larger cycles containing two or more Sn atoms are formed. Second, high molecular weight polymers containing Sn-O bonds cannot exist above room temperature. Third, "polycondensations" of Bu$_2$SnX$_2$ compounds with long diols will never yield polymers but the smallest thermodynamically stable macrocycles. These conclusions represent nothing extraordinary from the thermodynamical point of view, because for reasons of entropy almost all polymers will break down into cycles when the energy of activation is sufficiently lowered.

The experimental evidence for these conclusions was obtained from "polycondensations" of Bu$_2$Sn(OMe)$_2$ with various monodisperse and polydisperse poly(ethylene-glycol)s (eq. 4)(*15*). The marcocyclic structure (11) of the reaction products was proven by viscosity and SEC measurements, by mass spectrometry and

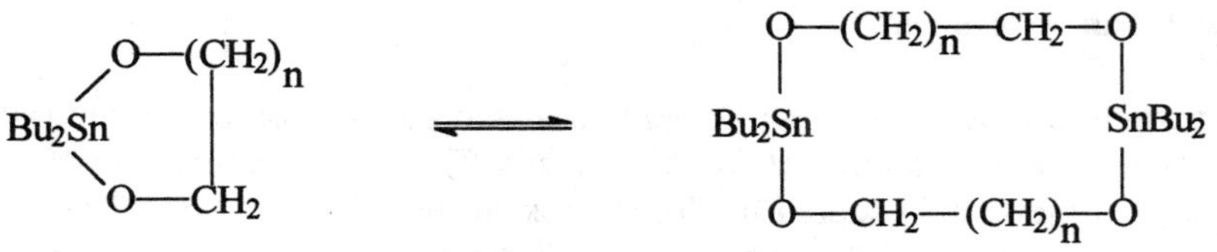

$$\begin{array}{ccc} & & (1) \end{array}$$

1a : n = 1   **1b** : n = 1

2a : n = 2   **2b** : n = 2

3a : n = 3   **3b** : n = 2

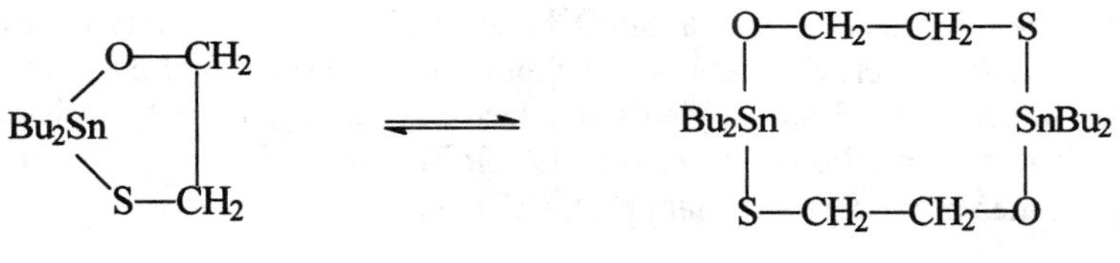

(2)

4a   **4b**

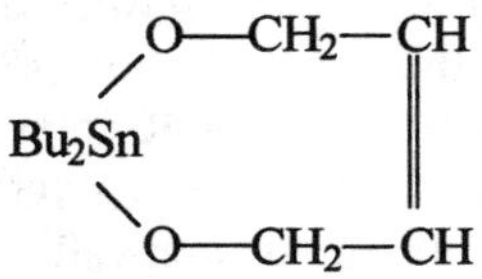

5 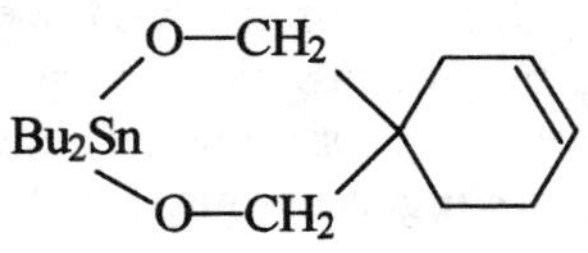   **6**

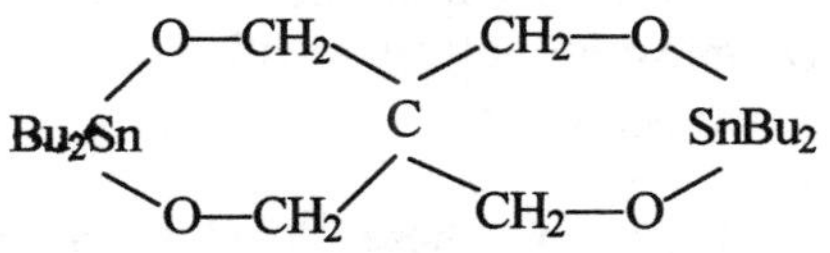

7

(3)

(4)

8

9

10

^{119}S NMR spectroscopy. Analogous results were obtained by polycondensation of Bu$_2$Sn(OMe)$_2$ with commercial poly(tetrahydrofuran) diols (16). Using commercial poly(propylene glycol)s, poly(styrene) diols, hydrogenated poly(butadiane) diols or poly(siloxane) diols a broad variety of tin-containing macrocycles became accessible by this simple methods. As examplified by eq. (5) all these macrocyclic oligomers or polymers can be used as "macroinitiators" for the polymerization of lactones. In this way various macrocyclic block copolymers can be prepared which, as described below, can be transformed into an even greater variety of tin-free polymers. A first interesting example of such a transformation is the ring-opening with dimercapto ethane (eq. 6). At room temperature a selective elimination of the Bu$_2$Sn group takes place without cleavage of ester groups, and telechelic A-B-A triblock copolymers having free OH endgroups can be isolated (16).

Telechelic Polyesters

As illustrated by equation (6) the treatment of tin-contaning macrocycles with dimercaptoethane is a simple method to prepare telechelic oligo - or polyesters having two OH endgroups which can further be modified in various ways, for instance by chain extension with diisocyanates. When the removal of Bu$_2$Sn groups with dimercaptoethane can be applied to the macrocycles 9, telechelic block copolyesters of two lactones are obtained. When applied to the "nanopretzels" (11) four armed stars having free OH endgroups are the reaction products (8).

The Sn-O bonds of the tin-containing macrocycles react easily with carboxylic acid chloride (eq. 7). The reaction conditions need to be optimized depending on the reactivity of the acid chloride to obtain 97 - 100 % functionalization (as evidenced by 400 MHz ^{1}H NMR spectra). Almost any kind of functional acid chloride can be used, so that a broad variety of telechelic polylactones can be synthesized and used for chain extension and other modifications of the endgroups (11). The tin containing macrocycles also react with carboxylic acid anhydrides. Functionalization with methacrylic anhydride (eq. 8) (11) yields telechelic polylactones which by copolymerization with suitable comonomers (e.g. hydroxy-ethyl methacrylate) may be useful as bone cements. Unfortunately acid chloride or anhydrides of N-protected amino acids or peptides are unstable and difficult to prepare. However, telechelic polylactones having amino acid or peptide endgroups can be synthesized by the reaction of tin-containing macrocycles with N-protected aminoacid thioarylesters (eq. 9) (19), which are easy to prepare from thiophenols and dicyclohexyl carbodiimide. This new coupling reaction is free of racemization and can also be used to bind penicilline derivatives to polylactones. Finally, it should be mentioned that the Boc protecting group was removed by anhydrous trifluoroacetic acid in CH$_2$Cl$_2$ without cleavage of the polyester chain (19). The liberated amino endgroups (stabilized as trifluoroacetates) can, in turn, serve as initiators for the polymerization of α-amino acid N-carboxyanhydrides or for stepwise peptide syntheses.

Ring-opening Polycondensations:

Ring-opening polymerization is usually the only synthetic strategy which allows to transform cyclic monomers into polymers (with Ziegler-Natta type polymerizations of a few cycloolefins being the only exception). We have recently demonstrated that the Sn-O bonds of tin-containing macrocycles, such as 3a or 8 are reactive enough to use them as bifunctional monomers for polycondensations (*12, 13, 16*). This new synthetic approach illustrated in eqs. (10) and (11) is called ring-opening polycondensation . In addition to dicarboxylic acid dichlorides the bisthioaryl esters of dicarboxylic acids may be used as electrophilic reation partners of the tin-containing macrocycles (*20*). The number of small tin-containing heterocycles which can serve as monomers for the ring-opening polycondensation is relatively small, but they can be used as initiators for ring-opening polymerizations (eq. 2), and the resulting tin-containing macrocyclic polymers (e.g. 8) can be used for in situ polycondensations (eq. 11). In other words, ring-opening polymerizations and ring-opening polycondensations can be combined in an "one-pot procedure" and this combination is called ROPPOC method. When macrocyclic block copolymers such as 9 or 12 are used as monomers for the polycondensation, multiblock copolymers are the reaction products. Therefore, the ROPPOC method is particularly useful for the synthesis of a broad variety of multiblock copolyesters with tailored properties. An interesting group of these multiblock copolymers are biodegradable thermoplastic elastomers composed of a flexible soft segment (e.g. poly(ethylene glycol)s or poly (THF)diols) and crystalline polyester segments (16).

Conclusion

Difunctional tin alkoxides ($R_2Sn(OR')_2$) offer several new synthetic strategies in the field of degradable materials: first, the macrocyclization of preformed polymeric diols, second, the macrocyclic polymerization of lactones, third, the synthesis of telechelic oligo- and polylactones in an "one pot procedure" and fourth, the ring-opening polycondensation possibly in direct combination with the macrocyclic polymerization of lactones.

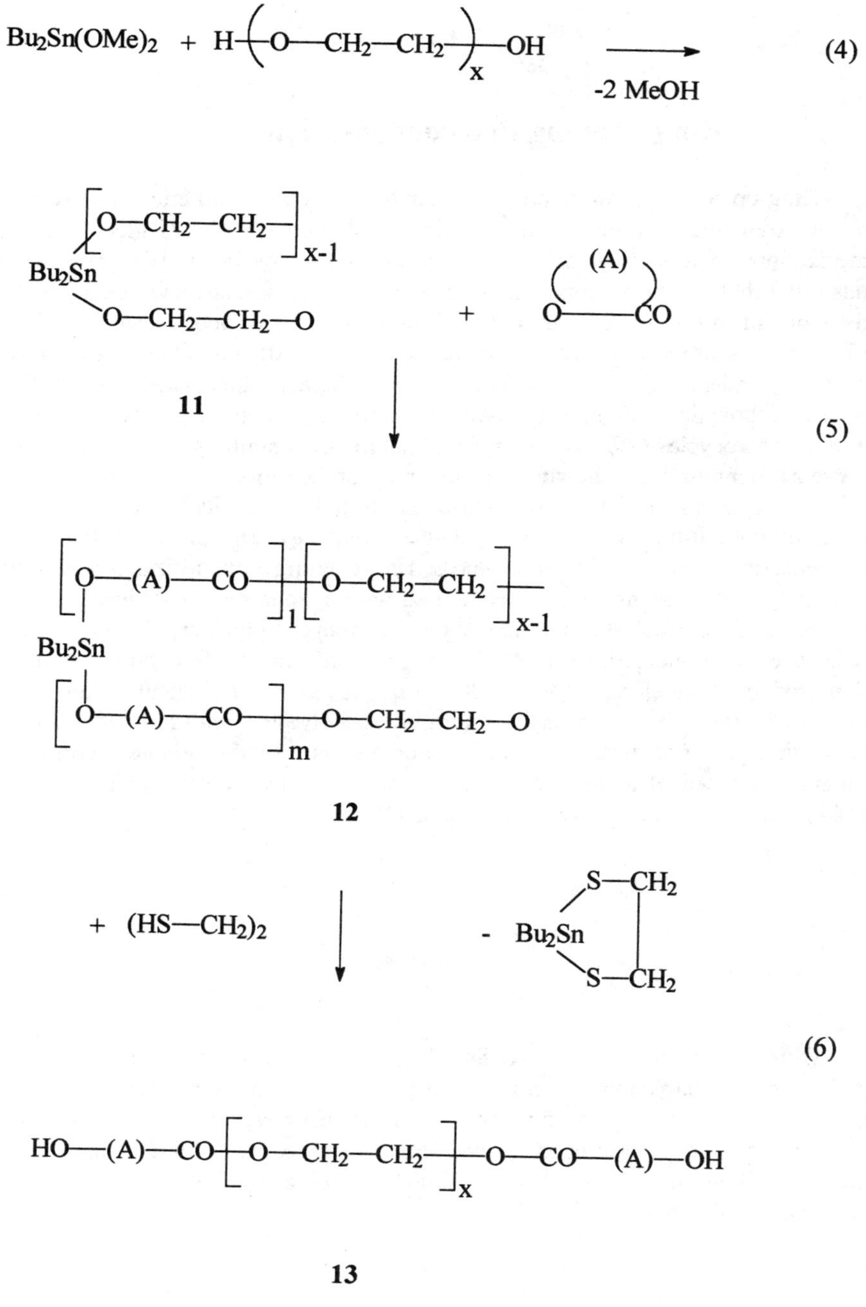

$$\text{Bu}_2\text{Sn(OMe)}_2 \ + \ \text{H}\left(\text{O}-\text{CH}_2-\text{CH}_2\right)_x-\text{OH} \quad \xrightarrow[\text{-2 MeOH}]{} \qquad (4)$$

$$\begin{array}{c} \text{Bu}_2\text{Sn} \underset{\text{O}-\text{CH}_2-\text{CH}_2-\text{O}}{\overset{\left[\text{O}-\text{CH}_2-\text{CH}_2-\right]_{x-1}}{\Big\langle}} \end{array} \qquad + \qquad \underset{\text{O}-\!\!\!-\!\!\!-\text{CO}}{\overset{(A)}{\frown}}$$

11

$$(5)$$

$$\begin{array}{c} \text{Bu}_2\text{Sn} \\ \left[\underset{|}{\text{O}}-(A)-\text{CO}\right]_1 \left[\text{O}-\text{CH}_2-\text{CH}_2\right]_{x-1} \\ \left[\underset{|}{\text{O}}-(A)-\text{CO}\right]_m \text{O}-\text{CH}_2-\text{CH}_2-\text{O} \end{array}$$

12

$$+ \ (\text{HS}-\text{CH}_2)_2 \qquad\qquad - \ \text{Bu}_2\text{Sn} \overset{\text{S}-\text{CH}_2}{\underset{\text{S}-\text{CH}_2}{\Big\langle}}$$

$$(6)$$

$$\text{HO}-(A)-\text{CO}\left[\text{O}-\text{CH}_2-\text{CH}_2\right]_x\text{O}-\text{CO}-(A)-\text{OH}$$

13

$$R\text{—}CO\left[O\text{—}(CH_2)_5\text{—}CO\right]O\text{—}(CH_2)_4\left[O\text{—}OC\text{—}(CH_2)_5\text{—}O\right]CO\text{—}R$$

$$+ 2\ R\text{-}COCl \qquad \uparrow \qquad - Bu_2SnCl_2 \qquad\qquad (7)$$

$$Bu_2Sn\ \begin{matrix}\left[O\text{—}(CH_2)_5\text{—}CO\right]O\text{—}CH_2\text{—}CH_2 \\[2mm] \left[O\text{—}(CH_2)_5\text{—}CO\right]O\text{—}CH_2\text{—}CH_2\end{matrix}$$

$$+ 2\quad (CH_2=CMeCO)_2\,O \qquad \downarrow \qquad - Bu_2Sn\,(O_2C\text{—}CMe=CH_2)_2 \qquad (8)$$

$$CH_2=\overset{Me}{\underset{}{C}}\text{—}CO\left[O\text{—}(CH_2)\text{—}CO\right]O\text{—}(CH_2)_4\text{—}O\left[CO\text{—}(CH_2)_5\text{—}O\right]CO\text{—}\overset{Me}{\underset{}{C}}=CH_2$$

$$\mathbf{8}\quad + 2\ BocNH\text{—}\overset{Me}{\underset{}{CH}}\text{—}CO\text{—}S\text{-}\langle\bigcirc\rangle\text{-}Cl$$

$$\downarrow \qquad - Bu_2Sn\,(S\text{-}C_6H_4Cl)_2 \qquad\qquad (9)$$

$$Boc\text{—}NH\text{—}\overset{Me}{\underset{}{CH}}\text{—}CO\left[O\text{—}(CH_2)_5\text{—}CO\right]O\text{—}CH_2\text{—}CH_2$$

$$Boc\text{—}NH\text{—}\underset{Me}{\overset{}{CH}}\text{—}CO\left[O\text{—}(CH_2)_5\text{—}CO\right]O\text{—}CH_2\text{—}CH_2$$

$$\text{Bu}_2\text{Sn}\underset{\text{O—CH}_2\text{—CH}_2}{\overset{\text{O—CH}_2\text{—CH}_2}{\big<}} \quad + \quad \text{ClCO—(CH}_2)_{\overline{n}}\text{—COCl}$$

$$\downarrow \quad -\text{ Bu}_2\text{SnCl}_2 \qquad\qquad (10)$$

$$\left[\text{—O—(CH}_2)_4\text{—O—CO—(CH}_2)_{\overline{n}}\text{CO—}\right]$$

$$\text{Bu}_2\text{Sn}\underset{\left[\text{O—(CH}_2)_5\text{—CO}\right]\text{O—CH}_2\text{—CH}_2}{\overset{\left[\text{O—(CH}_2)_5\text{—CO}\right]\text{O—CH}_2\text{—CH}_2}{\big<}} \quad + \quad \underset{\overset{|}{\text{CO—S—C}_6\text{H}_5}}{\overset{\text{CO—S—C}_6\text{H}_5}{\text{(A)}}}$$

$$\downarrow \quad -\text{ Bu}_2\text{Sn(S-C}_6\text{H}_5)_2 \qquad\qquad (11)$$

$$\left(\text{—CO—(A)—CO}\left[\text{O—(CH}_2)_5\text{—CO}\right]\text{O—(CH}_2)_4\text{—O}\left[\text{CO—(CH}_2)_5\text{—O}\right]\right)$$

Literature Cited

(1) Bornstein, J., Laliberté, B.R., Andrews, T.M., Montermoso,
J.C., J. Org. Chem. 1959, 24, 886

(2) Mehrotra, R.C.; Gupta, V.D., J. Organomet. Chem. 1965, 4, 145

(3) Considine, W.J., J. Organomet. Chem. 1966, 5, 263

(4) Pommier, J.C., Valade, J., J. Organomet. Chem. 1968, 12, 433

(5) Smith, P.J., White, R.F.M., J. Organomet. Chem. 1972, 40, 341

(6) Kricheldorf, H.R., Lee S.-R., Macromolecules 1995, 28, 6715

(7) Kricheldorf, H.R., Lee, S.-R., Bush, S., Macromolecules 1996,29, 1375

(8) Kricheldorf, H.R., Lee, S.-R, Macromolecules 1996, 29, 8689

(9) Kricheldorf, H.R., Lee, S.-R., Schittenhelm, N., Macromol.Chem. Phys.
1998, 199, 273

(10) Kricheldorf, H.R., Eggerstedt, S., Macromol. Chem. Phys. 1998, 199,
283

(11) Kricheldorf, H.R., Hauser, K., Macromolecules 1998, 31, 614

(12) Kricheldorf, H.R., Eggerstedt, S., J. Polym. Sci. Part A Polym. Chem.
1998, 36, 1373

(13) Kricheldorf, H.R., Eggerstedt, S., Macromolecules 1998, 31, in press

(14) Kricheldorf, H.R., Eggerstedt, S., Krüger, R., Macromol. Chem.
Phys. 1999 (Macrocycles 6.)

(15) Kricheldorf, H.R., Langanke, D., Macromol.Chem.Phys. in press
(Macrocycles 7.)

(16) Kricheldorf, H.R., Langanke, D., Macromol. Chem. Phys.
1999 in press (Macrocycles 8.)

(17) Kricheldorf, H.R., Lorenc, A., Damrau, D.-O., manuscript in preparation

(18) Davies, A.G. in "Organotin Chemistry" VCH, Weinheimn New York,
p.169

(19) Kricheldorf, H.R., Hauser, K., Macromol.Rapid Commun.1999 in press
("Polylactones 46.")

(20) Kricheldorf, H.R., Hauser, K., Krüger R., manuscript in preparation

Chapter 12

Polymerization and Copolymerization of Lactides and Lactones Using Some Lanthanide Initiators

Nicolas Spassky and Vesna Simic

Laboratoire de Chimie des Polymères, UMR 7610, Université Pierre et Marie Curie, 4, place Jussieu, 75252 Paris Cedex 05, France

A series of lanthanide alkoxides were used as initiators for the polymerization of lactides and several lactones (ε-caprolactone, δ-valerolactone, β-butyrolactone). Most of these initiators allow a controlled polymerization in mild conditions, i.e., reaction in solution (dichloromethane, toluene) at room temperature. Yttrium tris-(alkoxyethoxides) appear to be among the most satisfactory initiators as concerning reactivity and selectivity for the polymerization of lactide, ε-caprolactone and also trimethylene carbonate. Block copolymers were prepared from ε-caprolactone and (D) or (L)-lactides. The blending of these block copolymers leads to the formation of stereocomplexes.

Polyesters derived from lactides and lactones are interesting materials for biomedical, pharmacological and packaging applications (*1-5*). They are usually obtained from the corresponding monomers by ring opening polymerization using different types of initiators as decribed in a recent review (*6*). In order to obtain polymers with well defined characteristics and to be able to synthesize block copolymers, initiators leading to a living type process are desirable. Aluminum alkoxides were found to exhibit a controlled polymerization of some lactones and lactides (*7-9*) and the livingness of the polymerization systems was discussed in terms of the selectivity of the involved elementary reactions (*10*). Lanthanide alkoxides exhibit a much higher reactivity than other metal alkoxides for the polymerization of lactones and lactides and a living type behavior is observed for some of them (*11-18*). In this paper we want to discuss the results taken from the literature on the use of lanthanide initiators in the polymerization of cyclic esters and to present our recent investigations in this field.

Lanthanide alkoxide initiators

Although other derivatives of lanthanides, such as halogenides (*19*) cyclopentadienyl (*16,20*) and bimetallic complexes (*21*), are studied in the polymerization and copolymerization of cyclic esters, lanthanide alkoxides appear to give up to now the more reproducible results. The structure of the latter initiators, their functionality has been described in the literature.

In our studies we have mainly used two families of lanthanide alkoxides, i.e. the oxoalkoxide clusters of $Ln_5(\mu\text{-O})OiPr)_{13}$ type and the yttrium tris-alkoxyethoxides of $Y(OCH_2CH_2OR)_3$ type. Some results obtained with a bimetallic aluminum-yttrium alkoxide will be also reported in this paper. Almost all of the lanthanide initiators used in this work were prepared in the laboratory of Prof. L.G. Hubert-Pfalzgraf (Université Sofia-Antipolis, Nice). The oxoalkoxide clusters $Ln_5(\mu\text{-O})(OiPr)_{13}$ where Ln is La, Sm, Y and Yb were prepared according to the procedure described by Poncelet et al for yttrium derivative (*22*). In these compounds five lanthanide atoms are linked to a single central oxygen atom. The isopropoxy ligands of these pentanuclear oxoaggregates are distributed in three different groups corresponding to the formula $Ln_5(\mu_5\text{-O})(\mu_3\text{-OiPr})_4(\mu_2\text{-OiPr})_4(OiPr)_5$. All lanthanide atoms are octahedrally surrounded by six isopropoxide groups. The overall functionality (number of active metal-alcoholate groups) of these initiators is 2.6 (13/5). These pentanuclearoxoaggregate structure determined by ^1H and ^{13}C NMR and X-ray diffraction remains in solution (*23*). Yttrium tris-alkoxyethoxides initiators $[Y(OC_2H_4OR)_3]n$ were prepared from the reaction of yttrium with the corresponding alcohol according to the procedure described by Poncelet et al (*24*). The nuclearity (n) of these compounds depends on the R substituent and was found to be equal to n=2 for R=iPr, but reaches n = 10 for R=Me.

The heterometallic derivative $Y[Al(OiPr)_4]_3$ was obtained from exchange reaction between yttrium cluster compounds and aluminum (*25*).

Homopolymerization of lactides with lanthanide alkoxides.

The polymerization of lactides using lanthanide alkoxides as initiators was recently reported in several works.

In order to compare the reactivity of these compounds and kinetic results, a careful attention must be devoted to the experimental conditions used, e.g., monomer concentration $[M]_0$, monomer over initiator $[M]/[I]$ ratio, temperature, solvent used etc. In fact, only general trends can be established since experimental conditions usually differ between groups. Concerning cluster compounds the commercially available $Y_5(O)(OiPr)_{13}$ was used for the first time in the group of Feijen (*15*) for the polymerization of L-lactide. Later, a novel catalyst system derived from the reaction of tris (2,6-di-tert-butyphenoxy) yttrium and 2-propanol in situ was used and the kinetic data of both systems were compared (*18*). The latter in situ system was much more reactive and conversion of 98% can be obtained in 5 min in dichloromethane solution at 22°C (with $[M]_0 = 0175$ M and $[M]/[I] = 166$). Living type behavior with a narrow MWD (1.14) is observed.

In our work recently reported (26) we have compared the polymerization of (D,L)-lactide with different lanthanide oxoalkoxide clusters where Ln was La, Sm, Y and Yb in dichloromethane solution at room temeprature. In the same experimental conditions $[M]_0 = 0.9M$, and $[M]/[I] = 68$) the La initiator was found to be the most reactive and Yb initiator the less reactive.

Based on the time of half reaction a comparison of reactivities gives the following range :

Ln	La	Sm	Y	Yb
t 1/2 (min)	<1	30	50	700

The kinetic data obtained with these systems were recently reported (27). Linear Ln $[M]_0/[M]$ versus time plots were found as well as linear increase of molecular weight (Mn) with conversion. Both of these findings are in favor of a living type behavior. While the molecular weight distribution (MWD) remains narrow up to high conversion for Sm, Y and Yb systems, which is the sign of a good selectivity, it is not the case for La initiator for which a substantial broadening of MWD is observed at the end of the reaction. This broadening is due to the occurence of inter and intra transesterification reactions as revealed by MALDI-TOF MS analysis (27,28). It is also worthy to mention that the number of active sites (alkoxide initiating groups) per metal varies with the nature of the metal of the initiator and was found to be close to the expected value 2.6 for Y and Sm initiator, but was lower, close to 2.0, for La initiator.

The second type of initiators we have used are yttrium trisakoxides initiators derived from alcohols of ROC_2H_4OH type. Mc Lain et al. were the first to report on the very high reactivity of $Y(OCH_2H_4NMe_2)_3$ initiator in the polymerization of L-lactide (12,14), a conversion of 97% being obtained in 15 min ($[M]_0=1.2$ $[M]/[Y] = 360$). However very rapidly, e.g., after 20 min, transesterification reactions occur.

We have used $[Y(OCH_2H_4OiPr)_3]_2$ initiator and we have polymerized (D,L)-lactide in the same conditions as in the case of cluster $Y_5(O)(OiPr)_{13}$ initiator. We have observed a much faster polymerization since the time of half-reaction was less than two minutes and 94% of conversion were obtained in one hour (29). MWD was narrow (<1.20) up to high conversions and for several hours of polymerization. However after a very long time of polymerization, e.g., several days, a broadening of MWD occured. Kinetic data showed some differences with the cluster type initiators. The semi logarithmic plots were not linear from the beginning of polymerization which is typical for slowly initiated reactions. But after at different conversions a living type behavior was observed (27).

In THF solution the plots were linear from the beginning and the overall kinetic behavior was very similar in THF and DCM solvents. No differences in behavior were found between $[Y(OCH_2H_4OiPr)_3]_2$ and $[Y(OCH_2H_4OMe)_3]_{10}$ initiators. They show the same reactivity inspite of different degree of aggregation and different end groups. The number of active sites per metal was found to be 3.

As last initiator we have tried the bimetallic compound $Y[Al(OiPr)_4]_3$. In this initiator, the yttrium atom is hexacoordinated surrounded by six OiPr groups and each aluminum atom bears four OiPr groups, two of which are external and the two others shared between yttrium and aluminum atoms (25). The maximal functionality of this initiator is 12.

The polymerization of (D,L)-lactide performed with this initiator in the same conditions as with other Y initiators was found to be slow, 50% of conversion being reached in 1250 min at room temperature in DCM solution. The MWD was narrow (1.24) and molecular weight corresponded to a functionality of 12. However after prolongation of the polymerization a substantial broadening of MWD was observed. In bulk at 140°C a 95% conversion was reached in less than two hours with this initiator and the MWD was broad (1.56).

In table I are compared the results of homopolymerization of (D,L)-lactide with the three types of yttrium alkoxide initiators. The tris-alkoxides $[Y(OC_2H_4OR)_3]_n$ exhibit the higher reactivity and the best selectivity.

Table I. Polymerization of lactides using yttrium alkoxide initiators

Initiator	[M]/[Y]	t (min)	Conv (%)	Mn [b] (x10³)	MWD [b]
$Y_5(O)(OiPr)_{13}$[a]	68	90	62	1.9	1.20
	68	100	73	2.6	1.20
	170	120	41	2.5	1.17
$Y(OC_2H_4OiPr)_3$	70	25	89	2.9	1.13
	400	70	93	14.3	1.23
	400	12960	98	14.0	1.70
$Y[Al(OiPr)_4]_3$	70	1250	50	1.4	1.24
	70	2620	73	2.0	1.82

Polymerizations carried out at room temperature in DCM solution $[M]_0=0.9$ M ;

[a] in DCM/Toluene (80/20) solution [b] SEC in THF ; corrected values for poly LA.

It is worthy to mention that for all of these inititators the polymerization proceeds through acyl-oxygen cleavage of the lactide ring, as revealed from ^{1}H NMR analysis and there is no racemization as deduced from some runs performed with L-lactide with a subsequent control of the optical activity of prepared polymers.

Homopolymerization of different cyclic esters using $Y(OC_2H_4OiPr)_3$ as initiator system.

Poly(ε-caprolactone) and its copolymers are among the most studied synthetic biodegradable polymers and they have been commercialized. Poly (δ-valerolactone), also
a biodegradable polymer, was much less studied. ε-Caprolactone (ε-CL), d-valerolactone (δ-VL) and lactide (LA) are cyclic esters involving the same functional group. For this reason the initiators used for their polymerization are similar. ε-Caprolactone has been already polymerized by lanthanide initiators. Mc Lain and Drysdale (11) have used as initiators $Y_5(O)(OiPr)_{13}$ and $Y(OC_2H_4OEt)_3$ and some other derivatives. With the first initiator the polymerization in toluene solution at room temperature was almost complete in 5 min giving a polymer with a narrow MWD, but after 30 min a broadening of MWD appeared. With the second initiator the transesterification occured even more rapidly. The activity of lanthanide catalysts for ester interchange reactions was discussed by Okano et al (30). Using the in situ prepared yttrium alkoxide as initiator Stevels et al (31) have found that the polymerization of ε-CL with this system was the faster one observed up to now. Suprisingly the rate of polymerization was less faster than that of L-lactide using the same initiator. Living type behavior was also assumed by Yamashita et al (16) for ε-CL and δ-VL polymerization with different organolanthanide cyclopentadienyl complexes. Recently Deng et al (32) reported on the polymerization of ε-CL by some bimetallic initiators containing Y or Sm, Al and other metals including the $Y(Al[OiPr]_4)_3$ derivative. A fast polymerization was observed but the living behavior is not assumed.
We have polymerized ε-CL and δ-VL in DCM solution at room temperature with $Y(OC_2H_4OiPr)_3$ initiator system in the same conditions as (D,L)-LA. The results of polymerization are given in Table II. Another cyclic monomer, the trimethylene carbonate (TMC), was also polymerized in the same conditions. The polymerization of ε-CL proceeds in a much faster way than that of (D,L)-LA since 90% of conversion are reached in 10 min. A narrow MWD is observed. A slight broadening of MWD occurs after 60 min of reaction at complete conversion.
These results are different from those reported by Mc Lain and Drysdale (11) with $Y(OC_2H_4OEt)_3$ working in toluene solution. In their case the broadening of MWD occured much faster, which can be due to the nature of the solvent used. According to the authors the molecular weights are consistent with initiation by every alkoxide group, i.e., 3 alkoxide groups per Y atom. We have observed for $Y(OC_2H_4OiPr)_3$ an initiation with only two alkoxide groups. The kinetic data, e.g. semi-logarithmic plots Ln $[M]_0/[M]_t$ versus time, are similar for ε-CL and δ-VL. A short period of induction is observed followed by a linear relationship. This induction period can be related to structural rearrangements of the initiator

to form in the presence of monomer the working active site as suggested previously by Wurm et al (33) in the case of polymerization of ε-CL and diMe-TMC with aluminum alkoxides. Induction period was also found by Stevels and al in the case of polymerization ε-CL with in situ prepared Y alkoxide initiator.

Recently lanthanide monoalkoxide initiators carrying a bulky coordinate group, e.g. isopropoxylanthanide diethyl acetoacetate $(EA)_2LnOiPr$ with Ln=Nd, Y were reported to lead to a living type polymerization of ε-CL with successful preparation of different block copolymers (34). A narrow MWD was maintained up to complete conversion. It is believed that the steric effect due to the bulky groups surrounnding the active center of initiator hinders the polymer chain from the access to the

Table II. Homopolymerization of different cyclic monomers with $Y(OCH_2CH_2OiPr)_3$ as initiator system.

Monomer	Time (min)	Conv (%)	Mn calc[d]	Mn SEC[e] (corr)	MWD[e]	n[f]
VL	11	72[a]	3360	7700	1.33	3.5
	60	100	4700	10800	1.44	3.4
CL	10	90[b]	5300	13700 (7100)	1.13	2.1
	60	100	5300	15000 (8000)	1.33	2.1
(D,L)-LA	10	26[c]	1750	4000 (2200)	1.25	2.8
	60	76	5100	9300 (5200)	1.35	3
TMC	10	100[c]	4760	15250 (11450)	1.91	1.25
	60	100	4760	12800 (9600)	1.90	/

Polymerizations carried out at room temperature in dichloromethane solution ;
$[M]_0/[Y]_0 = 140$; $[M]_0 = 1M$
VL = δ-valerolactone, CL = ε-caprolactone, (D,L)-LA = racemic lactide, TMC = trimethylene carbonate.
Conversion determined by : [a] UV spectroscopy [b] gravimetric analysis [c] from H^1 NMR ;
[d] on the basis of 3 initiating OiPr group per yttrium atom [e] in THF, polystyrene standards ; in brackets corrected values for the corresponding polymers [f] number of active sites per yttrium atom.

active center and reduces the transesterification reactions. We have also observed a similar process in the case of polymerization of (D,L)-lactide with a bulky chiral aluminum alkoxide initiator (*35*).

Working with ε-CL in THF solution we have determined an apparent rate constant

k_{app} = 83 L. mol^{-1}min^{-1} ([M]$_0$=1.9M [M]$_0$/[I]$_0$ = 1540). Stevels et al have reported for the in situ Y alkoxide initiator kp=99 L. mol^{-1}min^{-1} (*31*). The differences observed in rate constants of polymerization with aluminum and yttrium alkoxides are much higher in the case of LA monomer, than in the case of ε-CL.

δ-Valerolactone exhibits an intermediate behavior between LA and ε-CL with a MWD somewhat broader than for both other monomers. For all of these cyclic esters a controlled polymerization with a living type behavior can be assumed, this allowing the possibility of preparation of block copolymers.

The polymerization of TMC was the faster of the series of cyclic monomers since complete conversion is obtained in much less than 10 min. According to MW determined by SEC the number of polymer chains initiated is only 1.25 per yttrium atom. In addition a broad MWD is observed. A very fast polymerization of TMC was recently reported by Soum et al (*36*) using La(OiPr)$_3$ as initiator. Polymerization in bulk at 80°C of TMC using anhydrous rare earth halides was studied by Shen et al (*37*). It was found that these initiators were much more active for the polymerization of TMC than that of ε-CL. Copolymerization of TMC with ε-CL and (D,L)-LA using the same initiators were reported recently by the same authors (*38,39*). A substantial study of the polymerization of aliphatic cyclic carbonates and particularly TMC by cationic and insertion initiators, e.g., Sn containing compounds, was performed in the group of Kricheldorf (*40-42*). The polymerization and copolymerization of 2,2 diMe-TMC with various cyclic esters and other cyclic compounds by a variety of initiators (anionic, Mg, Al, Zn derivatives) were extensively studied by Höcker and Keul (*33, 43-45*). Cyclic carbonates were shown to undergo volume expansion on polymerization (*46*).

The polymerization of racemic β-butyrolactone was successfully performed using [Y(OC$_2$H$_4$OMe)$_3$]$_{10}$ initiator system at room temperature in dichloromethane solution (*14*). The reaction is fast at the beginning (half-time of reaction close to 20 min) but then a slowing down is observed around 70% of conversion. A living type behavior is assumed (narrow MWD distribution (<1.2), linear $\overline{DPn}$ versus [M]/[I] relation, possibility to prepare block-copolymers). An O-acyl ring-opening occurs according to end groups NMR analysis and the three alkoxide groups are initiating polymeric chains.

The polymerization of racemic β-butyrolactone was successfully performed using [Y(OC$_2$H$_4$OMe)$_3$]$_{10}$ initiator system at room temperature in dichloromethane solution (*14*). The reaction is fast at the beginning (half-time of reaction close to 20 min) but then a slowing down is observed around 70% of conversion. A living type behavior is assumed (narrow

MWD distribution (<1.2), linear $\overline{DP}n$ versus [M]/[I] relation, possibility to prepare block-copolymers). An O-acyl ring-opening occurs according to end groups NMR analysis and the three alkoxide groups are initiating polymeric chains.

Table III. Polymerization of different cyclic esters using yttrium alkoxide initiators.

Initiator	*Mon*	*[M]/[Y]*	*t (min)*	*Conv (%)*	*Mn(x10³) (SEC)*	*MWD (SEC)*	*n* [g]
$Y(OC_2H_4OiPr)_3$	D,L-LA [a]	140	60	75	5.2 [e]	1.35	3.0 (3)
	CL [a]	140	10	90	7.1 [e]	1.13	2.0 (3)
	VL [a]	140	11	72	2.8 [f]	1.33	3.5 (3)
$Y(OC_2H_4OMe)_3$	BL [b]	142	44	65	3.2 [f]	1.18	3.0 (3)
$Y[Al(OiPr)_4]_3$	D,L-LA [a]	70	1350	50	1.4 [e]	1.23	12 (12)
	D,L-LA [c]	70	110	95	2.3 [e]	1.56	12 (12)
	CL [d]	70	50	52	3.6 [e]	1.51	4.8 (12)
$Y_5(O)(OiPr)_{13}$	D,L-LA [a]	68	80	62	2.0 [e]	1.20	2.6 (2.6)
	VL [d]	130	180	60	8.8 [f]	1.40	1 (2.6)
	BL [d]	133	1380	60	8.0 [f]	1.08	1 (2.6)

LA = lactide CL = ε-caprolactone VL = δ-valerolactone

BL =β-butyrolactone DCM = dichloromethane Tol = Toluene.

Polymerization conditions : [a] DCM sol., t amb. [b] Tol. sol., t amb. [c] in bulk, 140°C [d] in bulk, t amb [g] number of active sites ; in brackets maximum available OiPr groups [e] in THF ; corrected values [f] in THF ; polystyrene standards.

The polymerization of racemic β-butyrolactone was successfully performed using $[Y(OC_2H_4OMe)_3]_{10}$ initiator system at room temperature in dichloromethane solution (*14*). The reaction is fast at the beginning (half-time of reaction close to 20 min) but then a slowing down is observed around 70% of conversion. A living type behavior is assumed (narrow MWD distribution (<1.2), linear $\overline{DP}n$ versus [M]/[I] relation, possibility to prepare block-copolymers). An O-acyl ring-opening occurs according to end groups NMR analysis and the three alkoxide groups are initiating polymeric chains.

Polymerization with the cluster initiator $Y_5(\mu\text{-}O)(OiPr)_{13}$ was also attempted in bulk.

After a rapid initiation the reaction proceeds slowly in bulk, a narrow MWD (<1.2) is found and a linear correlation between $\overline{DP}n$ and [M]/[I] is observed. Only one alkoxide group over the 2.6 available seems to be active (*47*).

Polymerization of ε-CL was tried with bimetallic Y[Al(OiPr)$_4$]$_3$ initiator in bulk at room temperature. The time of half-reaction is approximatively 50 mn, MWD is not narrow (1.5) and only 4.8 active sites of the 12 available are used.

In the polymerization in bulk at 140°C of (D,L)-LA with the same initiator all of the alkoxide groups are active.

In table III are summarized the main results we have obtained with different cyclic esters using various yttrium alkoxide initiators.

Synthesis of block copolymers with cyclic esters using yttrium alkoxide initiators.

The living character of the polymerization of cyclic esters using yttrium alkoxide initiators make it possible the synthesis of block copolymers. In this review we will only discuss block copolymers containing polylactide sequences. Moreover, only block copolymers obtained by sequentiel addition of two cyclic esters will be considered but not those in which one block, already available, is used as macroinitiator, e.g., functionalized polyethylene oxide. The nature of the monomer first polymerized is highly important and it was reported by Feng and al (48) that in the case of ε-CL - LA blocks, ε-CL must be first monomer to be polymerized. Indeed ε-CL is not at all polymerized when added to living polylactide. Feng and al have used Teyssié bimetallic Zn,Al-μ-oxoalkoxide initiator which leads to a living type polymerization of ε-CL (49,50). Teyssié group was also the first to report on the possibility to use aluminum isopropoxide for living type polymerization of ε-CL (51). On this basis, Jacobs et al (52) prepared block copolymers of ε-CL and lactides using commercially available Al(OiPr)$_3$. The latter compound is in fact a mixture of trimeric and tetrameric species which polymerize these cyclic esters at different rates. High rank studies performed in Liège and Lodz groups have presently fully clarified mechanisms of homopolymerisations of ε-CL and LA with aluminum alkoxide initiators (53-55). On the other hand, in a recent report Duda et al revealed, that the initiation of ε-CL on living PLA is possible, but extremely slow and after complete incorporation of CL the obtained copolymer exhibits a random type structure (56). It must be mentioned that stereoblocks of L-and D-lactides were prepared in the group of Feijen (57) by sequential addition of both enantiomers using Al(OiPr)$_3$ as initiator.

Block copolymers using lanthanide initiators were first prepared by Mc Lain et al (12) with Y(OCH$_2$CH$_2$NMe$_2$)$_3$ system as initiator. In contrast to previous described synthesis which were performed, at least for lactide sequence, at elevated temperatures (70-90°C), this initiator allows the synthesis of block copolymers at room temperature in dichloromethane solution in a very short time. We have prepared in our group β-butyrolactone-b-L-lactide copolymers using Y(OCH$_2$CH$_2$OMe)$_3$ as initiator system (14). Commercial Y$_5$(O)(OiPr)$_{13}$ was used in the group of Feijen (*15*) for the systhesis of ε-CL-*b*-LA copolymers at room temperature, but this system is much less reactive than the previously described ones. In contrast, the in situ prepared yttrium alkoxide initiator by the same group (*17*) is a very reactive system for the synthesis of block copolymers with a lactide sequence. We have recently used Y(OCH$_2$CH$_2$OiPr)$_3$ initiator system for the synthesis of ε-CL-*b*-(D or L)-LA block copolymers in dichloromethane solution at room temperature. This system shows a good reactivity and copolymers with different sequential lengths can be obtained. A monofunctional yttrium initiator was also used by Shen et al (*34*) for the synthesis of the same block copolymers.

The main experimental results obtained with yttrium alcoholate initiators are reported in table IV.

Table IV. Block copolymers with cyclic esters using yttrium initiators

Initiator	M1	[M1]/[Y]	M2	t^a (min)	T^b (°C)	Mn^c (x10^3)	MWD	Ref
Y$_5$(O)(OiPr)$_{13}$	CL	150	L-LA	1152	amb	11400	1.20	(15)
	CL	150	L-LA	10800	amb	15200	1.27	(15)
Y(OiPr)$_3$ in situ	CL	166	L-LA	8	amb	13900	1.16	(17)
Y(OC$_2$H$_4$OMe)$_3$	BL	71	L-LA	120	20/70	5300	1.17	(14)
Y(OC$_2$H$_4$OiPr)$_3$	CL	110	L-LA	135	amb	27500	1.40	Our
	CL	120	D-LA	90	amb	28800	1.38	group
Y(OC$_2$H$_4$NMe$_2$)$_3$	CL	87	L-LA	30	amb	n.i.^{d}	1.36	(13)
(EA)$_2$YOiPr ^{e}	CL	200	DL-LA	750	20/70	52000	1.31	(34)

^{a} overall time for block copolymerization ^{b} two values when the two stages were run at different temperatures ^{c} SEC in THF, polystyrene standards ^{d} not indicated ^{e} EA =diethyl acetoacetate ether

CL = ε-caprolactone   BL = β-butyrolactone   LA = lactide

Diblock copolymers of δ-VL and LA, and ε-CL and LA, were also prepared using respectively as initiator MeONa (*58*) or tBuOLi (*59*). Some racemization reaction occured in the case of the first initiator.

Formation of stereocomplexes

It is known that is some cases the blending of two polyenantiomeric chains leads to materials that have enhanced properties as compared to those of polyenantiomers. These polyenantiomeric blends cocrystallise in particular racemic lattices and form the so-called "stereocomplexes". Polylactides are among examples of macromolecules forming stereocomplexes and were extensively studied during the last years (*60-*

63). The formation of stereocomplexes is not limited to the blending of optically pure polyenantiomers : it can also occur with enriched stereocopolymers and stereoblock copolymers *(64)*. For example the already cited stereoblock copolymer prepared by sequential addition of D- and L-lactide using Al(OiPr)$_3$ as initiator leads to a material which exhibits a melting point of 205°C *(57)*, much higher than that found for pure poly-L-lactide (170°C). We have prepared the same lactide stereoblock L-*b*-D using Y(OC$_2$H$_4$OiPr)$_3$ initiator (Mn = 17 000 (correct. value) ; Mw/Mn = 1.19 ; [a]$^{25}_D$ = - 1.0 (c= 1, CHCl$_3$)) and we have found a higher melting point (218°C) which can be due to a better balance between L- and D-blocks in the copolymers.

Another interesting concept consists to prepare blends of block copolymers containing enantiomeric lactide sequences of opposite configurations. Stevels et al *(65)* have synthesized block copolymers of ε-CL and D-or-L-lactide by sequential polymerization using Y$_5$(O)(OiPr)$_{13}$ as initiator. Binary blends of these block copolymers in (1 : 1) ratio provided materials which showed a melting point close to 60°C corresponding to ε-CL block and a melting point around 203-212°C assigned to the formation of PLA stereocomplex. The formation of stereocomplexes and the corresponding melting points are depending on the composition of block copolymers. We have prepared the same type of block copolymers using Y(OC$_2$H$_4$OiPr)$_3$ initiator system (see table IV) and after blending we have observed also the formation of stereocomplex materials with the following thermal characteristics : T$_{PCL}$ = 57°C, T$_{PLA}$ = 229°C (stereocomplex).

Further studies on block-copolymers and stereocomplexes are in progress.

Acknowledgements

We are highly indebted to Prof. Liliane H. Hubert-Pfalzgraf for providing us with most of the lanthanide derivatives used in this work as initiators and for fruitful discussions.

Literature cited

(1) - Holland, S.J. ; Tighe, B.J. ; Gould, L.P. J. *Controlled Release* **1986**, *4*, 155

(2) - Pierre, T.St. ; Chiellini, E. *J. Bioactive and Compatible Polymers* **1987**, 2, 4

(3) - Vert, M. ; Li, S.N. ; Spenlehauer, G ; Guérin, P. *J. Mater Sci., Mater. Med.* **1992**, *3*, 432

(4) - Nieuwenhuis, *J. Clin. Mater.* **1992,** *10*, 59

(5) - Sinclair, R.J. *J. Macromol. Sci. Chem. Int. Ed., Engl.* **1996**, *A33*, 585

(6) - Löfgren, A. ; Albertsson, A.C. ; Dubois, P. ; Jérôme, R.J. *Macromol. Sci., Rev. Macromol. Chem. Phys.* **1995**, *C35*, 379

(7) - Kricheldorf, H.R. ; Berl, M. ; Scharnagl, N. *Macromolecules* **1988**, *21*, 286

(8) - Dubois, P. ; Jacobs, C. ; Jérôme, R. ; Teyssié, P. *Macromolecules* **1991**, *24*, 2266

(9) - Penczek, S. ; Duda, A. ; Slomkowski, S. *Makromol. Chem. Macromol. Symp.* **1992**, *54/55*, 31

(10) - Penczek, S. ; Duda, A. *Macromol. Symp.* **1996**, 107, 1

(11) - Mc Lain, S.J. ; Drysdale, N.E. *Polym. Prepr. (Am. Chem. Soc. , Div. Polym. Chem.)* **1992**, *33*, 174

(12) - Mc Lain, S.J. ; Ford, T.M. ; Drysdale, N. E. *Polym. Prep. (Am. Chem. Soc., Div. Polym. Chem.)* **1992**, 33, 463

(13) - Mc Lain, S.J. ; Ford, T.M. ; Drysdale, N. E.; Jones, N. ; Mc Cord, E.F. *Polym. Prepr. (Am. Chem. Soc. Div. Polym. Chem.)* **1994**, *35*, 534

(14) - Le Borgne, A. ; Pluta, C. ; Spassky, N. *Makromol. Rapid Commun.* **1995**, *15*, 955.

(15) - Stevels, W. M. ; Ankoné, M.J.K. ; Dijkstra ; P.J. ; Feijen, J. *Macromol. Chem. Phys.* **1995**, *196*, 1153

(16) - Yamashita, M. ; Takemoto, Y. ; Ihara, E. ; Yasuda, H. *Macromolecules* **1996**, *29*, 1978

(17) - Stevels, W. M. ; Ankoné, M.J.K. ; Dijkstra, P.J. ; Feijen, J. *Macromolecules* **1996**, *29*, 3332

(18) - Stevels, W. M. ; Ankoné, M.J.K. ; Dijkstra, P.J. ; Feijen, J. *Macromolecules* **1996**, *29*, 6132

(19) - Shen, Y. ; Zhu, K.J. ; Shen, Z. ; Yao, K.M. *J. Polym. Sci., Pt. A, Polym. Chem.* **1996**, *34*, 1799

(20) - Evans, W.J. ; Katsumata, H. *Macromolecules* **1994**, *27*, 2330

(21) - Shen, Z. ; Chen, X. ; Y. ; Zhang, Y. *J. Polym. Sci., Pt.. A, Polym. Chem.,* **1994**, *32*, 597

(22) - Poncelet, O. ; Sartrain, W.J. ; Hubert-Pfalzgraf, L.G., Folting, K. ; Caulton, *K. G. Inorg. Chem.* **1989**, *28*, 263

(23) - Hubert-Pfalzgraf, L.G. New *J. Chem.* **1995**, *19*, 727

(24) - Poncelet, O. ; Hubert-Pfalzgraf, L.G.; Daran, J.C. ; Astier, R.J *J. Chem. Soc. Chem. Commun.* **1989,** 1849

(25) - Bradley D.C. ; Merhotra, R.C. ; Gaur D.P. *Metal Alkoxides, Academic, London,* **1978.**

(26) - Simic, V.; Spassky, N. ; Hubert-Pfalzgraf, L.G. *Macromolecules* **1997**, *30*, 7338

(27) - Spassky, N. ; Simic, V. ; Hubert-Pfalzgraf, L.G. ; Montaudo, M.S. *Macromol. Symp.* in press.

(28) - Spassky, N. ; Simic, V. ; Montaudo, M.S.; Hubert-Pfalzgraf, L.G. *Macromol. Chem. Phys.,* in press

(29) - Simic, V. ; Girardon, V. ; Spassky, N. ; Hubert-Pfalzgraf, L.G. ; Duda, A. *Polym. Degrad. Stab.* **1998**, *59*, 227

(30) - Okano, T. ; Hayashizaki, Y. ; Kiji, J. *Bull. Chem. Soc. Jpn.* **1993**, *66*, 163

158

(31) -   Stevels, W.M. ; Ankoné, M.J.K. ; Djikstra, P.J. ; Feijen, J.
          Polym. Prep. (Am. Chem. Soc. Div. Polym. Chem.),**1996,** *37,*
          190
(32) -   Deng, X. ; Zhu, Z. ; Xiong, C. ; Zhang, L. *J. Appl. Polym.* Sci.
          1997, *64,* 1295
(33) -   Wurm, B. ; Keul, H. ; Höcker, H. *Macromol. Chem. Phys.*
          1994, *195,* 3489
(34) -   Shen, Y. ; Shen, Z. ; Zhang, Y. ; Yao, K. *Macromolecules*
          1996, *28,* 8289
(35) -   Spassky, N. ; Wisniewski, M. ; Pluta, C. ; Le Borgne, A.
          Macromol. Chem. Phys. **1996,** *197,* 2627
(36) -   Onfroy, V. ; Schappacher, M. ; Soum, A. *Preprints Int. Symp.*
          on Ionic Polym., IP'97, Paris **1997,** p.246
(37) -   Shen, Y. ; Shen, Z. ; Zhang, Y. ; Hang, Q. *J. Polym. Sci., A,*
          Polym. Chem. **1997,** *35,* 1339
(38) -   Shen, Y. ; Shen, Z. ; Zhang, Y. ; Hang, Q. ; Shen, L. ; Yuan,
          H. *J. Appl. Polym. Sci.* **1997,** *64,* 2331
(39) -   Huang, Q. ; Shen, Z. ; Zhang, Y. ; Shen, Y. ; Shen, L. ; Yuan,
          H. *Polym. J.* **1998,** *30,* 1668
(40) -   Kricheldorf, H.R. ; Jensen, J. ; Kreiser-Saunders I. *Makromol.*
          Chem. **1991,** *192,* 2391
(41) -   Kricheldorf, H.R. ; Weegen-Schulz, B. ; Jenssen, J. *Makromol.*
          Chem., Macromol. Symp. **1992,** *60,* 119
(42) -   Kricheldorf, H.R. ; Weegen-Schulz, B. *J. Polym. Sci. Pt. A,*
          1995, *33,* 2193
(43) -   Keul, H. ; Bächer, R. ; Höcker, H. *Makromol. Chem.,* **1986,**
          187, 2589
(44) -   Keul, H. ; Müller, A.J. ; Höcker, H. *Makromol. Chem.,*
          Macromol. Symp. **1993,** *67,* 289
(45) -   Gerhard-Abozari, E. ; Keul, H. ; Höcker, H. *Macromol. Chem..*
          Phys. **1994,** *195,* 2371
(46) -   Takata, T. ; Sanda, F. ; Ariga, T. ; Endo, T. *Macromol. Rapid*
          Comm.. **1997,** *18,* 461
(47) -   Thiam, M. ; Spassky, N. to be published
(48) -   Teyssié, P. ; Ouhadi, T. ; Bioul, J.P. *Int. Rev. Sci., Phys.*
          Chem. Ser. **1975,** *118, 191*
(49) -   Hamitou, A. ; Ouhadi, T. ; Jérôme R. ; Teyssié, P. *J. Polym.*
          Sci., Polym. Chem. Ed., **1977,** *15,* 865
(50) -   Feng, X.D. ; Song, C.X. ; Chen, W.Y. *J. Polym. Sci., Polym.*
          Lett Ed. **1983,** *21,* 593
(51) -   Ouhadi, T. ; Stevens, C. ; Teyssié, P. *Makromol. Chem.,*
          Suppl. I **1975,** *191*
(52) -   Jacobs, C. ; Dubois, P. ; Jérôme, R. ; Teyssié, P.
          Macromolecules, **1991,** *24,* 3027
(53) -   Ropson, N. ; Dubois, P. ; Jérôme, R. ; Teyssié, P.
          Macromolecules **1995,** *28,* 7589
(54) -   Duda, A. *Macromolecules,* **1996,** *29,* 1399
(55) -   Kowalski, A. ; Duda, A. ; Penczek, S. *Macromolecules,* **1998,**
          31, 2114

(56) -   Duda, A.; Biela, T. ; Libiszowski, J. ; Penczek, S. ; Dubois, P.
          ; Mecerreyes, D.; Jerôme, R. *Polym. Degr. Stab.* **1998**, *59*, 215
(57) -   Yui, N. ; Dijkstra, P.J. ; Feijen, J. *Makromol. Chem.* **1990**,
          191, 481
(58) -   Kurcok, P. ; Penczek, J. ; Franek, J. ; Jedlinski, Z.
          Macromolecules **1992**, *25*, 1992
(59) -   Bero, M. ; Adamus, G. ; Kasperczyk, J. ; Janeczek, H. *Polym.*
          Bull. **1993**, *31*, 9
(60) -   Ikada, Y. ; Janshidi, K. ; Tsuji, H. ; Hyon, S.H.
          Macromolecules **1987**, *20*, 906
(61) -   Loomis, G.L. ; Murdoch, J.R. ; Gardner ; K.H. *Polym. Prep.*
          (Am. Chem. Soc., Div. Polym. Chem.) **1990**, *31*, 55 ; US Patent
          4, 800, 2189 **1989**, jan. 24
(62) -   Brochu, S. ; Prud'homme, R.E. ; Barakat, I. ; Jérôme, R.
          Macromolecules **1995**, *28*, 5230
(63) -   Brizzolara, D. ; Cantow, H.J. ; Diederichs, K. ; Keller, E. ;
          Domb, A.J. *Macromolecules*, **1996**, *29*, 191
(64) -   Spassky, N. ; Pluta, C. ; Simic, V. ; Thiam, M. ; Wisniewski,
          M. *Macromol. Symp.* **1998**, *128*, 39
(65) -   Stevels, W.M. ; Ankoné, M.J.K. ; Dijkstra, P.J. ; Feijen, J.
          Macromol. Symp. **1996**, *102*, 107

Chapter 13

Thermodynamics, Kinetics, and Mechanisms of Cyclic Esters Polymerization

Andrzej Duda and Stanislaw Penczek

Center of Molecular and Macromolecular Studies, Polish Academy of Sciences, Sienkiewicza 112, PL–90–363 Lodz, Poland

Recent advances in pseudoanionic ring-opening polymerization (ROP) of cyclic esters initiated with covalent metal alkoxides and carboxylates are reviewed. General aspects of covalent ROP are discussed: first, thermodynamics, particularly of monomers exhibiting low ring strain. Then, structures of covalent initiators, their behavior in initiation, formation of active species, kinetics, and mechanism of propagation. Polymerization with reversibly aggregating species is analyzed in more detail. Mechanism of initiation and propagation, in the polymerizations induced with metal carboxylates (tin(II) octoate ($Sn(Oct)_2$)) is discussed. It is shown that the actually initiating species are formed by converting $Sn(Oct)_2$ into OctSnOR and/or $Sn(OR)_2$ in reactions with ROH present in the system. Chain transfer reactions (transesterifications, both intra- and intermolecular), taking place in the polymerization of lactones and lactides, are quantitatively analyzed.

Ring-opening polymerization (ROP) of cyclic esters is one of the preferred methods for a synthesis of high molar mass aliphatic polyesters (*1-3*) and more recently has even been extended to the enzyme catalyzed processes (*4-7*). Although polycondensation is at the basis of the major industrial aromatic polyesters (e.g. polyterephthalates), ROP has already been used in industrial production of two aliphatic polyesters, namely poly(ε-caprolactone) (*1*) and poly(L-lactide) (*3*). These are the ecologically degradable polymers and may play an important role in future polymer industry. Particularly poly(L-lactide), made from renewable starting materials (carbohydrates of various agricultural origin), can become attractive for countries that do not have their own sources of olefins.

On the other hand, ROP of cyclic esters became an efficient tool in studies of the mechanism of anionic and pseudoanionic (covalent) ROP. This is because in many cyclic ester/initiator systems the termination could be excluded. There are, however, two well documented chain transfer reactions. Both are based on transesterification, taking place also in polycondensation: back- and/or end-to-end-biting and chain transfer to foreign macromolecules followed by the chain rupture.

Intention of the present paper is to review basic principles of thermodynamics, kinetics, and mechanisms of cyclic esters polymerization. Although this paper is based mostly on the work of our own, it describes general phenomena related directly to the controlled synthesis of poly(aliphatic ester)s. Elementary chain-growth reactions: initiation and propagation are discussed from the point of view of the livingness of polymerization. Eventually, the molecular structures of the growing species are compared with their activity in transesterification.

Thermodynamics

Aliphatic cyclic esters of various ring sizes provide high molar mass polyesters. In Table I we compare the thermodynamic data of polymerization: enthalpy (ΔH_p), entropy (ΔS°_p), and a monomer concentration at equilibrium ($[M]_{eq}$), at 298 K for oxirane and five cyclic esters of various ring sizes (8-13).

In ROP, conducted at constant pressure, the change of enthalpy is mostly due to the monomer ring strain energy, if monomer-polymer-solvent specific interactions can be neglected (14-17). The major contributions to the ring strain come from: deviation from the non-distorted bond angle values (e.g.: for cyclic esters: 110.5° (C-C-C), 109.5° (C-O-C(O)) or 110° (O-C(O)-C)), bond stretching and/or compression, repulsion between eclipsed hydrogen atoms, and non-bonding interactions between substituents (angular, conformational, and transannular strain, respectively) (14, 17).

Moreover, polymerization of majority of monomers is accompanied by the entropy decrease. Thus, polymerization is thermodynamically allowed when the enthalpic contribution into free energy prevails (thus when $\Delta H_p < 0$ and $\Delta S_p < 0$, the inequality $|\Delta H_p| > -T\Delta S_p$ is required; cf. eq 1). Therefore the higher is the ring strain the lower is the resulting monomer concentration at equilibrium (eq 2).

$$\Delta G_p = \Delta H_p - T\Delta S_p \qquad (1)$$

$$\ln[M]_{eq} = \Delta H_p / RT - \Delta S^\circ_p / R \qquad (2)$$

(where T is the absolute temperature and R the gas constant)

Although formation of the three-membered α-lactone intermediates in the nucleophilic substitution reaction of α-substituted carboxylate anions was postulated on the basis of quantum mechanical calculations (18), these cyclic esters have never been isolated. Therefore, we decided to give for comparison in Table I thermodynamic parameters for another highly strained three-membered oxacyclic monomer - oxirane. The four-membered β-propiolactone also belongs to the most strained cyclic monomers and its equilibrium monomer concentration is unmeasurably low, namely about 10^{-10} mol·L^{-1} at room temperature.

Table I. Standard Thermodynamic Parameters of Polymerization for some Selected Oxacyclic Monomers (298 K)

Monomer[a]	EO	PL	BL	VL	LA	CL
Ring size	3	4	5	6	6	7
Conditions[b]	lc	lc	lc	lc	ss	lc
$\dfrac{\Delta H_p}{kJ \cdot mol^{-1}}$	-140	-82.3	5.1	-27.4	-22.9	-28.8
$\dfrac{\Delta S^o{}_p{}^c}{J \cdot mol^{-1} \cdot K^{-1}}$	-174	-74	-29.9	-65.0	-25.0	-53.9
$\dfrac{[M]_{eq}{}^d}{weight\ fraction}$	$3.6 \cdot 10^{-16}$	$2.8 \cdot 10^{-11}$	$2.9 \cdot 10^{2}$	$3.9 \cdot 10^{-2}$	$1 \cdot 10^{-3}$	$6 \cdot 10^{-3}$
Ref	8	9	10	11	12	13

[a] EO = ethylene oxide (oxirane), PL = β-propiolactone, BL = γ-butyrolactone, VL = δ-valerolactone, LA = L,L-dilactide, CL = ε-caprolactone. [b] The first letter denotes the state of monomer, the second that of polymer: l = liquid, c = condensed, s = solution. [c] Standard states: weight fraction = 1 (lc) or 1.0 $mol \cdot L^{-1}$ (ss). [d] Calculated from eq 2 (lc) or determined experimentally at the monomer - polymer equilibrium (ss); in order to obtain the approximate value in $mol \cdot L^{-1}$ the weight fraction value should be multiplied by 10.

However, both the six- and seven-membered monomers, namely L,L-dilactide (LA) and ε-caprolactone (CL) have relatively high equilibrium monomer concentrations, that can not be neglected in the practical considerations: in handling of the final polymer and in studies of polymerization, particularly at elevated temperatures. It has to be remembered, that whenever in the final polymer objects (films, injecting molded pieces) one or more macromolecules are ruptured, the active species may emerge, from which the unzipping will take place until the equilibrium monomer concentration is reached. Fortunately, at least in the case of L,L-dilactide its final product of hydrolysis, lactic acid, is an ecologically friendly compound.

More recently, we have determined the standard thermodynamic parameters for polymerization of L,L-dilactide (12). Its equilibrium concentration appeared considerably high, particularly at elevated temperatures (at which LA is usually polymerized). For a temperature range from 80 to 133°C [LA]$_{eq}$ changes from 0.058 to 0.151 $mol \cdot L^{-1}$. This six-membered cyclic diester assumes irregular skew-boat conformation, in which two ester groups can adopt planar conformation and has therefore a relatively high enthalpy of polymerization equal to 22.9 $kJ \cdot mol^{-1}$. This is very close to the ring strain of δ-valerolactone and CL, equal to 27.4 and 28.8 kJ·mol^{-1}, respectively. Strain comes in these compounds from C-H bonds interactions and from distortion of the bond angles. In contrast, high ring strain in the four-membered β-propiolactone is mostly because of the bond angles distortion and bond stretching.

In the five-membered cyclics ring strain comes almost exclusively from the conformational interactions (*14, 16*). It is known, however, that the five-membered esters are not strained because of the reduced number of the C-H bond oppositions, caused by the presence of the carbonyl group in the monomer ring. Indeed, for γ-butyrolactone (BL) we have ΔH_p = 5.1 kJ·mol^{-1} and ΔS°_p = -65 J·mol^{-1}·K^{-1}. These give $[BL]_{eq} \approx 3\cdot10^3$ mol·L^{-1}, whereas the monomer concentration in bulk does not exceed 13 mol·L^{-1}. Therefore in the majority of the polymer chemistry textbooks it is stated that BL is not able to polymerize (e.g. one finds in Odian's textbook: "...the 5-membered lactone (γ-butyrolactone) does not polymerize..." (*19*)). BL is indeed not able to give a high molar mass homopolymer, but this feature sometimes is incorrectly identified with an inability of BL to undergo the ring-opening reaction at all. The equilibrium constants for the first few monomer additions (K_0, K_1, ...) are not the same, as for an addition to a high polymer (K_n) due to an influence of the head- and tail-end-groups and usually K_0 and K_1 are larger than K_n:

$$X + BL \xrightleftharpoons{K_0} X\text{-bl}^*$$

$$X\text{-bl}^* + BL \xrightleftharpoons{K_1} X\text{-bl-bl}^*$$

$$\text{---} \qquad (3)$$

$$X\text{-(bl)}_n\text{-bl}^* + BL \xrightleftharpoons{K_n} X\text{-(bl)}_{n+1}\text{-bl}^*$$

$$\quad (PBL)_n \qquad\qquad\qquad\qquad (PBL)_{n+1}$$

(where X denotes initiator, bl the polyester repeating unit derived from BL, and bl* the pertinent active center)

Moreover, the concentration term contribution into the Gibbs energy, usually presented as RTln [M], because -(m)$_n$- ≈ -(m)$_{n+1}$-, should be given differently for short chains, when this equality may not hold, namely:

$$\Delta G_n = \Delta H_n - T\Delta S^\circ_n + RT \ln\{[(PBL)_{n+1}]/[BL][(PBL)_n]\} \qquad (4)$$

at the very beginning of the BL ring-opening, when $[(PBL)_n] \gg [(PBL)_{n+1}]$ may outweight a sum of the enthalpic and entropic contributions (both positive) and eventually gives the negative value of ΔG_n. Therefore, even when $\Delta H_p \approx 0$ as for reactions 3, formation of short BL oligomers is not thermodynamically forbidden, because apparently $-\Delta S^\circ_p < |R\ln\{[(PBL)_{n+1}]/[BL][(PBL)_n]\}|$.

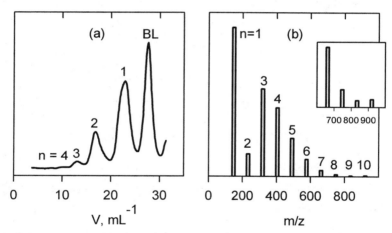

Thus, we studied polymerization (actually, oligomerization) of BL initiated with aluminum *tris*-isopropoxide (Al(O^{i}Pr)$_3$) (eq 5) (*20-22*). Both size exclusion chromatography (SEC) (Figure 1 (a)) and mass spectrometry (MS) with chemical ionization (Figure 1 (b)) show, that indeed oligomers are formed - up to decamers as detected by MS, whereas only the tetramer was seen in the SEC traces.

Figure 1. Oligomerization of γ-butyrolactone (BL) initiated with Al(O^{i}Pr)$_3$. SEC trace (a) and mass spectrum (chemical ionization) (b) of the isolated series of the linear oligomers: H-[O(CH$_2$)$_3$C(O)]$_n$-OiPr . Conditions of oligomerization: [BL]$_0$ = 3.8 mol·L^{-1}, [Al(O^{i}Pr)$_3$]$_0$ = 0.2 mol·L^{-1}, THF solvent, 80 °C (20).

Moreover, BL was found to copolymerize effectively with CL (*21, 22*). Molar masses (the number-average values, M_n) of the resulting poly(CL-*co*-BL)'s were controlled by the consumed comonomers to the initiator (Al(O^{i}Pr)$_3$) in the feed concentrations ratio. Figure 2 gives the ^{13}C NMR spectrum of CL/BL copolymers with $M_n \approx 10^4$. The presence of several new peaks in the >C=O group absorption range, differing from εεε and γγγ sequences (where ε and γ denote the copolymer repeating units derived from CL and BL, respectively) indicates formation of copolymers; for molar fraction of the γ-oxybutyryl units (F_{BL}) equal to 0.11 this is mostly a random one, whereas for $F_{BL} = 0.43$ closer to pseudoperiodic copolymer is formed.

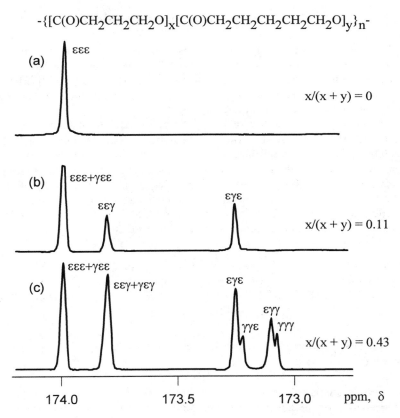

Figure 2. $^{13}C\{^{1}H\}$ (100 MHz ^{13}C, INVGATE) spectra, in the >C=O group absorption range, of the isolated poly(CL-co-BL) containing (in mole fractions) 0 (a), 0.11 (b), and 0.43 (c) γ-oxybutyryl units. Chloroform-d solvent (Reprinted with permission from reference 21. Copyright 1996 Wiley-VCH).

Thus, the thermodynamic polymerizability can not be taken as a direct measure of a monomer reactivity. For instance, the rate constants of basic hydrolysis of BL and CL at comparable conditions are close one to another ($1.5 \cdot 10^{-4}$ and $2.6 \cdot 10^{-4}$ mol$^{-1} \cdot$L$\cdot$s^{-1}, respectively) (23) although the corresponding ring strains differ considerably (cf. Table I). For another pair of monomers, β-propiolactone (PL) and CL the rate constants of propagation, in the anionic process, differ about 10^{3} times in favor of the less strained CL (24). These difference of reactivities can, however, be related to the structure of active species: PL propagates on the caboxylate, whereas CL on the much more reactive alkoxide anions and the ring strain of monomers appears to be, in this case, of a minor importance in relation to the rate of polymerization.

Polymerization Mechanism

In the present review almost exclusively the covalent (pseudoanionic) polymerization is discussed. Since mid 1970-ies our laboratory has been engaged, however, in the studies of ionic polymerization of lactones (24-29). Then we found out, that ionic polymerization is much less selective than the covalent one (30-32) and studied the latter process in a more detail (33-42).

Some facts related to the ionic polymerization have to be mentioned. Particularly important was the first observation of a living polymerization[*] of cyclic esters. In 1976 two papers were published, showing, that introduction of the crown ether (25) or cryptand (43) in the anionic polymerization of β-propiolactone initiated with potassium acetate gives indeed a living polymerization (eq 6). M_n of the growing polyester became a linear function of monomer conversion and the semilogarithmic kinetic plot was also a linear function of time (25). Thus, the two criteria (44) were simultaneously fulfilled, as required for a living process.

$$
\underset{\text{CH}_3\text{CO}^{\ominus}\,\text{K}^{\oplus}\cdot\text{DBC}}{\overset{\text{O}}{\underset{\|}{\parallel}}} + n\;\; \underset{\text{H}_2\text{C}\text{---}\text{O}}{\overset{\text{H}_2\text{C}\text{---}\text{C}\diagup\!\!\diagup^{\text{O}}}{}} \xrightarrow[\text{propagation, } k_p]{\text{initiation, } k_i} \text{living polymerization} \qquad (6)
$$

(where DBC denotes dibenzo-18-crown-6 ether, k_i and k_p the respective rate constants)

Living character of the anionic polymerization of β-propiolactone enabled determination of the absolute rate constants of propagation on macroions ($k_p^{(-)}$) and macroion pairs ($k_p^{(\pm)}$). Depending on the temperature of polymerization macroions are more or less reactive than macroion pairs. For example, this inversion of reactivities was observed at $T_i = 293$ K in DMF solvent; namely, $k_p^{(\pm)} > k_p^{(-)}$ below T_i (28).

Later it was shown, on the basis of more detailed studies, that already at room temperature there is some measurable transfer to monomer (cf. eq 7).

---

[*] There is a number of ways of understanding of the expression "living polymerization", coined by Szwarc in 1956. We follow here the recommendation of IUPAC, where living polymerization requires $k_t = 0$ and $k_{tr} = 0$. Since "never" these rate constants are in reality equal to 0, we think that whenever k_t or k_{tr} are the measurable ones, their values have to be determined. To avoid the internal contradiction one may say, for example: living with $k_p / k_{tr} \approx 10^5$. In our opinion, in any system for which the expression "living" is used an attempt has to be made to at least to estimate k_p and k_{tr}.

Namely, $k_p/k_{tr} = 4 \cdot 10^4$ for β-propiolactone (CH_2Cl_2 solvent, 20°C), and $k_p/k_{tr} \le 2.0 \cdot 10^2$ for β-butyrolactone (THF solvent, 20°C) propagation to transfer rate constants ratios were determined (29). Both electronic and steric effects of the methyl group are responsible for a lower value of the k_p/k_{tr} ratio in β-butyrolactone, when compared to β-propiolactone and this is at least partially because of the differences in the rate constants of propagation ($4 \cdot 10^{-3}$ and 10^{-6} $mol^{-1} \cdot L \cdot s^{-1}$ at 20°C in THF solvent, respectively).

(where R stands for H (β-propiolactone) or CH_3 (β-butyrolactone), k_{tr} denotes the rate constant of transfer; $K^{\oplus} \cdot$ DBC counterion omitted)

These k_p/k_{tr} values set also the limits of the polymerization degree for the corresponding polyesters. On the basis of these measurements the limit of M_n, for poly(β-propiolactone) is of the order of 10^6, whereas for poly(β-butyrolactone) it would be, at the same conditions, about $1.7 \cdot 10^4$. Claims of the synthesis of poly(β-butyrolactone) of higher molar mass ($M_n = 3.8 \cdot 10^4$) in anionic polymerization were published (45). On the other hand, it is known for a long time that alumoxane initiators give much higher molar masses ($M_n \le 10^5$), but molar masses distributions of thus prepared poly(β-butyrolactone) are polymodal (46-48). Lowering the temperature could help, but then polymerization could become exceedingly slow.

Similar reactions can take place at the stage of initiation. Then, transfer influences exclusively the structure of the end-groups and usually does not change the number of macromolecules formed. Therefore, at this stage transfer is less important, since eventually growing species are quantitatively formed. Moreover, independently of the extent of these side reactions they do not break the material chains. Usually, both initiation by addition and transfer coexist, and their relative importance depends on the ionic initiator used: more important for alcoholates, less important for carboxylates (24, 49, 50). Sometimes the reversibility of transfer may minimize its importance. Actually, Dale has shown, that alcoholates react reversibly with higher lactones, including ε-caprolactone, giving enolates (51), e.g.:

Although the quantitative data are not yet available, the importance of transfer may depend on the counterion structure and the extent of pairing. Transfer processes are even more important for the polymerization of other oxacyclic monomers, such as propylene oxide, where transfer in anionic polymerization does not allow forming of high polymers (*52*).

Moreover, it has been known that in anionic polymerization of cyclic esters extensive back biting takes place (*26, 53*), although it was observed that with covalent initiators conditions could be found, allowing preparation of polymers (e.g. poly(ε-caprolactone)) free of cyclics, under the kinetic control (*30-32, 54*). Therefore, we mostly concentrated our efforts on understanding of the covalent polymerization.

Covalent (Pseudoanionic) Initiators and Initiation

There are three major groups of the covalent initiators mostly used in polymerization of cyclic esters and which we studied in our work. These are: alkylmetal alkoxides, metal alkoxides, and metal carboxylates.

Alkylmetal alkoxides

In this group of initiators the most extensively studied were dialkylaluminum alkoxides (R_2AlOR') introduced over 30 years ago (*55, 56*) but studied quantitatively and understood only recently (*30, 33, 34, 36, 39*).

Depending on the size of substituents (R and R') these initiators are known to exist in solution mostly as dimeric or trimeric species (*57-59*). More complicated structures have also been proposed. Aggregates are usually in equilibria with unimeric structures. The rate of interconversion depends on the number of factors (e.g. solvent, temperature, R and R' substituents). Diethylaluminum ethoxide (Et_2AlOEt) used in our studies gives, however, simple ^1H NMR spectrum, indicating that either unimeric species are present or that there is a fast exchange between the aggregated and nonaggregated (unimeric) forms (*32*). The same dialkylalkoxy compound may assume in one solvent the aggregated structure and deaggregated in another one. Deaggregation proceeds easier in polar and nucleophilic solvents, able to interact with Al atoms. If solvent is too strong as a complexing agent, stronger than monomer itself, the "initiator" may be exclusively unimeric, but initiation may not take place at all or become very slow. This kind of behavior was observed in our attempts of initiating the CL polymerization by Et_2AlOEt in hexamethylphosphorous triamide (HMPT) (*60*).

$$\text{R} \diagdown \atop \text{R} \diagup \!\!\!\!\!\!> \text{Al}\!-\!\text{OR'} \ + \ (n+1)\ \underset{\displaystyle \smile}{\text{O}}\!-\!\overset{\displaystyle \overset{O}{\|}}{\text{C}} \ \longrightarrow \ {\text{R} \diagdown \atop \text{R} \diagup}\!\!\!\!\!\!> \text{Al}\!-\!\text{O}\ \underset{\displaystyle \smile}{}\ \overset{\displaystyle \overset{O}{\|}}{\text{C}}\!\!-\!\!\Big(\!\text{O}\ \underset{\displaystyle \smile}{}\ \overset{\displaystyle \overset{O}{\|}}{\text{C}}\!\Big)_{\!\!n}\!\!-\!\text{OR'} \qquad (9)$$

The ^1H NMR spectrum of a low molar mass oligomer, formed when a high enough ratio $[R_2AlOR']_0/[\text{monomer}]_0$ is used, shows exclusively signals expected for the stoichometric reaction product, responsible for further polymerization. The alkyl groups stay intact on Al atom and only the alkoxide group is involved in initiation, as it is schematically shown in eq 9 above (*32, 33*).

Metal alkoxides

Metal alkoxides give stronger aggregates in comparison to alkylmetal alkoxides. This results from the presence of a larger proportion of oxygen atoms, forming bridges between the metal atoms. Therefore, whenever steric factors permit, relatively stable aggregates are formed.

Aluminum *tris*-isopropoxide has been the most often used in polymerization. For a long time it was not clear, why various authors observed different numbers of chains supposedly growing from one aluminum atom. Then, it was found that the two known aggregates, namely a trimer (A_3) and a tetramer (A_4), do not only exchange slowly, in comparison with the rate of polymerization (eq 10):

$$4\ A_3 \underset{}{\overset{K_{34},\ \text{slow}}{\rightleftharpoons}} 3\ A_4 \tag{10}$$

(where K_{34} denotes the equilibrium constant)

but react with monomers with rates, that differ from 10^2 times (LA) to 10^5 times (CL), at least at moderate temperatures (*37, 38, 40*). Similar difference of reactivities between A_3 and A_4 was already reported for the Meerwein-Ponndorf-Verley reaction (*64*). However, these differences of reactivities are not observed when $Al(O^iPr)_3$ is used in the presence of an alcohol, since the exchange between A_3 and A_4 is then much faster (*35*).

Figure 3 shows the ^1H NMR spectra of both $Al(O^iPr)_3$ aggregates. Detailed assignments of peaks in this spectra is given in ref *37*. As it follows from Figure 3, A_3 and A_4 have areas of different chemical shifts for the same (chemically) groups, but being in different magnetic environment. Therefore, it was possible to observe directly in ^1H NMR simultaneous consumption of the actual initiator. The methine protons absorption range was particularly useful for this purpose.

This is illustrated in Figure 4, for the polymerization of CL (eq 11), showing that at the complete consumption of A_3 in initiation and polymerization of CL, A_4 is left almost intact (*36, 37*).

$$A_3 + A_4 + n \left[\text{(caprolactone)}\right] \xrightarrow[\text{propagation, } k_p]{\text{initiation, } k_i}$$

$$\xrightarrow{\hspace{2cm}} Al\{[O(CH_2)_5\overset{O}{\overset{\|}{C}}]_x OCH(CH_3)_2\}_3 + A_4$$

(11)

Similar behavior of the $Al(O^iPr)_3$ initiator we observed also in the polymerization of the much less reactive monomer - L,L-dilactide (*40*). In Table II are given kinetic parameters revealing difference of A_3 and A_4 reactivities in their reactions with CL and LA monomers.

Table II. Comparison of the Rate Constants of Propagation (k_p) and Initiation (k_i) for Polymerization of CL and LA Initiated with A_3 and A_4[a]

Monomer	ε-caprolactone	L,L-dilactide
$\dfrac{k_p}{mol^{-1} \cdot L \cdot s^{-1}}$	0.5	7.5×10^{-5}
$\dfrac{k_i(A_4)}{mol^{-1} \cdot L \cdot s^{-1}}$	$5 \cdot 10^{-6}$	$2.7 \cdot 10^{-8}$
$k_i(A_3)/ k_i(A_4)$	10^5	$4.1 \cdot 10^3$

[a] Conditions of polymerization (concentrations given in $mol \cdot L^{-1}$): $[CL]_0 = 2.0$, $[LA]_0 = 1.0$, $3[A_3]_0 = 4[A_4]_0 = [Al(O^iPr)_3]_0 = 0.01$; THF solvent, 20°C (Data taken from ref 40).

Fortunately, A_3 and A_4 can be prepared as individual compounds. Moreover A_3 is soluble in some solvents in which A_4 is practically insoluble at all (e.g. in pyridine). Therefore, as it is seen in Figure 3, almost pure A_3 and A_4 were obtained and applied then in polymerizations. The rate of $A_3 \rightleftharpoons A_4$ interconversion is relatively low; kinetic data are given in Figure 5. Time required for conversion of a few percent of A_4 (unreactive) into A_3 (reactive) is, at room temperature, over 10^3 times longer than time needed for complete (over 90%) polymerization of CL or LA, however, reactivities of both aggregates tend to converge with increasing temperature (*40*).

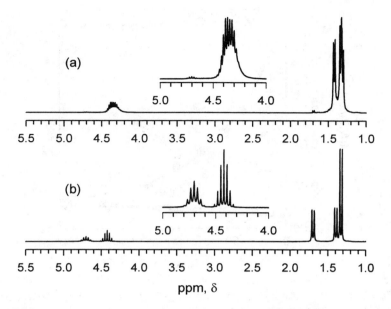

Figure 3. ^1H NMR (300 MHz) spectra of (a): 98% trimeric Al(O^{i}Pr)$_3$ (A$_3$) and (b): tetrameric Al(O^{i}PrO)$_3$ (A$_4$). Conditions measurements: [Al(O^{i}Pr)$_3$]$_0$ = 0.1 mol·L^{-1}, benzene-d$_6$ solvent, 23°C (Reprinted with permission from reference 37. Copyright 1995 Wiley-VCH).

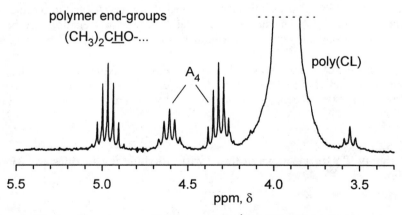

Figure 4. The methine protons fragment of the ^1H NMR (200 MHz) spectrum of the living poly(CL). Conditions of polymerization (concentrations given in mol·L^{-1}): [CL]$_0$ = 2.0, [Al(O^{i}Pr)$_3$]$_0$ = 3[A$_3$]$_0$ + 4[A$_4$]$_0$ = 0.07; 3[A$_3$]$_0$ = 0.037; benzene-d$_6$ solvent, 25°C. Spectrum recorded 20 min. after "complete" CL consumption (≈ 5 min.) (Reprinted with permission from reference 37. Copyright 1995 Wiley-VCH).

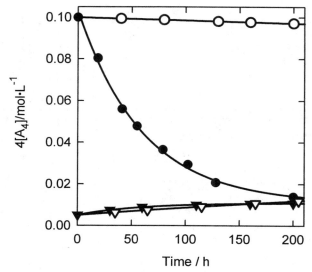

Figure 5. Kinetics of $A_3 \rightarrow A_4$ and $A_4 \rightarrow A_3$ conversions. ($\triangledown$): $A_3 \rightarrow A_4$, 25 °C; ($\blacktriangledown$): $A_3 \rightarrow A_4$, 70 °C; ($\bigcirc$): $A_4 \rightarrow A_3$, 25 °C; ($\bullet$): $A_4 \rightarrow A_3$, 70 °C. Reaction conditions: $3[A_3]$ + $4[A_4]$ = 0.1 mol·L^{-1}, benzene-d_6 solvent (Reproduced with permission from ref 38. Copyright American Chemical Society 1995).

A_4 is thermodynamically more favorable at room and lower temperatures, whereas at higher temperatures equilibrium 9 shifts into the side of A_3. The respective equilibrium constants (K_{34}, cf. eq 4) are equal to 3.0 and $2.7 \cdot 10^{-2}$ mol·L^{-1} at 25 and 70°C, respectively (38). Thus, some energy is needed to break A_4 and to produce less stable (less favorable) A_3 form. Since A_4 has a higher aggregation number its formation is driven by enthalpy, whereas conversion into A_3 is helped by increasing entropy.

The behavior of some other initiators, like μ-oxoalkoxides of lanthanium, samarium or yttrium also looks today bizzare, and perhaps their closer studies will reveal that their strange behavior (not covered by this review) has its origin in aggregation.

Propagation in Polymerizations Initiated with Alkylmetal Alkoxides and Metal Alkoxides.

The number of chains growing from one initiator molecule.

Before discussing the peculiarities of propagation a question should be answered: how many chains grow from one metal atom ?

Figure 6 illustrates an evolution of molar masses (M_n) with conversion for the poly(LA) formed in the polymerization of LA initiated with various alkoxymetallic initiators: $R_nMt^{(x)}(OR')_{x-n}$. Lines in Figure 6 (a) are calculated on the assumption that every alkoxide substituent starts growth of one polyester chain; points are experimental. In Figure 6 (b) the data of Figure 6 (a) are normalized in such a way, that give calculated and experimental M_n for identical starting concentrations of alkoxxide groups (and not molecular initiators, as in Figure 6 (a)).

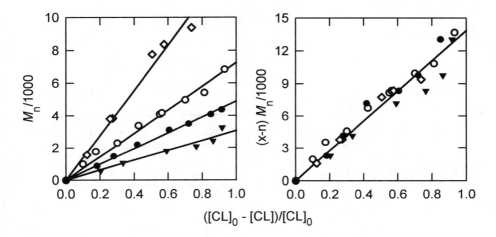

Figure 6. Dependence of M_n *of poly(L-lactide) on the monomer conversion degree. Polymerization of L,L-dilactide initiated with metal alkoxides* $(R_nMt^{(x)}(OR')_{x-n})$: *Bu₃SnOEt (◇), Sn(OBu)₂ (○), Al(OⁱPr)₃ trimer - A₃ (●), and Ti(OⁱPr)₄ (▼). Conditions: [LA]₀ = 1.0 mol·L⁻¹, [R_nMt^{(x)}(OR')_{x-n}]₀ =10⁻² mol·L⁻¹, THF solvent, 80°C (except Sn(OBu)₂ for which polymerization was conducted at 25°C) (65).*

There is a good agreement, for Al, Ti, Sn(II), and Sn(IV) alkoxides, between theoretically predicted dependencies and experimental points, meaning that each alkoxide group starts one chain for all of these initiators differing in the numbers of alkoxide groups attached to one metal atom.

A number of chains growing from one metal atom was also established from a direct determination of molar mass. First, for the growing chains attached to the metal atom (e.g. three arm "star" for Al.(OⁱPr)₃) and then after hydrolysis of the alkoxide bonds. Then, as in the case of aluminum *tris*-alkoxides, the growing species should have a molar mass three times larger than the finally detached macromolecules. Results of such measurements are collected in Table III. Since indexes of the polydispersity (M_w/M_n) of the killed polymers are close to 1, similar values of M_w/M_n could be assumed for the living polymers, and therefore $M_w(\text{MALLS}) \approx M_n$.

As it follows from Table III, the predicted and measured values of M_n for the killed polymers are close one to another, and the respective values for the living polymers are approximately three times higher. Thus, it means, that the growing macromolecules form three-arm branched structures, i.e. three chains grow on one Al atom, as presented schematically by eq 12.

Table III. Comparison of M_w of Living Poly(CL) ((...-CH$_2$O)$_3$Al) with M_n of Deactivated Poly(CL) (...-CH$_2$OH)^a

$\dfrac{[CL]_0}{mol \cdot L^{-1}}$	$\dfrac{3[A_3]_0}{10^{-4}\,mol \cdot L^{-1}}$	$10^{-5}M_n$				
		calcd ^{b} living	MALLS ^{c} living	calcd ^{d} deactd	SEC ^{e} deactd	$\dfrac{M_w}{M_n}^f$ deactd
1.47	4.9	3.57	1.19	1.19	1.08	1.03
2.00	12.9	1.77	0.59	0.59	0.65	1.05

^{a} Initiator: A$_3$, THF solvent, 25°C. ^{b} M_n(calcd, living) = 114.14([CL]$_0$/3[A$_3$]$_0$); ^{c} M_w(MALLS) measured by the multiangle laser light scattering in THF solvent at 20°C; ^{d} M_n(calcd, deactivated) = (1/3)·M_n(calcd, living); ^{e} absolute values, measured by light scattering and RI detectors in series. ^{f} measured by SEC. (Reproduced with permission from ref 38. Copyright American Chemical Society 1995).

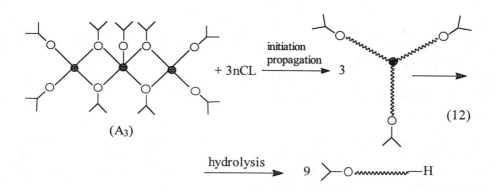

(where ○ and ⬤ denote the oxygen and aluminum atoms, respectively;

>— the isopropyl group; ∿∿∿ the PCL chain: —(O(CH$_2$)$_5$C(O)—)$_n$)

Kinetics of Propagation

First-order kinetic plots for CL and LA polymerization in a few systems, involving Al(O^{i}Pr)$_3$, Sn(OBu)$_2$, Et$_2$AlOEt, Bu$_3$SnOEt, Ti(OiPr)$_4$, and Fe(OEt)$_3$ are illustrated in Figure 7. Their linearity from the very beginning of polymerization means that the rate of initiation is at least comparable to the rate of propagation and that termination is practically absent.

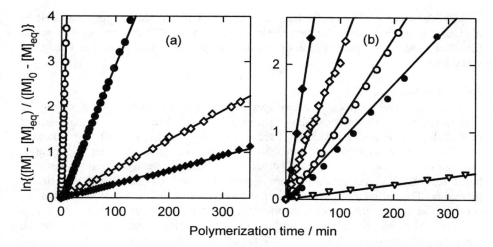

Figure 7. Kinetics of ε-caprolactone (a) and L,L-dilactide (b) polymerization initiated with metal alkoxides $(R_nMt^{(x)}(OR')_{x-n})$. [CL] and [LA] determined dilatometrically and polarimetrically, respectively. Conditions (concentrations given in $mol·L^{-1}$): (a, ○) $3[A_3]_0 = 3·10^{-3}$, 25 °C; (a, ●) $[Sn(OBu)_2]_0 = 3.3·10^{-3}$, 80 °C; (a, ◇) $[Et_2AlOEt]_0 = 8·10^{-3}$, 25 °C; (a, ◆) $[Bu_3SnOEt]_0 = 9.9·10^{-2}$, 80 °C; (b, ◆) $[Ti(O^iPr)_4]_0 = 10^{-2}$, 80 °C; (b, ◇) $[Fe(OEt)_3]_0 = 1.4·10^{-3}$, 80 °C; (b, ○) $3[A_3]_0 = 2·10^{-2}$, 80 °C; (b, ●) $[Sn(OBu)_2]_0 = 3·10^{-3}$, 50 °C; (b, ▽) $[Bu_3SnOEt]_0 = 5·10^{-2}$, 80 °C; $[CL]_0 = 2$, $[LA]_0 = 1.0$; THF solvent (65).

Propagation, taking an example of R_2AlOR', can be visualized in the simplest form as shown in eq 13.

$$(13)$$

(where R denotes the alkyl or alkoxide group)

and proceeds, the most probably, similarly to the $B_{AC}2$ mechanism, according to the Ingold's classification (66).

On the basis of Figure 7 some comparison of reactivities is possible. This is discussed in more detail in the next section, since the "total" reactivity and the actual (absolute) constant of propagation do not have to give the same dependencies on the metal structure.

(a) dialkylaluminum alkoxides.

Structures and behavior of active species in propagation reflect the properties of initiators, described in the preceding section. The "only" difference, when an initiator like dialkylaluminum alkoxide is used, is related to the size of the alkoxide substituent, becoming during propagation a polymer chain (cf. eq 9). Indeed, propagation initiated with R_2AlOR' proceeds exclusively on the aluminum-oxygen bond R_2Al-OR', therefore during propagation the size of the alkoxy substituent increases. The influence of the size of the polymer chain on the rate constant of propagation is not yet known, neither for dialkylalkoxy nor trialkoxyaluminum initiators. The ester bond appear in the polymer repeating units, with their electron donating oxygen atoms from carbonyl groups, providing an additional source of coordination of aluminum atoms. Information on the state of aggregation of the growing species comes from kinetics (*34, 36, 39*), whereas the information on the degree of solvation was obtained from the ^{27}Al NMR studies (*56*).

It was assumed in further analysis that the total concentration of active species ($[P_i^*]$) is equal to the starting concentration of alkoxide groups in initiator. This assumption, in turn, is justified on the basis of the data in Figure 6. Thus, in the present case $[P_i^*]$ is equal to the initiator concentration in the feed ($[I]_0$).

It was observed that the rate of CL polymerization initiated with R_2AlOR' is directly proportional to the first power of concentration of the active species $[P_i^*]$ only at the low enough $[P_i^*]$. At higher concentration the rate is either proportional to $[P_i^*]^{1/2}$ when R = ^iBu or to $[P_i^*]^{1/3}$, when R = Et (*36*). These conclusions are derived from data given in Figure 8, where relative polymerization rates ($r_p = [CL]^{-1}\cdot d[CL]/dt = \{\ln([CL]_0/[CL])\}/t$) are plotted against the starting concentrations of initiator in the bilogarithmic coordinates. Such kinetic behavior can be explained by kinetic scheme given below.

$$x \ldots\text{-}P_n^* \quad \underset{k_{dag}}{\overset{k_{ag}}{\rightleftharpoons}} \quad (\ldots\text{-}P_n^*)_x \qquad (K_{ag} = k_{ag}/k_{dag})$$

$$\ldots\text{-}P_n^* + M \quad \underset{k_d}{\overset{k_p}{\rightleftharpoons}} \quad \ldots\text{-}P_{n+1}^* ; \quad (\ldots\text{-}P_n^*)_x + M \xrightarrow{\quad\times\quad}$$

(14)

(where k_p, k_d, k_{ag}, and k_{dag} denote the rate constants of propagation, depropagation, aggregation, and deaggregation, respectively; M the monomer molecule; $\ldots\text{-}P_n^*$ the growing macromolecule: $R'O[C(O)(CH_2)_5O]_n$-AlR_2 in the case of CL polymerization; x = 2 for R = ^iBu and x = 3 for R = Et)

It follows from these changes of the kinetic propagation order in initiator, that aggregation of active species takes place into dimers: $[P_i^*]_0^{1/2}$ or trimers: $[P_i^*]_0^{1/3}$ (where $[P_i^*] = [I]_0$) and that only the deaggregated species propagate, whereas the aggregated ones are dormant. At these conditions only a certain fraction of chains are instantaneously active; they become dormant upon aggregation and restore their reactivity when deaggregate. The rates of this aggregation - deaggregation processes are not yet known.

Using the analytical method elaborated by us, solution of the corresponding kinetic scheme (eqs14) and thus determination of k_p and the k_{ag}/k_{dag} ratio was eventually possible (*34, 36, 39*). The pertinent data are collected in Table IV.

Both k_p and K_{ag} decrease with increasing solvating power of the solvent; this is in agreement with the proposed coordinaton-insertion mechanism of the propagation (scheme 13). Moreover, in spite of the fact that degree of aggregation (x depends on the size of the alkyl substituent at the Al atoms, k_p remains practically the same (at least in THF as a solvent).

We have also successfully applied this approach to determine k_p and K_{ag} not only in the polymerization of CL, studied by our own, but also to analyze the experimental results of the other research groups, reported for polymerizations proceeding with a fast reversible deactivation by aggregation of active centers. These involved anionic polymerizations of *o*-methoxystyrene, methyl methacrylate, ethylene oxide, and hexamethylcyclotrisiloxane (*67*).

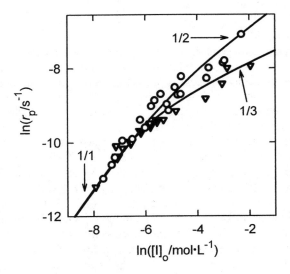

Figure 8. Bilogarithmic dependencies of the relative polymerization rates (r_p) against the starting concentrations of initiator ([I]$_0$). Polymerization of ε-caprolactone initiated with dialkylaluminum alkoxides. Initiators: (▽)-(C$_2$H$_5$)$_2$AlOC$_2$H$_5$, (○)-[(CH$_3$)$_2$CHCH$_2$]$_2$AlOCH$_3$, (▼)-(C$_2$H$_5$)$_2$AlOCH$_2$CHCH$_2$. Conditions: [CL]$_0$ = 2.0 mol·L^{-1};THF solvent, 25 °C. Points experimental; curves plotted on the basis of the computer simulation of aggregation-deaggregation equilibrium (scheme 14) with experimentally determined K$_{ag}$ (Reproduced with permission from reference 36. Copyright Wiley-VCH 1994).

Table IV. Propagation Rate Constants (k_p) and the Aggregation Equilibrium Constants (K_{ag}) for Polymerization of ε-Caprolactone Initiated with Dialkylaluminum Alkoxides (R_2AlOR')[a]

Solvent	ε[b]	R_2AlOR'	x[c]	$\dfrac{k_p}{mol^{-1} \cdot L \cdot s^{-1}}$	$\dfrac{K_{ag}}{(mol \cdot L^{-1})^{1-x}}$
CH$_3$CN	37	Et$_2$AlOEt	3	$7.5 \cdot 10^{-3}$	77
THF	7.3	Et$_2$AlOEt	3	$3.9 \cdot 10^{-2}$	$4 \cdot 10^4$
C$_6$H$_6$	2.3	Et$_2$AlOEt	3	$8.6 \cdot 10^{-2}$	$2.4 \cdot 10^5$
THF	7.3	$^iBu_2AlOMe$	2	$3.9 \cdot 10^{-2}$	77

[a] Conditons: $[CL]_0 = 2.0$ mol·L^{-1}, 25°C. [b] Solvent dielectric constant [c] Aggregation degree (scheme 14) (Data taken from references 36 and 39).

(b) Tris-alkoxyaluminum

Propagation on *tris*-alkoxyaluminum active species, was a matter of controversy, as discussed in the section describing initiation. Eventually, it has been revealed that, indenpedently on the monomer used, three polyester chains grow from one Al atom (eq 12). However, the structure of the living poly(CL) given in eq 12 is an oversimpliftication because the ^{27}Al NMR spectra of the growing species show that additional coordination takes place *(32, 38, 65)*.

Chemical shifts in the ^{27}Al NMR spectra are very sensitive to the coordination number of the Al atoms *(68)*. In Figure 9 ^{27}Al NMR spectrum of *tris*-isopropoxyaluminum trimer (A$_3$) is compared with spectra of the living poly(ε-caprolactone), living poly(γ-butyrolactone), and living poly(L-lactide) - all three polyesters prepared with A$_3$ as an initiator. In initiator itself the tetra and pentacoordinated Al atoms are present, according to the formula given in eq 9. Spectrum of A$_3$ exhibits signals at δ ≈ 60 and 30 ppm in expected proportion (2:1), marked in Figure 9 (a) as Al(4) and Al(5). There is also small signal present in the vicinity of δ = 0 ppm - it is due to the presence of ≈1.5 mol-% admixture of the tetramer (A$_4$). In the poly(CL) spectrum a strong peak due to the hexacoordinated Al atoms (Al(6)) prevail (at δ ≈ 4 ppm) but a small shoulder in the field characteristic to Al(5) is also apparent (Figure 9 (b)). Spectrum of poly(BL) shows, apart from the sharp peak due to Al(6), the broad signal coming from Al(4) (Figure 9 (c)), whereas in poly(LA) the Al(4) atoms are present (Figure (9 (d)).

The observed additional coordination of Al atoms in growing species could result from the intramolecular complexation by the acyl oxygen atoms from the polyester repeating units, as it was originally proposed by the the Liège group *(69)*. Some of the structures are illustrated schematically below (Al(4)1, Al(4)2, Al(5), and Al(6)).

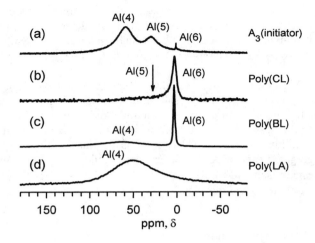

Figure 9. ^{27}Al NMR (75 MHz ^{27}Al, 70 °C, benzene-d_6 solvent) spectra of the aluminum tris-isopropoxide trimer (A_3) and the living polyesters prepared with A_3. (a): A_3, (b): living poly(ε-caprolactone), (c): living poly(γ-butyrolactone), and (d): living poly(L-lactide). Polymerization conditions (concentrations given in mol·L^{-1}): $3[A_3]_0 = 0.1$, $[CL]_0 = 2.0$, $[BL]_0 = 3.8$, $[LA]_0 = 1.0$, 80 °C, benzene-d_6 solvent (65).

Al(4)1   Al(4)2

Al(5)   Al(6)

(where R: ($CH_2)_5$ (poly(CL)), ($CH_2)_3$ (poly(BL)) or $CH(CH_3)$ (poly(LA))

Polymerization of CL exhibited approximately first order of propagation both in monomer (internally) and in active species (*38*).Thus, determination of the absolute rate constant of propagation k_p seems to be straightforward. The average k_p value (for both five- and six-coordinated species) is equal to 0.62 $mol^{-1} \cdot L \cdot s^{-1}$ (25°C, THF solvent) (*38*).

In poly(LA), according to the ^{27}Al NMR spectrum (Figure 9 (d)), hexacoordinated species are not present. This may result from the steric hindrance caused by the presence of methyl groups. Moreover, the fractional orders in active species, observed by us in the polymerization of LA initiated with A_3 (*40*) suggest that an aggregation - deaggregation equilibrium: 2 Al(4)1 ⇌ Al(4)2 takes place and propagation proceeds on the non-aggregated species Al(4)1. Therefore kinetics of LA polymerization was analyzed in terms of scheme 14. $K_{ag} = 92 \; mol^{-1} \cdot L$ and $k_p = 8.2 \cdot 10^{-3} \; mol^{-1} \cdot L \cdot s^{-1}$ (80°C, THF solvent) was determined this way.

Metalcarboxylates - Tin Octoate

Tin octoate $(Sn(OC(O)CH(C_2H_5)C_4H_9)_2$, tin(II) 2-ethylhexanoate, denoted further as $Sn(Oct)_2$) is probably the most often used initiating compound in the polymerization of cyclic esters (*41, 42, 70-84*). However, it is also the least understood and several "mechanisms" of initiation with $Sn(Oct)_2$ were proposed (e.g., ref *71, 74, 80, 82*).

One of the most developed ideas is based on the assumption that $Sn(Oct)_2$ is not bound to the polyester chain during the chain growth, and that $Sn(Oct)_2$ merely acts as a catalyst, by activating monomer before the actual propagation step. We shall call this mechanism the "activated monomer mechanism" (AMM) by analogy with formally similar concept developed earlier. This expression was coined by Szwarc for anionic polymerization of N-carboxy anhydrides (*86*). The AMM can be presented schematically, according to the papers cited above, in in the following way:

$$(15 \; (a))$$

(where K_{ac} denotes the equilibrium constant of activation, k_p the rate constant of propagation; ROH stands for water and/or alcohol molecule, either present as adventitious impurities or intentionally added)

There are variations of this scheme, not changing its basic feature: there is no Sn atoms covalently bonded to the growing chains and propagation step involves the monomer⋯Sn(Oct)$_2$ complex. For example, in reference 80 the poly(LA) chain growth reaction is written as follows:

$$R = H \text{ or Bzl} \qquad (15\,(b))$$

This mechanism would lead to the simple kinetics (independently whereas it is scheme 15 (a) or 15 (b)), and the rate of polymerization (propagation) (R_p) is than given by the following equation:

$$R_p = k_p K_{ac} \cdot [...-OH]_0 \cdot [Sn(Oct)_2]_0 \cdot [M] \qquad (16)$$

(where $[...-OH]_0$ denotes starting concentration of the hydroxy group containing compounds and $[M]$ the instantaneous monomer concentration).

The work of our own, to be discussed below, shows that AMM does not operate in the Sn(Oct)$_2$ / alcohol (or water) / cyclic ester system (41, 42). We also proposed a general mechanism of initiation with Sn(Oct)$_2$ (and presumably all of the metal carboxylates) in the polymerization of cyclic esters. There was no systematic kinetic work before we started our measurements, although a work of Zhang et al. (71, 72) has to be mentioned, since these authors came earlier to conclusions we describe in the present paper.

In order to check whether eq 16 holds we performed two series of kinetic experiments with butyl alcohol (BuOH) as a coinitiator. In Figure 10 dependencies of the relative rate of polymerization ($r_p = R_p/[M]$ on $[BuOH]_0$ (for constant $[Sn(Oct)_2]_0$) and on $[Sn(Oct)_2]_0$ (for constant $[BuOH]_0$) are given. These results show, that in both cases the rate first increases and then levels off. Such kinetic behavior suggests strongly that Sn(Oct)$_2$ is not a catalyst or initiator exclusively by itself. On the contrary, these results indicate that Sn(Oct)$_2$ has to react with alcohol added in order to produce the true initiator. The rate of polymerization increases with added alcohol as long, as there is enough of Sn(Oct)$_2$ in the system to give the initiator (Figure 10 (a)). The same argument holds when $[BuOH]_0$ is kept constant and Sn(Oct)$_2$ is being added (Figure 10 (b)). When all of the BuOH is already used in reaction with Sn(Oct)$_2$ its further addition is no more increasing the rate.

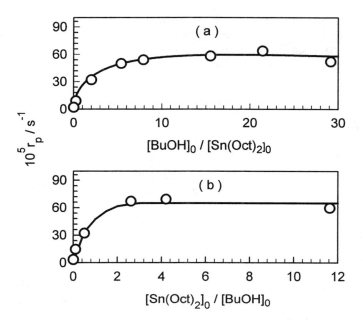

Figure 10. Dependencies of the rate of polymerization (r_p = -d[CL]/[CL]dt = (1/t)·ln([CL]$_0$/[CL])) of ε-caprolactone (CL) initiated with tin octoate (Sn(Oct)$_2$) on: (a) the ratio of starting concentrations of butyl alcohol (BuOH) to Sn(Oct)$_2$ ([BuOH]$_0$/[Sn(Oct)$_2$]$_0$) with [Sn(Oct)$_2$]$_0$ = 0.05 mol·L^{-1}, and kept constant through this series of experiments; (b) the ratio of starting concentrations of BuOH to Sn(Oct)$_2$ ([Sn(Oct)$_2$]$_0$/[BuOH]$_0$) with [BuOH]$_0$ = 0.10 mol·L^{-1} constant through this series of experiments (THF solvent, 80 °C) (Reproduced with permission from reference 42. Copyright Wiley-VCH 1998).

Thus, the only possibility is, that the following reactions take place:

$$Sn(Oct)_2 + ROH \rightleftharpoons OctSn-OR + OctH$$

$$(OctSn-OR + ROH \rightleftharpoons RO-Sn-OR + OctH) \qquad (17)$$

$$...-Sn-OR + nM \longrightarrow ...-Sn-O-(m)_n-R$$

(where R denotes H or alkyl group, OctH the 2-ethylhexanoic acid, M the CL or LA molecules, and m the polyester repeating unit)

Several authors proposed already a long time ago (*71*), as well as more recently (*79, 84*) that Sn(Oct)$_2$ reacts either with H$_2$O or ROH present in the system and gives the tin alkoxide acting as initiator. However, the clear cut evidence was lacking and therefore in the most recent and comprehensive studies the AMM was advanced and became more popular (80-82). Besides, AMM was also adopted by other authors describing merely their synthetic work and not being involved in any mechanistic studies (73, 74).

If the chemical changes described by eqs 17 take place indeed, we should be able to observe OctSn-O-(m)$_n$-R species directly in the polymerizing mixture. MALDI-TOF mass spectrometry gave final answer - the macromolecules containing OctSnO-... end-groups were directly observed (41), ending this way the long lasting mechanistic controversy and discussing the AMM in which Sn(Oct)$_2$ acts exclusively as a catalyst, activating monomer before the actual chain propagation step.

In Figure 11 (a) the corresponding mass spectrum is given and the peaks are identified. A fragment of the complex signal A could also be due to the stannoxane OctSnOSnO[(CH$_2$)$_5$C(O)O]$_n$(CH$_2$)$_5$C(O)OBu, H$^\oplus$ species. A' comes probably from the stannoxane, Na$^\oplus$ adduct. When this same sample was hydrolyzed at mild conditions in order to break the ...Sn-O-... bonds, peaks related to macromolecules with OctSnO-... moieties disappeared (Figure 11 (b)).

In this kind of MALDI-TOF experiments high starting concentration of Sn(Oct)$_2$ has to be used since otherwise population with Sn atoms in the chains is to low to be observed. Indeed, for the low starting concentration of Sn(Oct)$_2$ essentially two populations of the poly(CL) macromolecules were observed, namely similar to populations C and F from Figure 11 (i.e. H-(cl)$_n$-OBu and (CL)$_n$ cyclics, respectively). However, when at least the starting concentration of Sn(Oct)$_2$ was increased, then we observed an additional population of Sn-containing species. Another important prerequisite is to apply both delayed extraction and reflector modes in order to obtain high enough resolution of the spectra, enabling observation of the superfine structure of the MALDI signals (as e.g. in Figure 12).

Peaks related to macromolecules containing Sn atoms are very particular, because of the Sn isotopes distribution. Therefore, these peaks cannot be mistaken and taken for peaks of species not containing Sn. Moreover, comparison of the experimentally observed and computed "isotopic multiplets" show their identity. For example, this was shown for polymerizations of CL with Sn(Oct)$_2$/H$_2$O system (65). In the low molecular part of MALDI-TOF spectrum of the resulting polymer a series of cyclics were detected. One series is due to the SnO[(CH$_2$)$_5$C(O)O]$_n$ macrocyclics. Below, in Figure 12 we compare the enlarged signal of a cyclic decamer, namely OSn[O(CH$_2$)$_5$C(O)]$_{10}$ with the simulated peak of this particular species - they are identical. Tin-containing cyclics can be formed from the respective linear derivative by the end-to-end-biting reaction:

$$OctSn[O(CH_2)_5C(O)]_nOH \longrightarrow OSn[O(CH_2)_5C(O)]_n + OctH \qquad (18)$$

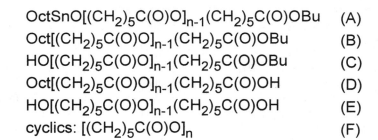

$$\text{OctSnO}[(CH_2)_5C(O)O]_{n-1}(CH_2)_5C(O)OBu \qquad (A)$$
$$\text{Oct}[(CH_2)_5C(O)O]_{n-1}(CH_2)_5C(O)OBu \qquad (B)$$
$$\text{HO}[(CH_2)_5C(O)O]_{n-1}(CH_2)_5C(O)OBu \qquad (C)$$
$$\text{Oct}[(CH_2)_5C(O)O]_{n-1}(CH_2)_5C(O)OH \qquad (D)$$
$$\text{HO}[(CH_2)_5C(O)O]_{n-1}(CH_2)_5C(O)OH \qquad (E)$$
$$\text{cyclics: } [(CH_2)_5C(O)O]_n \qquad (F)$$

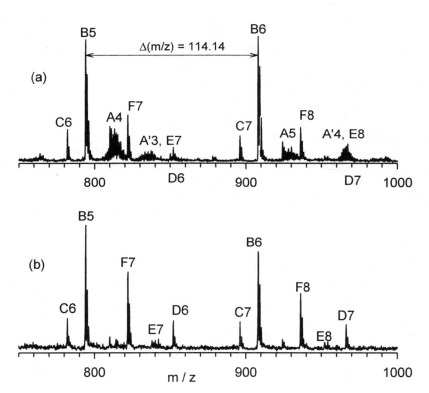

Figure 11. 750-1000 m/z fragment of the MALDI-TOF mass spectra of the reacting mixture: ε-caprolactone/butanol/Sn(Oct)₂ recorded before (a) and after hydrolysis, removing Sn from macromolecules (b). Numbers put after the structure symbol stand for n given in the formula above; all marked peaks come exclusively from the Na⊕ adducts of the respective structures. Conditions (concentrations given in mol·L⁻¹): [CL]₀ = 1.0, [BuOH]₀ = 0.13, [Sn(Oct)₂]₀ = 1.0, THF solvent, 80 °C (Reproduced with permission from reference 41).

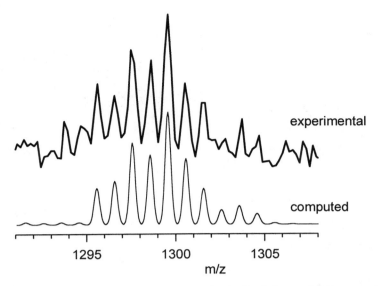

Figure 12. Comparison of the MALDI-TOF "multiplet" of the cyclic Sn-containing poly(CL) decamer (n = 10 in eq 18) from the experimental mass spectrum determined for the reacting mixture: ε-caprolactone/H_2O/$Sn(Oct)_2$), with the computed spectrum (program IsoPro 3.0) of the same macromolecule (both $Na^{\oplus}$ adducts). Conditions of polymerization (concentrations in mol·L^{-1}): $[CL]_0$ = 2.0, $[H_2O]_0$ = 0.03, $[Sn(Oct)_2]_0$ = 1.0; THF solvent, 80 °C (65).

The ways of formation of all of the macromolecules which are observed in mass spectrum in Figure 11 (a) are described by reactions shown in scheme 19. The actual monomer addition proceeds as a multicenter concerted reaction, similarly to the propagation step we have illustrated for Al-alkoxides in scheme 13.

Some of the reactions in scheme 19 are less important when a low starting concentration of $Sn(Oct)_2$ is used, but all of the macromolecular structures shown above should be present, at least in tiny concentrations. Although this scheme is using CL as a model monomer Sn-containing polyester chains in polymerization of LA initiated with $Sn(Oct)_2$ were also observed (65).

All of these data show that the Sn atom is present in the growing chains, and polymerization proceeds on the tin(II)-alkoxide bond. Thus, metal carboxylates have to be first converted into their corresponding alkoxides. Either by intentionally added alcohols or water (coinitiators) or by interaction with the adventitiously present coinitiators. In this case their concentration can be found "backwards", by counting the number of macromolecules formed. Structure of these adventitiously present coinitiators (e.g. H_2O, hydroxyacids - "open chain monomers" etc.) can be judged from the MALDI-TOF. The fragments of these coinitiators would constitute the end-groups of macromolecules.

$$Sn(Oct)_2 + ROH \rightleftharpoons OctSnOR + OctH$$

$$(and/or \longrightarrow OctSnOH + OctOR)$$

$$OctSnOR + \underset{(CH_2)_5}{\overset{O}{\underset{\|}{C}}-O} \longrightarrow OctSn-O(CH_2)_5C(O)-OR$$

$$(A1)$$

$$A1 + (n-1) \underset{(CH_2)_5}{\overset{O}{\underset{\|}{C}}-O} \longrightarrow OctSn-[O(CH_2)_5C(O)]_n-OR$$

$$(A)$$

$$A + ROH \rightleftharpoons OctSnOR + H-[O(CH_2)_5C(O)]_n-OR$$

$$(C)$$

$$OctH + C \xrightarrow{Sn(Oct)_2} Oct-[(CH_2)_5C(O)O]_{n-1}-(CH_2)_5C(O)OR + H_2O$$

$$(B)$$

$$Sn(Oct)_2 + H_2O \rightleftharpoons OctSnOH + OctH \qquad\qquad (19)$$

$$OctSnOH + n \underset{(CH_2)_5}{\overset{O}{\underset{\|}{C}}-O} \longrightarrow OctSn-[O(CH_2)_5C(O)]_n-OH$$

$$C + OctSn-[O(CH_2)_5C(O)]_n-OH \rightleftharpoons A + H-[O(CH_2)_5C(O)]_n-OH$$

$$(E)$$

$$OctH + E \xrightarrow{Sn(Oct)_2} Oct-[(CH_2)_5C(O)O]_{n-1}-(CH_2)_5C(O)OH + H_2O$$

$$(D)$$

$$OctSn-O(CH_2)_5C(O)-[O(CH_2)_5C(O)]_{n-1}-OR \rightleftharpoons$$

$$\rightleftharpoons OctSn-[O(CH_2)_5C(O)]_{n-x}-OR + [O(CH_2)_5C(O)]_x$$

$$(F)$$

Chain Transfer

Whenever the polymer repeating units contain the same heteroatoms as these present in monomers and reacting in the propagation step, chain transfer to polymer (macromolecules) with chain scission occurs. Either uni- or/and bimolecular reactions take place:

$$...\text{-(m)}_n\text{-m*} + M \underset{k_d}{\overset{k_p}{\rightleftharpoons}} ...\text{-(m)}_{n+1}\text{-m*} \qquad \text{(propagation)} \quad (20\,(a))$$

$$...\text{-(m)}_n\text{-m*} + M \underset{k_{p(x)}}{\overset{k_{tr(1)}}{\rightleftharpoons}} ...\text{-(m)}_{n-x}\text{-m*} + (m)_x \qquad \begin{array}{l}\text{(unimolecular}\\ \text{chain transfer)}\end{array} \quad (20\,(b))$$

$$...\text{-(m)}_n\text{-m*} + ...\text{-(m)}_p\text{-m*} \underset{k_{tr(2)}}{\overset{k_{tr(2)}}{\rightleftharpoons}} ...\text{-(m)}_{n+q}\text{-m*} + ...\text{-(m)}_{p-q}\text{-m*} \quad (20\,(c))$$
$$\text{(bimolecular chain transfer)}$$

(where m denotes the polymer repeating unit, m* - the involved active species, M - the monomer molecule; $k_{tr(1)}$, $k_{tr(2)}$ - the rate constants of the uni- and bimolecular chain transfer, respectively; k_p and $k_{p(x)}$ - the rate constants of propagation of the monomer and the cyclic x-mer, respectively; k_d - the rate constant of depropagation)

These reactions of transfer are of various intensities and have been observed in the polymerizations of cyclic ethers, acetals, esters, amides, and sulfides (26, 53, 86-91).

Intramolecular Chain-Transfer

Initially, we elaborated methods of measuring the $k_p/k_{tr(1)}$ ratio in the polymerization of ε-caprolactone and we correlated this ratio with k_p, determined from the polymerization kinetics. The $k_{tr(1)}$ was determined from the rates of appearance of cyclics measured by SEC (30-32, 92). On the basis of eq 20 (a) and 20 (b), assuming $k_p \gg k_d$ and introducing the selectivity parameter $\beta = k_p/k_{tr(1)}$ we have:

$$\beta = \frac{k_p}{k_{tr(1)}} = \frac{\ln([M]_0 / [M])}{[M(x)]_{eq} \cdot \ln\{[M(x)]_{eq} / ([M(x)]_{eq} - [M(x)])\}} \qquad (21)$$

(where [M] denotes concentration of monomer, $[M]_0$ the concentration of monomer in the feed, [M(x)] the concentration of a cyclic x-mer, $[M(x)]_{eq}$ the equilibrium concentration of a cyclic x-mer)

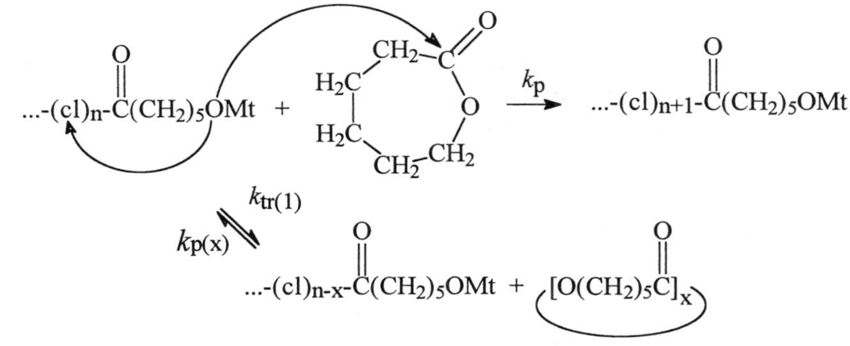

...-$(cl)_n$-$\overset{O}{\overset{\|}{C}}$$(CH_2)_5OMt$ + [ε-caprolactone ring] $\xrightarrow{k_p}$ ...-$(cl)_{n+1}$-$\overset{O}{\overset{\|}{C}}$$(CH_2)_5OMt$

$k_{tr(1)}$

$k_{p(x)}$

...-$(cl)_{n-x}$-$\overset{O}{\overset{\|}{C}}$$(CH_2)_5OMt$ + $[O(CH_2)_5\overset{O}{\overset{\|}{C}}]_x$

(22)

(where cl denotes the poly(CL) repeating unit and Mt = (e.g.) Na, Sm<, Al.<)

The parameters β (for x = 2), determined in the polymerization of ε-caprolactone (eq 22) conducted with ionic and covalent active species, are compared with the respective propagation rate constants in Table V.

Table V. Propagation rate constants (k_p) and the selectivity parameters ($\beta = k_p/k_{tr(1)}$) for the polymerization of ε-caprolactone [a]

Active species	$\dfrac{k_p}{mol^{-1}\cdot L\cdot s^{-1}}$	$\beta = \dfrac{k_p/k_{tr(1)}}{mol^{-1}\cdot L}$
...-$(CH_2)_5O^{\ominus} Na^{\oplus}$	≥1.70	$1.6\cdot10^3$
...-$(CH_2)_5O$-$Sm[O(CH_2)_5$-...$]_2$	2.00	$2.0\cdot10^3$
...-$(CH_2)_5O$-$Al(C_2H_5)_2$	0.03	$4.6\cdot10^4$
...-$(CH_2)_5O$-$Al[CH_2CH(CH_3)_2]_2$	0.03	$7.7\cdot10^4$
...-$(CH_2)_5O$-$Al[O(CH_2)_5$-...$]_2$	0.50	$3.0\cdot10^5$

[a] Polymerization conditions: 20°C, THF solvent (Reprinted with permission from reference 92. Copyright Wiley-VCH 1998).

Analysis of these data shows that there are two factors influencing the $k_p/k_{tr(1)}$ ratio, namely the intrinsic reactivity of the growing species and the steric hindrance. The Reactivity-Selectivity Principle has already been discussed in our previous work in terms of the early and late transition states formation (31). In the most simplified way it could be said, that the less reactive species are more discriminating; therefore with increase of k_p the $k_p/k_{tr(1)}$ ratio is decreasing.

In order to see clearly the influence of the steric hindrance the actual structures of the growing species in polymerization initiated with dialkylaluminum alkoxides - with less steric hindrance and trialkoxyaluminum - more sterically hindered are compared (structures G and H).

$$..-\overset{\overset{O}{\|}}{C}(CH_2)_4CH_2OAl\overset{CH_2CH_3}{\underset{CH_2CH_3}{\diagdown}}$$

$$..-\overset{\overset{O}{\|}}{C}(CH_2)_4CH_2OAl\overset{OCH_2(CH_2)_4\overset{O}{\overset{\|}{C}}-...}{\underset{OCH_2(CH_2)_4\underset{O}{\overset{\|}{C}}-...}{\diagdown}}$$

(one polyester chain at one Al atom)   (three polyester chains at one Al atom)

(G)   (H)

There is only one chain growing from the Al atom in G structure and three chains grow from one Al atom in structure H. The state of solvation of the actually growing species is not known at present, however, whatever is the solvation, the structure H is more sterically hindered. Reaction with a monomer molecule is not very much perturbed, because of its small size. Reaction with a polymer chain requires, however, penetration of the bulky structure H through a polymer coil in order to find a convenient conformation for the transfer reaction to proceed. Therefore for the more bulky growing species k_p may not be affected although $k_{tr(1)}$ is decreasing (cf. Table V, entries $-Al(CH_2CH_3)_2$ and $-Al[CH_2CH(CH_3)_2]_2$).

Intermolecular Chain-Transfer

As we already mentioned in the previous paragraphs, both intramolecular and intermolecular chain transfer usually coexist. In the polymerization of CL we could measure *exclusively* the $k_{tr(1)}$ (intramolecular) by observing the rate of cyclics formation, whereas for the polymerization of LA intramolecular transfer seems to be less important and therefore we were able to determine $k_{tr(2)}$ (intermolecular). In the studied systems we secured quantitative and fast initiation (in comparison with propagation). Thus, we had to exclusively consider propagation and the intermolecular transfer. In the absence of any transfer Poisson distribution should result; when, however, transfer to polymer proceeds, the molar masses distribution (MMD) broadens with conversion. At the first approximation we assumed, that this change of MMD is exclusively due to intermolecular transfer. Further work will introduce a correction, taking into account the intramolecular transfer that can also influence the M_w/M_n ratio.

In the respective kinetic scheme (eqs 23) propagation is accompanied by intermolecular chain transfer (bimolecular segmental exchange). Although we were unable to find an analytical solution for the corresponding set of differential equations, relating eventually M_w/M_n and conversion to the $k_p/k_{tr(2)}$ ratio (*84-86*), the numerical solution gave the required dependencies of M_w/M_n on the monomer conversion. From these plots the $k_p/k_{tr(2)}$ ratio could be determined. Actually, computation was made for various arbitrarily assumed $k_p/k_{tr(2)}$ and monomer to initiator starting concentration ($[M]_0/[I]_0$) ratios. Then, the experimentally determined plots of M_w/M_n = f(conversion) were compared with the set of computed dependencies and the best fit gave the $k_p/k_{tr(2)}$ ratio for the studied system. An example of such computed dependencies related to eqs 23 is given in Figure 13. Thus, as expected, M_w/M_n increases with monomer conversion reaching eventually the value predicted by the theory of the segmental exchange (*96*). This is M_w/M_n = 2, characteristic for the most probable distribution. Moreover, for a given α the higher M_w/M_n are obtained for the lower $k_p/k_{tr(2)}$ ratios, as it is intuitively expected.

$$(23)$$

(where la denotes poly(LA) repeating unit; Mt = Al.<, Fe<, La<, Sm<, -Sn<, -Ti<)

The results obtained for the polymerization of LA initiated with Al(O^{i}Pr)$_3$ trimer or Fe(OEt)$_3$ are shown in Figures 14 (a) and (b), respectively. The experimental points were obtained from SEC measurements based on the poly(LA) standards prepared in our laboratory.

Data collected in Figure 15 summarize results obtained in this way for L,L-dilactide polymerization with a series of initiators. The selectivity parameters (γ =

$k_p/k_{tr(2)}$) for the intermolecular transfer are compared with the corresponding propagation rate constants (k_p). It is remarkable that, at least for the metal alkoxides used as initiators (viz. the resulting active species), this dependence conforms to the rules of the Reactivity-Selectivity Principle, i.e. the high selectivities result in lower reactivities.

The highest selectivities are provided by tin(II) derivatives (e.g., tin octoate ($Sn(Oct)_2$) or tin dibutoxide ($Sn(OBu)_2$), for which $k_p/k_{tr(2)} \approx 200$ was determined. Unfortunately, for the $LA/Sn(Oct)_2$ and $LA/Sn(OBu)_2$ polymerizing systems concentrations of the actually (momentary) propagating active species are not yet known and the absolute rate constants of propagation (k_p) could not therefore be determined. Therefore those points are not introduced to the Figure 15.

It is also interesting to note, that the alkylmonoalkoxides, such as diethylaluminum or tributyltin alkoxide, deviate from this order of reactivities and selectivities.

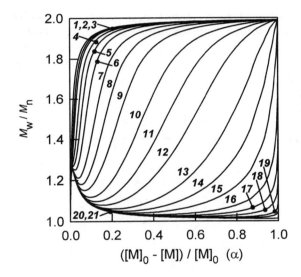

*Figure 13. Numerically simulated plots of M_w/M_n versus monomer conversion degree (α). Assumptions: $[M]_0 = 1.0\ mol \cdot L^{-1}$, $[I]_0 = 10^{-2}\ mol \cdot L^{-1}$; $k_p/k_{tr(2)} = 10^{-3}$ (**1**), 10^{-2} (**2**), $2 \cdot 10^{-2}$ (**3**), $5 \cdot 10^{-2}$ (**4**), 10^{-1} (**5**), $2 \cdot 10^{-1}$ (**6**), $5 \cdot 10^{-1}$ (**7**), 10^0 (**8**), $2 \cdot 10^0$ (**9**), $5 \cdot 10^0$ (**10**), 10^1 (**11**), $2 \cdot 10^1$ (**12**), $5 \cdot 10^1$ (**13**), 10^2 (**14**), $2 \cdot 10^2$ (**15**), $5 \cdot 10^2$ (**16**), 10^3 (**17**), $2 \cdot 10^3$ (**18**), $5 \cdot 10^3$ (**19**), 10^4 (**20**), 10^5 (**21**); $k_p \gg k_d$. (Reproduced with permission from reference 94. Copyright Wiley-VCH 1998)*

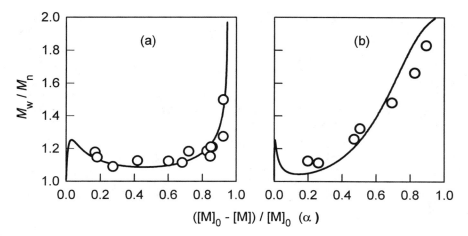

Figure 14. Dependencies of M_w/M_n on degree of monomer conversion (α) determined for L,L-dilactide polymerization initiated with: (a) $Al(O^iPr)_3$ and (b) $Fe(OEt)_3$. Conditions (concentrations given in $mol \cdot L^{-1}$): $[LA]_0 = 1.0$, $[Al(O^iPr)_3]_0 = 10^{-2}$, $[Fe(OEt)_3]_0 = 1.4 \cdot 10^{-3}$; THF solvent, 80 °C. Points experimental, lines computed assuming $k_p/k_{tr(2)} = 100$ (a) and 60 (b). (Reproduced with permission from reference 94. Copyright Wiley-VCH 1998)

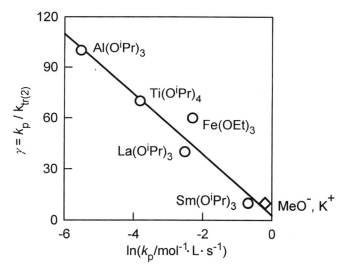

Figure 15. Dependence of $\gamma = k_p/k_{tr(2)}$ on ln k_p determined in polymerizations of L,L-dilactide initiated by metal alkoxides. Covalent alkoxides: (○); ionic alkoxide - MeOK: (◇). Conditions: $[LA]_0 = 1.0$ $mol \cdot L^{-1}$, THF solvent, 80 °C (MeOK initiated polymerization - at 20 °C)). (Reproduced with permission from reference 94. Copyright Wiley-VCH 1998).

Controlled Synthesis of Poly(ϵ-Caprolactone) and Poly(L,L-Dilactide)

It is not our intention to comprehensively review the synthetic aspects of the poly(aliphatic ester)s chemistry. In this area there is over one hundred exact papers published, devoted particularly to poly(CL) and poly(LA). The large majority of these papers were published before some of the initiation mechanisms described in this paper became apparent. Therefore it was not understood, that structures of the polymer end groups were less uniform than believed. Several other polyesters (apart from poly(CL) and poly(LA)), such as poly(β-lactones) or polycarbonates, are not covered by the present review. Besides, a plethora of various block and graft copolymers or end-functionalized homo- and copolymers was prepared. Some of these products were prepared with $Al(O^iPr)_3$ before the difference in reactivities of the corresponding trimer and tetramer was revealed, as described in this review. Thus, the resulting structures could not in fact correspond to what was published. On the other hand, when $Sn(Oct)_2$ was used it was not known to some authors that octanoic acid esters end-groups are formed. Therefore, at least some of these works should be rectified.

We first used dialkylaluminum alkoxides in our quantitative synthetic studies. This was at the time, when the difference of aluminum *tris*-isopropoxide trimer (A_3) and tetramer (A_4) reactivites were not yet appreciated in studies of polymerization. Thus, "puzzling results" (expression taken from reference 69) of other authors working with $Al(O^iPr)_3$ forced us to look into initiators giving a known number of macromolecules per one molecule of initiator. This criteria were met indeed by dialkylaluminum alkoxides, shown already in early sixties to act as initiators *(55)*. Thus, we prepared with these initiators a number of well defined macromolecules with well controlled size, architecture, and the end-groups *(30, 33, 97)*. Simple macromolecules with one desired end were synthetised first *(33)* Then initiator $R_2Al-OR'O-AlR_2$, giving macromolecules growing in two directions was prepared and used in the syntehsis of "two ended" macromolecules *(97)*. Later on, these and similar initiators were used a number of times by others (e.g. *54, 98-102*).

After understanding a difference between A_3 and A_4 reactivities, the isolated A_3 has become the most versatile one for the controlled polymerization of cyclic esters, particularly of poly(CL). It provides a relatively fast and quantitative initiation, moderately fast propagation and relatively good selectivity *(21, 37, 38, 92, 94)*.

Thus, $Al(O^iPr)_3$ (in the form of A_3 *(40)*) looks to be ideally suited also for the synthesis of poly(LA), since apart from a good selectivity it provides, in contrast to $Sn(Oct)_2$, a direct control of the molar mass of the resulting polyester, by simply adjusting the $([LA]_0-[LA]_{eq})/[Al(O^iPr)_3]_0$ ratio *(40, 102, 103)*. On the other hand, it looks that there is a certain limit of M_n of poly(LA), which can be obtained with $Al(O^iPr)_3$. Reasons are not yet well-understood *(103)*.

The highest controlled masses of poly(LA) have recently been obtained with $Sn(OBu)_2$ as an initiator *(65)*. As already mentioned, this initiator provides relatively high $k_p/k_{tr(2)}$ ratio, equal to 200. Both low and high molar mass poly(LA)'s (up to $M_n \approx 10^6$) were prepared in a controlled way. Some molar mass data of poly(LA) prepared with $Sn(OBu)_2$ are given in Figure 16.

194

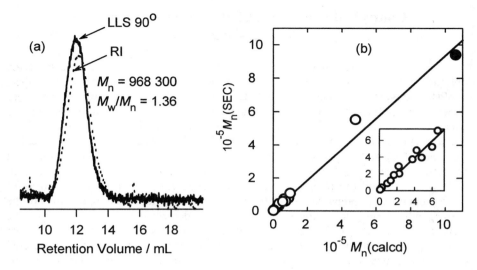

Figure 16. Molar mass data for poly(L-lactide) prepared with Sn(OBu)₂ as an initiator. (a) SEC trace of a high molar mass polymer prepared in bulk at 120 °C (6.8·10⁻³ mol-% of initiator). Conditions of measurement: TSK-Gel G 3000 HR and G 6000 HR columns, CH₂Cl₂ as an eluent, 0.8 mL·min⁻¹. (b) Dependence of M_n measured by SEC on M_n calculated from the feed composition; $M_n(calcd) = 144.13([LA]_0 - [LA]_{eq})/2[Sn(OBu)_2]_0$. Polymerization conditions: (○) $[LA]_0$ from 1.0 to 3.0 mol·L⁻¹, THF solvent, 80 °C; (●) as in a caption for figure (a) (65).

Nevertheless, other Sn(II) derivatives, including Sn(Oct)₂, can advantageously be used, taking into account its real role and necessity of using a coinitiator. Preparation of the very high molar mass poly(LA) was already reported a long time ago (71). Control of molar masses can be achieved by dosing the respective amount of a coinitiator. It has to be remembered that a certain proportion of macromolecules would bear the octanoic acid ester end groups. Otherwise, Sn(OR)₂, described above, and Sn(Oct)₂/ROH system are close one to another.

Acknowledgement

This work was supported financially by the Polish State Committee for Scientific Research (KBN) grant 3 T09B 105 11.

References

1. Pitt, C. G. *"Poly-ε-Caprolactone and Its Copolymers"*, in *Biodegradable Polymers as Drug Delivery Systems;* Chasin, M.; Langer, R., eds., Marcel Dekker, Inc.: New York 1990; p.71.
2. Kharash, G. B.; Sanchez-Riera, F.; Severson, D. K. *"Polymers of Lactid Acid"* in *Plastics from Microbes;* Mobley, D. P., ed., Hanser Publishers: Munich, New York 1994; p. 93.
3. Lunt, J. *Polymer Degrad.Stab.* **1998**, *59*, 145.
4. Kobayashi, S.; Uyama, H.; Namekawa, S. *Polymer Degrad.Stab.* **1998**, *59*, 191.
5. Kobayashi, S.; Uyama, H.; Namekawa, S.; Hayakawa H. *Macromolecules* **1998**, *31*, 5655.
6. Bisht, K. S.; Svirkin, Y.Y.; Henderson, L. A.; Gross, R. A.; Kaplan, D. L.; Swift, G. *Macromolecules* **1997**, *30*, 2705.
7. Bisht, K. S.; Henderson, L. A.; Gross, R. A.; Kaplan, D. L.; Swift, G. *Macromolecules* **1997**, *30*, 7735.
8. Ostrovskii, V. E.; Khodzhemirov, V. A.; Barkova, A.; P. *Dokl.Akad.Nauk SSSR* **1970**, *191*, 1095.
9. Evstropov, A. A.; Lebedev, B. V.; Kulagina, T. G.; Lyudvig, Ye. B.; Belenkaya, B. G. *Vysokomol.Sedin., Ser.A* **1979**, *21*, 2038.
10. Evstropov, A. A.; Lebedev, B. V.; Kiparisova E. G.; Alekseev, V.A.; Stashina, G. A. *Vysokomol.Sedin., Ser.A* **1980**, *22*, 2450.
11. Evstropov, A. A.; Lebedev, B. V.; Kulagina, T. G.; Lebedev, N. K. *Vysokomol.Sedin., Ser.A* **1982**, *24*, 568.
12. Duda, A.; Penczek, S. *Macromolecules* **1990**, *23*, 1636.
13. Lebedev, B. V.; Evstropov, A. A.; Lebedev, N. K.; Karpova E. A.; Lyudvig, Ye. B.; Belenkaya, B. G. *Vysokomol.Sedin., Ser.A* **1978**, *20*, 1974.
14. Sawada, H. *Thermodynamics of Polymerization;* Marcel Dekker, Inc.: New York and Basel, Switzerland, 1976.
15. Leonard, J.; Maheux, D. *J.Macromol.Sci.-Chem.* **1973**, *A7*, 1421.
16. Brown, H. C.; Brewster, J. M.; Shechter, H. *J.Am. Chem. Soc.* **1954**, *76*, 467.
17. Saiyasombat, W.; Molloy, W.; Nicholson T. M.; Johnson, A. F.; Ward, I. M.; Poshyachinda, S. *Polymer* **1998**, *39*, 5581.
18. Antolovic, D.; Shiner, V. J.; Davidson, E. R. *J.Am.Chem.Soc.* **1988**, *110*, 1375.
19. Odian, G. *Principles of Polymerization*, 3rd ed., Wiley-Interscience: New York, 1991; p. 570.
20. Duda, A.; Penczek, S.; Dubois, Ph.; Jérôme, R. *International Conference on Advanced Materials ICAM'97, European Materials Research Society Spring Meeting E-MRS'97, Strasbourg (France), June 1997, Book of Abstracts*, H-VII.4.
21. Duda, A.; Penczek, S.; Dubois, Ph.; Mecerreyes, D.; Jérôme, R. *Macromol.Chem.Phys.* **1996**, *197*, 1273.
22. Duda, A.; Penczek, S.; Dubois, Ph.; Mecerreyes, D.; Jérôme, R. *Polymer Degrad.Stab.* **1998**, *59*, 215.
23. Gol'dfarb, Ya. I.; Belen'kii, L. I. *Russ.Chem.Rev.* **1960**, *29*, 214.; Huisgen, R.; Ott, H. *Tetrahedron* **1959**, *6*, 253.
24. Sosnowski, S.; Slomkowski, S.; Penczek, S. *Makromol.Chem.* **1991**, *192*, 735.

25. Slomkowski, S.; Penczek, S. *Macromolecules* **1976**, *9*, 367; **1980**, *13*, 229.
26. Sosnowski, S.; Slomkowski, S.; Penczek, S.; Reibel, L. *Makromol.Chem.* **1983**, *184*, 2159.
27. Sosnowski, S.; Slomkowski, S.; Penczek, S. *J.Macromol.Sci., Chem.* **1983**, *A20*, 979.
28. Slomkowski, S. *Polymer* **1986**, *27*, 71.
29. Duda, A. *J.Polym.Sci., Part A: Polym.Chem.* **1992**, *30*, 21.
30. Hofman, A.; Slomkowski, S.; Penczek, S. *Makromol.Chem., Rapid Commun.* **1987**, *8*, 387.
31. Penczek, S.; Duda, A.; Slomkowski, S. *Makromol.Chem., Macromol.Symp.* **1992**, *54/55*, 31.
32. Penczek, S.; Duda, A. *Macromol.Symp.* **1996**, *107*, 1.
33. Duda, A.; Florjanczyk, Z.; Hofman, A.; Slomkowski, S.; Penczek, S. *Macromolecules*, **1990, 23**, 1640.
34. Penczek, S.; Duda, A. *Makromol.Chem., Macromol.Symp.* **1991**, *47*, 127.
35. Duda, A. *Macromolecules* **1994**, *27*, 577; **1996**, *29*, 1399.
36. Duda, A.; Penczek, S. *Macromol.Rapid Commun.* **1994**, *15*, 559.
37. Duda, A.; Penczek, S. *Macromol.Rapid Commun.* **1995**, *196*, 67.
38. Duda, A.; Penczek, S. *Macromolecules* **1995**, *28*, 5981.
39. Biela, T.; Duda, A. *J.Polym.Sci., Part A: Polym.Chem.* **1996**, *34*, 1807.
40. Kowalski, A.; Duda, A.; Penczek, S. *Macromolecules* **1998, 31**, 2114.
41. Kowalski, A.; Libiszowski, J.; Duda, A.; Penczek, S. *Polym.Prepr. (Am.Chem.Soc., Div.Polym.Chem.)* **1998**, *39(2)*, 74.
42. Kowalski, A.; Duda, A.; Penczek, S. *Macromol.Rapid Commun.* **1998**, *19*, 567.
43. Deffieux, A.; Boileau, S. *Macromolecules* **1976**, *9*, 369.
44. Penczek, S.; Kubisa, P.; Szymanski, R. *Makromol.Chem., Rapid Commun.* **1991**, *12*, 77.
45. Jedlinski, Z.; Kurcok, P.; Lenz, R. W. *J.Macromol.Sci.-Pure Appl.Chem.* **1995**, *A32*, 797.
46. Agostini, D. E.; Lando, J. B.; Shelton, J.R. *J.Polym.Sci., Polym.Chem.Ed.* **1971**, *6*, 2771.
47. Iida, M.; Araki, T.; Teranishi, K.; Tani, H. *Macromolecules* **1977, 10**, 275.
48. Gross, R.A.; Zhang, Y.; Konrad, G.; Lenz, R. W. *Macromolecules* **1988**, *21*, 2657.
49. Jedlinski, Z.; Kowalczuk, M.; Kurcok, P. *Macromolecules* **1991**, *24*, 1218.
50. Sosnowski, S.; Slomkowski, S.; Penczek, S. *Macromolecules* **1994, 26**, 5526.
51. Dale, J.; Schwartz, J.-E. *Acta Chem.Scand.* **1986, B40**, 559.
52. Gagnon, S. D. *"Propylene Oxide and Higher 1,2-Epoxide Polymer"*, in *Encyclopedia of Polymer Science and Engineering*, ed. Bikales, N. et al., J.Wiley & Sons: New York, 1986; vol. 6, p. 273.
53. Ito, K.; Yamashita, Y. *Macromolecules* **1978**, *11*, 68.
54. Ouhadi, T.; Stevens, Ch.; Teyssié, Ph. *Makromol.Chem., Suppl.* **1975, 1**, 191.
55. Cherdron, H.; Ohse, H.; Korte, F. *Makromol.Chem.* **1962, 56**, 187.
56. Hsieh, H. L.; Wang, I. W. *Am.Chem.Soc., Symp.Ser.* **1985**, *286*, 161.
57. Mole, T.; Jeffery, E. A. *Organoaluminium Compounds*, Elsevier Publishing Company: Amsterdam, London , New York, 1972; Chapter 8.

58. Bradley, D. C.; Mehrothra, R. C.; Gaur, D. P. *Metal Alkoxides,* Academic Press, London 1978, pp. 74, 122.
59. Eisch, J. J. *"Aluminium",* in *Comprehensive Organometallic Chemistry,* ed. G.Wilkinson et al., Pergamon Press: Oxford U.K. 1982; vol.1, p.583.
60. Duda, A.; Penczek, S. unpublished data.
61. Cox, F. E.; Hostetler F. (Union Carbide Corp.), *US Patent* **1962,** 3 021 313.
62. Kricheldorf, H. R.; Berl, M.; Scharnagl, N. *Macromolecules* **1988,** *21,* 286.
63. Jacobs, C.; Dubois, Ph.; Jérôme, R.; Teyssié, Ph. *Macromolecules* **1991,** *24,* 3027.
64. Shiner, V. J.; Whittaker, D. *J.Am.Chem.Soc.* **1969,** *91,* 394.
65. Baran, J.; Duda, A.; Kowalski, A.; Libiszowski, J.; Penczek, S., in preparation.
66. Jones, R. A. Y. *Physical and Mechanistic Organic Chemistry,* Cambridge University Press: Cambridge, UK, 1984; Chapter 12.
67. Duda, A.; Penczek, S. *Macromolecules* **1994,** *27,* 4867.
68. Benn, R.; Rufinska, A. *Angew.Chem., Int.Ed.Engl.* **1986,** *25,* 861.
69. Ropson, N.; Dubois, Ph.; Jérôme, R.; Teyssié, Ph. *Macromolecules* **1993,** *26,* 6378.
70. Kleine J.; Kleine, H.-H. *Makromol.Chem.* **1959,** *30,* 23.
71. Leenslag, J. W.; Pennings, A. J. *Makromol.Chem.* **1987,** *188,* 1809.
72. Jamshidi, K.; Eberhard, R. C.; Hyon, S.-H.; Ikada, Y. *Polym.Prepr. (Am.Chem.Soc.,Div.Polym.Chem.)* **1987,** *28(1),* 236.
73. Nijenhuis, A. J.; Grijpma, D. W.; Pennings, A. J. *Macromolecules* **1992,** *25,* 6419.
74. Doi, Y. J.; Lemstra, P. J.; Nijenhuis, A. J.; van Aert, H. A. M.; Bastiaansen, C. *Macromolecules* **1995,** *28,* 2124.
75. Dahlman, J.; Rafler, G.; Fechner, G.; Meklis, B. *Brit.Polym.J.* **1990,** *23,* 235.
76. Rafler, G.; Dahlman, J. *Acta Polym.* **1992,** *43,* 91.
77. Dahlman, J.; Rafler, G. *Acta Polym.* **1993,** *44,* 103.
78. Zhang, X.; Wyss, U. P.; Pichora, D.; Goosen, M. F. A. *Polym.Bull. (Berlin)* **1992,** *27,* 623.
79. Zhang, X.; MacDonald, D. A.; Goosen, M. F. A.; McCauley, K. B. *J.Polym.Sci., PartA: Polymer.Chem.* **1994,** *32,* 2965.
80. Kricheldorf, H. R.; Kreiser-Saunders, I.; Boettcher, C. *Polymer* **1995,** *36,* 1253.
81. In't Veld, P. J. A.; Velner, E. M.; van de Witte, P.; Hamhuis,J.; Dijkstra,P. J.; Feijen, J. *J.Polym.Sci., Part A: Polym.Chem.* **1997,** *35,* 219.
82. Schwach, G.; Coudane, J.; Engel, R.; Vert, M. *J.Polym.Chem., Part A: Polym.Chem.* **1997,** *35,* 3431.
83. Witzke, D. R.; Narayan R.; Kolstad, J. J. *Macromolecules* **1997,** *30,* 7075.
84. Storey, R. F.; Taylor, A. E. *J.Macromol.Sci.-Pure Appl.Chem.* **1998,** *A35,* 723.
85. Szwarc, M. *Pure Appl.Chem.* **1966,** *12,* 127.
86. Rozenberg, B. A.; Irzhak, V. I.; Enikolopyan, N.S. *Interchain Exchange in Polymers* (in Russian), Khimiya: Moscow 1975.
87. Penczek, S.; Kubisa, P.; Matyjaszewski, K. *Adv.Polym.Sci.* **1985,** *68/69,* 1.
88. Pruckmayr, G.; Wu, T. K. *Macromolecules* **1978,** *11,* 265.
89. Chwialkowska, W.; Kubisa, P.; Penczek, S. *Makromol.Chem.* **1982,** *183,* 753.
90. Wichterle, O. *Makromol.Chem.* **1960,** *35,* 127.

91. Goethals, E. J.; Simonds, R.; Spassky, N.; Momtaz, A. *Makromol.Chem.* **1980,** *181,* 2481.
92. Baran, J.; Duda, A.; Kowalski, A.; Szymanski, R.; Penczek, S. *Macromol.Symp.* **1998,** *128,* 241.
93. Baran, J.; Duda, A.; Kowalski, A.; Szymanski, R.; Penczek, S. *Macromol.Rapid Commun.* **1997,** *18,* 325.
94. Penczek, S.; Duda, A.; Szymanski, R. *Macromol.Symp.* **1998,** *132,* 441.
95. Szymanski, R. *Macromol.Theory Simul.* **1998,** *7,* 27.
96. Flory, P. J. *J.Am.Chem.Soc.* **1942,** *64,* 2205.
97. Sosnowski, S.; Slomkowski, S.; Penczek, S.; Florjanczyk, Z. *Makromol.Chem.* **1991,** *192,* 1457.
98. Dubois, Ph.; Jérôme, R.; Teyssié, Ph. *Polym.Bull.(Berlin)* **1989,** *22,* 475.
99. Dubois, Ph.; Ropson, N.; Jérôme, R.; Teyssié, Ph. *Macromolecules* **1993,** *26,* 2730.
100. Dubois, Ph.; Degée, Ph.; Jérôme, R.; Teyssié, Ph. *Macromolecules* **1996,** *29,* 1965.
101. Kricheldorf, H. R.; Kreiser-Saunders, I. *Polymer* **1994,** *35,* 4175.
102 Dubois, Ph.; Jacobs, C.; Jérôme, R.; Teyssié, Ph. *Macromolecules* **1991,** *24,* 2266.
103. Degée, Ph.;Dubois, Ph.; Jérôme, *Macromol.Chem.Phys.* **1997,** *198,* 1973.

Polymer Evaluation

Chapter 14

Can the Glass Transition Temperature of PLA Polymers Be Increased?

K. Marcincinova Benabdillah, M. Boustta, J. Coudane, and M. Vert

C.R.B.A. ESA C.N.R.S. 5473, Faculty of Pharmacy, Montpellier I University, 15, av. Charles Flahault, 34060 Montpellier, France

Hydrolyzable lactic acid-based aliphatic polyesters are among the most promising environment-friendly degradable polymers for temporary commodity applications. The properties of the semi-crystalline members of the family are comparable to those of many presently used, industrial thermoplastic polymers. However, a glass transition temperature below 60°C and a rather low viscosity above T_g are among the weaknesses of PLA polymers and stereocopolymers. In the past years, many copolymers have been studied. This paper reviews the main linear degradable copolymers of lactic acids reported in literature and discusses the influence of the comonomers on thermal properties, in particular glass transition temperature which is to be considered as one of the key characteristics for the development of PLA-based commodity devices.

Introduction

In the sixties US companies patented sutures based on poly(glycolic acid) (PGA), and glycolic acid-rich poly(lactic acid-co-glycolic acid) polymers (PLAXGAY where X=percentage of L-lactyl units and Y=percentage of glycolyl units, as suggested by Vert (1)). This was the beginning of the search for synthetic polymers which can degrade after use, a characteristic typical of natural biopolymers. Soon after, poly(lactic acid)s (PLAX) were proposed as bioresorbable matrices for biomedical and pharmaceutical uses (2, 3).

High molecular weight PLA are synthesized by ring-opening polymerization of lactides according to the following reaction :

200

$$n \quad \underset{CH_3}{\overset{H}{\underset{*}{C}}} \overset{O-C\overset{O}{\parallel}}{\underset{C-O}{\overset{\cdots}{\underset{O}{\parallel}}}} \underset{CH_3}{\overset{H}{\underset{*}{C}}} \quad \xrightarrow{\text{initiator}} \quad [-O-\underset{CH_3}{\overset{H}{\underset{*}{C}}}-\overset{O}{\overset{\parallel}{C}}-O-\underset{H}{\overset{CH_3}{\underset{*}{C}}}-\overset{O}{\overset{\parallel}{C}}-]_n$$

Many initiators can be used (4-6) and many mechanisms, namely cationic, anionic, coordination-insertion, etc. have been proposed. In most cases the exact mechanism of polymerization and corresponding chain structures are still questioned, especially for the industrially used initiating species, namely stannous octoate, zinc metal or zinc lactate (7).

Because of their excellent biocompatibility and mechanical properties semicrystalline polylactides are of particular interest for orthopaedic and maxillo-facial bone surgery (8-11). The osteosynthetic devices exposed to dynamic stresses in animal and human bodies require very pure polymers of high tacticity and crystallinity (12). On the other hand, matrices for drug delivery systems are generally prepared from amorphous PLA stereocopolymers and PLAGA copolymers (13-16). Micro- and nanoparticles, micro- and nanocapsules, films and implants were considered for the delivery of many bioactive compounds, namely anticancer agents, hormones, anesthetics and antibiotics (17-20). For all these therapeutic applications, the polymers work at body temperature i.e. at 37°C, a value reasonably lower than the glass transition temperature, c.a. 55-60°C, of most of the members of PLA family. Therefore, the glass transition is not a source of major problems for this type of applications, unless uptake of water decreases the glass transition below body temperature.

Therapeutic applications do not constitute the only field of interest of degradable and bioresorbable polymers. There are many other applications relevant of the concepts of temporary uses and elimination after use. Insofar as the temporary commodity applications are concerned, bacterial poly(hydroxy butyrate), PHB, was for a long time considered as the most promising degradable aliphatic polyester. Nowadays, PLA is supplanting PHB and is considered as the degradable polymer having the highest potential to be industrially developed after the pilot scale facilities set up by Cargill Dow Polymers in USA, Mitsui in Japan, and Neste in Europe. However, environmentally degradable polymers aimed at commodity applications have to fulfill rather severe thermal requirements related to the larger temperature range to be faced outdoor. With this respect, the glass transition temperature is one of the most important characteristics of a thermoplastic material. It must be high enough to allow a device to preserve its initial form as it is the case for poly(methyl methacrylate) and poly(ethylene terephtalate) whose T_g values are above 100°C.

The glass transition temperature of polymers is generally determined by differential scanning calorimetry (DSC). However broad-line NMR and mechanical relaxation measurements are also used. Jamshidi (21) compared the glass transition temperatures of different polylactides determined by the three methods and concluded to "a fairly good agreement of the results". T_g was found almost independent on the configurational structure although many other factors can affect this values (molecular weights, polydispersity, residual solvent, absorbed water, oligomers, etc.) (1, 21).

The melting temperature of PLA polymers depends also on many factors. Contrary to T_g, it can be modified by varying the contents in D- and L-lactides of the feed used for the synthesis. Stereocopolymers containing more than 90% of D- or L-lactyl units are semicrystalline. Melting temperature of PLA 100 approaches a constant value of 184°C as the M_w of the polymer increases (21). The melting temperature and the heat of fusion decrease with incorporation of D-units (1). The melting temperature of semicrystalline PLA polymers is also affected by the distribution of L- and D-lactyl units, which depends on the copolymerization initiation and propagation stages and on the composition of the feed in L-, D- and mesolactides. It is worth noting that L- and D-lactyl units can be redistributed with more or less randomization due to transesterification side-reactions (22).

Another means to modify the thermal properties of polymers is copolymerization combined with stereocopolymerization, occasionally. Several strategies have been developed to synthesize lactyl-containing copolymer chains : ring-opening copolymerization, polymerization initiated by macromonomers, and polymerization of asymmetric cyclic compounds consisting of two different units. Ring-opening copolymerization of lactides with other cyclic monomers results generally in random copolymers (23, 38). If the difference of comonomer reactivities is important and transesterification reactions are limited, copolymers with blocky structure are prepared (41). Real block copolymers can be obtained by successive addition of comonomers or by using a prepolymer as macroinitiator (55). Polymerization of dioxane-2,5-diones composed of two different hydroxy acids or morpholine-2,5-diones containing one lactyl unit and one amino acid unit can give alternating copolymers (25, 64). Two main conditions must be fulfilled : one-side attack of the monomer ring and absence of transesterification reactions. These different strategies led to many copolymers (Table I).

Table I : Main copolymers of lactides

Chemical nature	Comonomer	Comonomer formula	Distribution of comonomers	References
Polyester	Glycolide		random alternating	23, 29 26
Polyester	β-butyrolactone		random block	32 30, 31

Table 1. Continued

Polyester	γ-butyrolactone		random	33, 34
Polyester	δ-valerolactone		block	35
Polyester	ε-caprolactone		random block	36, 38, 39 47
Polyester	1,4-dioxane-2,5-dione		random	52
Polyesterether	Ethylene oxide		block	55, 59
Polyesterether	1,5-dioxepane-2-one		random	54
Polyesteramide	1,4-morpholine-2,5-diones		random alternating	66, 68 62, 64
Polyestercarbonate	Trimethylene-carbonate		random	70, 71
Polyestercarbonate	2,2-dimethyltri methylenecar-bonate		random block	72 72
Polyestercarbonate	2,2-[2-pentene-1,5-diyl]trimeth-ylenecarbonate		random	73
Polyesteranhydride	Adipic anhydride		random	74
Polyestersiloxane	Polydimethyl-siloxane	$HO{-}({-}\overset{\underset{CH_3}{|}}{\underset{|}{Si}}{-}O{-})_x{-}H$ with CH_3	block	75

The polymerization conditions influence directly the comonomer distribution in the resulting material and consequently the thermal properties. This paper reviews the most important linear degradable copolymers of lactic acid and their thermal properties. Glass transition temperature data are primarily discussed with respect to temperature stresses usually found outdoor.

Polyesters

Poly(lactide-co-glycolide).

Glycolide is the simplest cyclic dimer derived from α-hydroxy acids and the most common comonomer of lactides. Three methods are available to synthesize high molecular weight PLAXGAY. The first one is the ring-opening polymerization of a feed composed of lactide and glycolide. This method yields almost random copolymers. Initially, the difference of reactivities leads to preferential polymerization of glycolide. Later on, monomer units are usually redistributed and randomized, as a result of transesterification reactions (23).

The second method is based on the ring-opening polymerization of 3-methyl-dioxane-2,5-dione, a cyclic dimer composed of one lactic and one glycolic acid units (24) :

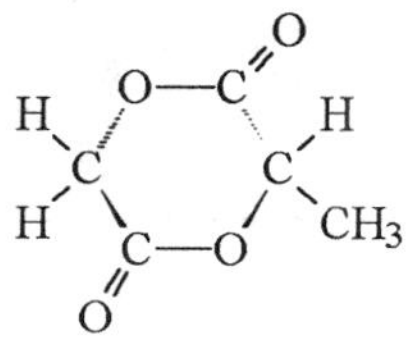

The polymerization was carried out at 140°C for 5h in the presence of stannous octoate as the initiator. Shen (25) showed by NMR spectroscopy that only one sequence existed in the polymer chain and suggested that the ring opening was uniform and thus led to alternating copolymers. The author did not comment on the possibility of randomization by transesterification reactions.

The third method consists in the polycondensation of 2-bromopropionyl glycolic acid at rather low temperature (about 80°C) :

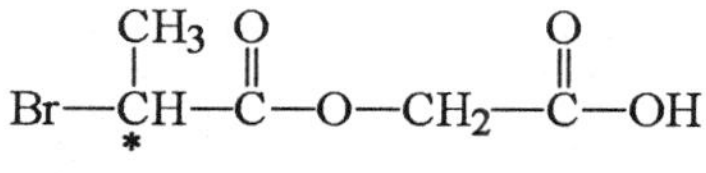

This method precluding transesterification reactions, the resulting high molecular weight PLA25GA50 chains were composed of 100% alternating units (26).

The introduction of a small proportion of glycolyl units into PLA100 chains decreases melting temperature, heat of fusion and glass transition temperature. The presence of more than 25 wt% of glycolyl units in random copolymers of L-lactide and glycolide depresses completely the crystallization of L-lactyl sequences, as shown by Gilding (23). The glass transition temperature of PLAXGAY copolymers can be as low as 36°C, depending on the content in glycolyl units (1, 29). For high proportions of glycolide (>70wt%) melting reappears, because the glycolyl sequences become long enough to crystallize. However, one must keep in mind that the morphology and the thermal properties of PLAXGAY depend not only on the ratio of incorporated comonomers but also on their distribution. For example, it was found that PLAXGA50 could crystallize when glycolyl sequences were long enough, i.e. $\overline{L}_{Glycolyl} > 10$ (27).

Several studies concerning the influence of the composition of PLAGA copolymers on thermal properties were found in literature (22, 28, 29). In contrast, the importance of the sequence lengths on the copolymer properties was rather neglected.

Variations of glass transition and melting temperature of random L-lactide/glycolide copolymers reported by Gilding (23) and Grijpma (29) are compared in Figure 1. One can observe some differences in the data published by the two authors. These differences can be due to different synthesis conditions (4 h at 200°C (23), 1 week at 110°C (29), both in the presence of stannous octoate) and/or different treatment of resulting copolymers (extraction (23), no purification (29)), and/or different unit distributions.

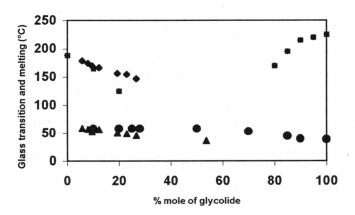

Figure 1 : Glass transition and melting temperatures of poly(L-lactide-co-glycolide) (■ T_m, ● T_g, adapted from (23), ◆ T_m, ▲ T_g, adapted from (29) with permission

The glass transition temperature of poly(3-methyl-1,4-dioxane-2,5-dione), alternating copolymer of D,L-lactic acid and glycolic acid, was reported to be 47°C (25), a value which is comparable with that of the random PLA50GA50 (about 48°C (23)).

The combination of three monomer units in PLAGA copolymers, namely glycolyl, D- and L-lactyl, enables to cover a broad spectrum of performance characteristics. In any case, the copolymerization of D- and L-lactides with glycolide leads to copolymers with lower glass transition temperature than that of PLA stereocopolymers.

Poly(lactide-co-β-butyrolactone).

Both block and random copolymers of lactide and β-butyrolactone have been prepared. Zhang et al. (30) and Reeve et al. (31) synthesized diblock PLA-PHB copolymers using natural low molecular weight poly(R-β-hydroxybutyrate) as a macroinitiator for lactides. The diblock structure of these copolymers was confirmed by ^1H NMR spectroscopy. Reeve et al. (31) analyzed different diblock PLA-PHB copolymers with constant PHB sequence length ($\overline{DP}=26$). The copolymers PHB-PLA50 and PHB-PLA100 with short PLA chains ($\overline{DP}=13$) had similar thermal behaviours. Under these conditions, PLA100 chains did not crystallize. For both copolymers, the glass transition temperatures of PHB ($\approx3°C$) and PLA ($\approx60°C$) sequences and the melting temperature of PHB sequence ($\approx148°C$) were observed. In the case of copolymers with longer PLA100 segments ($\overline{DP}=23$), the crystallizations of both blocks were detected by X-ray diffraction. For the same copolymer, miscibility was observed after quenching and resulted in a new T_g at approximately 20°C. Broad and complex crystallization exotherm and melting endotherm were observed when the quenched material was heated again. Therefore, the authors concluded to the presence of both PHB and PLA crystalline phases (42). Upon annealing, phase separation occured at 55°C for 24 h.

Random L-lactide/R-β-butyrolactone copolymer was prepared using distannoxane catalysts (32). Higher reactivity of L-lactide resulted in different comonomer molar ratio in the copolymer (R-β-butyrolactone/L-lactide=83/17) compared with that in the initial mixture (R-β-butyrolactone/L-lactide=90/10). The random PLAHB copolymer containing 17% of L-lactyl units was semicrystalline with a glass transition at 12°C and a melting point at 120°C (32). A small amount of lactide in the random copolymer affected much more the melting temperature of PHB than higher amounts in the block copolymers. Anyhow, the introduction of β-HB units into a PLA chain led to a decrease of the glass transition temperature here again.

Poly(lactide-co-γ-butyrolactone).

Nakayama et al. (33, 34) succeeded in copolymerizing γ-butyrolactone with L-lactide. Large differences between comonomer compositions in the feed and in copolymer chains were assigned to the poor polymerizability of γ-butyrolactone. For example, the highest content of γ-butyrolactone in the copolymer chain was 28 mole% to be compared to 66 mole% of γ-butyrolactone in the feed mixture. ^1H and ^{13}C NMR spectra showed random comonomer distributions. From a general viewpoint, T_g and T_m decreased as the content of γ-butyrolactone in the PLA-γ-BL copolymer increased (Fig. 2).

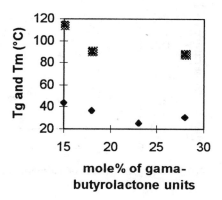

Figure 2 : Thermal properties of poly(L-lactide-co-γ-butyrolactone), T_m (■) and T_g (♦), adapted from (33) with permission

Very small amounts of γ-methyl and γ-ethyl γ-butyrolactone were also copolymerized with L-lactide (33). Less than 1 mole% of γ-substituted γ-butyrolactones depressed T_g of PLA100 from 56 to 47°C and T_m 177°C to 159°C.

Poly(lactide-co-δ-valerolactone).

Block copolymers of L-lactide and δ-valerolactone have been prepared (35). The diblock structure was obtained by addition of L-lactide to living poly(δ-valerolactone) (PVL) macroinitiator as shown by ^{1}H and ^{13}C NMR spectroscopy. All copolymers were semicrystalline and exhibited two melting temperatures corresponding to PLA and PVL sequences (Fig. 3). Unfortunately, the glass transition temperatures of the copolymers were not reported. It is likely that the T_g values were low because the T_g of PVL homopolymer was c.a. - 66°C (35).

Poly(lactide-co-ε-caprolactone).

PLACL random copolymers have been synthesized in order to produce materials with intermediate and adjustable properties (36-40). Because of a large difference of comonomer reactivities, the copolymers had a blocky structure in the first stages of polymerization. Later on, transesterification reactions occurred but their extent depended strongly on the initiator, and on the polymerization time and temperature (40-45). Vanhoorne (41) reported that randomization of comonomer units distribution was more important in the case of L-lactide than in the case of D,L-lactide. Block copolymers were obtained by successive addition of comonomers.

208

Transesterification reactions could be avoided by choosing convenient initiators and by controlling polymerization conditions carefully (46, 47).

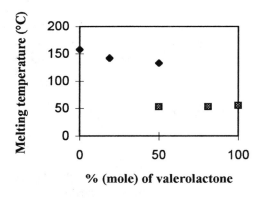

Figure 3 : Melting temperatures of PLA (◆) and PVL (■) segments in diblock copolymers, adapted from (35) with permission

Thermal properties depended strongly on the comonomer ratio and distribution. Random copolymers exhibited one glass transition temperature and one melting point only (38). Introduction of ε-oxycaproyl units into PLA depressed both T_g and T_m (Fig. 4). The effect of ε-oxycaproyl units on the melting temperature and the heat of fusion of PLA100 was found much smaller than that of D-lactyl units (48, 49). Analysis of copolymers by wide angle X-ray scattering (WAXS) suggested that there was no substantial incorporation of ε-oxycaproyl units into the crystalline lattice of PLA 100. In contrast, changes of the glass transition temperature were proportional to the ε-oxycaproyl units content and indicated their homogenous distribution in the amorphous phase. Decreases of T_g and T_m were associated with a decrease of copolymer melt viscosity, a very important parameter for the processing of heat sensitive drug delivery systems. For example, PLACL containing 30wt% of ε-caprolactone could be extruded at temperatures below 100°C (39).

Tapered copolymers were formed In the first stages of the ring-opening polymerization of lactide/ε-caprolactone mixtures, and three glass transitions were observed : two corresponding to homopolymers, and the third one at an intermediate value which was characteristic of a diffuse interface (41).

Diblock copolymers usually show two glass transitions. If any of the blocks is able to crystallize and the segments are long enough to allow crystallization to occur, one melting endotherm is also observed. If the two blocks can crystallize, two melting endotherms are then observed. Typically, a diblock PLA100-PCL showed the glass transition of PCL amorphous domains at about -60°C and two distinguished melting endotherms at 56°C (PCL) and 165°C (PLA 100). The behaviour of these copolymers was that of a material where microphase separation occured (47).

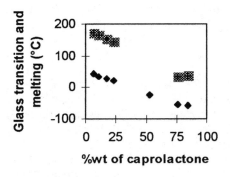

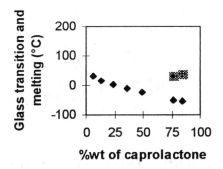

Figure 4 : Thermal properties of random : top : poly(L-lactide-co-ε-caprolactone) and bottom : poly(DL-lactide-co-ε-caprolactone), ■ T_m, ◆ T_g, adapted from (38) with permission

The largely different properties of PLA and PCL homopolymers make their combination by polymerization or blending of interest to generate a wide range of characteristics : permeability, hydrophilicity/hydrophobicity, degradation rate, thermal and mechanical properties, etc. (36, 44, 50). Most applications proposed for these materials concern drug delivery (36, 39). Nevertheless, PLACL copolymers can also be used as artificial skin, nerve guides or burn wound coverings (40). Applications in orthopaedy have been proposed for crystalline and stiff PLACL containing low amounts of ε-oxycaproyl units (39). A potential was attributed to PLACL copolymers with regards to uses as degradable devices, PCL being known as biodegradable under outdoor conditions (51).

Correlation between copolymer characteristics on one side and comonomer units distribution and homopolymer sequence lengths on the other side was attempted recently (38). Establishing such relationships is necessary if one wants to develop new copolymer materials with properties adjustable through copolymerization. So far, no PLACL copolymer has been shown to have a higher T_g than PLA homopolymers.

Poly (lactide- co- 3(S)-[(benzyloxycarbonyl) methyl]- 1,4-dioxane- 2,5-dione).

Hydrophobicity and slow hydrolysis rates of widely used degradable polyesters appear sometimes as serious shortcomings for the development of certain drug release systems. That is why the work of several research teams concentrated on the introduction of hydrophilic groups into PLA.

Kimura (52) succeeded in copolymerizing the six-membered cyclic diester, 3(S)-((benzyloxycarbonyl) methyl]-1,4-dioxane-2,5-dione, BMD, with L-lactide :

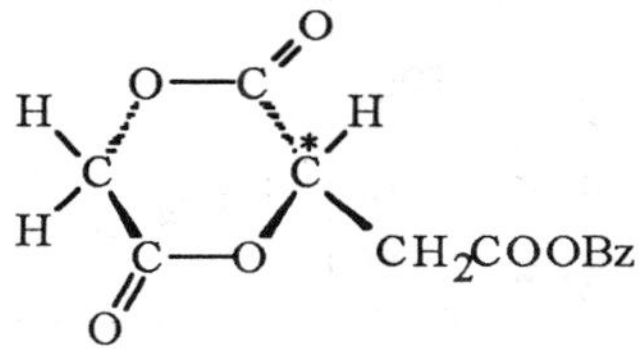

For small amounts of BMD, the copolymer compositions were almost identical to those of the monomer feed. Random incorporation of BMD units was confirmed by ^{13}C NMR spectrometry.

Complete deprotection of benzyl groups by hydrogenolysis resulted in copolyesters built of L-lactic acid, glycolic acid and L-α-malic acid repeat units with pendant carboxyl groups.

Crystallinity and melting temperature of the copolymers decreased with the increase in BMD unit content. After deprotection, slightly higher melting temperatures were observed and explained by an increase of the weigth fraction of L-lactyl units (Fig. 5). The glass transition temperatures of the synthesized copolymers were all close to 20°C.

The pendant carboxyl groups of the deprotected copolymers behaved as hydrophilic groups and as sites for chemical modification, such as coupling of a photo affinity labeling agent, for example (52). Basically, bioactive agents can also be attached to the copolymer chain and the hydrolysis rate can be adjusted through the comonomer ratio. Poly(L-lactide-co-BMD) polymers were intended to be used for the development of drug delivery systems as well as other biomedical materials and devices.

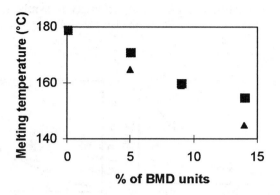

Figure 5 : Melting temperatures of protected (▲) and deprotected (■) poly(L-lactide-co-BMD), adapted from (52) with permission

Polyesterethers

Poly(lactide-co-1,5-dioxepane-2-one).

A heterocyclic compound, namely 1,5-dioxepane-2-one, DXO was synthesized and copolymerized with L- and DL-lactide (53, 54) :

No racemization reactions were observed during the polymerization as shown by ^{13}C NMR spectroscopy (54). The resulting copolymers exhibited one T_g value only located between the glass transition temperatures of the homopolymers ($T_{g\ (PLA)} \approx 55°C$, $T_{g\ (PolyDXO)} \approx -36°C$) and depending on the composition (Fig. 6). Melting endotherms were observed in poly(L-lactide-co-DXO) having contents in L-lactyl units as high as 30-40%. The presence of crystalline domains was explained by either blocky distribution of comonomer units or isomorphism (54).

Polylactide-co-poly(ethylene oxide).

Lactyl units were also combined with various types of polymers such as poly(ethylene oxide), PEO. The aim was generally the promotion of hydrophilicity. In order to synthesize AB- and ABA-type copolymers, mono- and bifunctional PEO polymers were used as macroinitiators or co-initiators of lactide (55-58). Chemical modifications of polymer end groups were necessary to synthesize multiblock PLA-PEO copolymers (59). PLA and PEO were also physically blended (55).

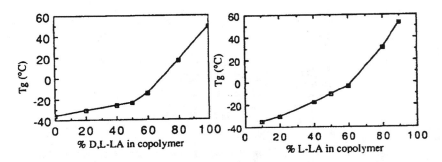

Figure 6 : Glass transition temperatures of PLADXO as a function of composition, from (54), with permission

Whether homopolymers are present in PLA-PEO block copolymers is one of the problems which can affect the thermal behaviour of these copolymers. PLA-PEO copolymers can be distinguished from corresponding homopolymer blends by differences in their physical behaviour. After water extraction, the IR spectra of blends and showed complete disappearance of PEO characteristic bands and resembled to that of pure PLA in agreement with the solubilization of the hydrophilic polyether compound (55). In contrast, no change was detected in the case of the block copolymer. Using small angle X-ray scattering (SAXS), a microphase separation could be observed in PLA-PEO copolymers (60) whereas the homopolymer blends showed a macrophase separation.

An increase of the oxyethylene fraction in PLA100-PEO-PLA100 triblock copolymers decreased the glass transition and the melting temperature as refered to PLA100. Crystallinity also decreases with increasing comonomer ratios. When more than 60% of oxyethylene units are incorporated, PLA100 segments become amorphous and PEO segments crystallize (Tab. II) (57).

A rather different behaviour was observed in the case of PLAGA-PEO diblock copolymers. By introducing glycolyl units into PLA100 segment, the crystallinity of the latter was destroyed. The disappearance of melting temperature of PEO segment was explained by an increase in the phase compatibility of PLAGA and PEO segments (60).

Despite the number of articles concerning PLA-PEO copolymers, comparison of the results is difficult because of different synthesis methods, segment lengths, molecular weights, etc. These multiphase amphiphilic materials would be worth of a complete behaviour study. Anyhow, the presence of PEO segments tends to lower the T_g temperature as refered to PLA homopolymers.

Table II : Thermal properties of PLA100-PEO-PLA100 copolymers (57)

% of oxyethylene units	T_g (°C)	T_m (°C)	ΔH_m (Jg^{-1})
30	20	153	38
33	29	151	33
38	18	145	32
52	9	137	19
62	-12	38	52

Polylactide-co-poly(ethylene oxide-co-propylene oxide).
Commercial poly(ethylene oxide-co-propylene oxide) or Pluronic® (PN) was copolymerized with L-lactide in an attempt to prepare fibres with improved flexibility and degradability (61). The authors concluded on the lowering of T_g and T_m of PLA segments because of both the decrease of molecular weight of the PN segment and the increase of its molar ratio in the copolymer. Actually, the comparison of the copolymers is almost impossible because of the large differences in molecular weights which are known to influence thermal properties.

Polyesteramides

Alternating high molecular weight polydepsipeptides were prepared by ring-opening polymerization of morpholine-2,5-dione derivatives (62-65) :

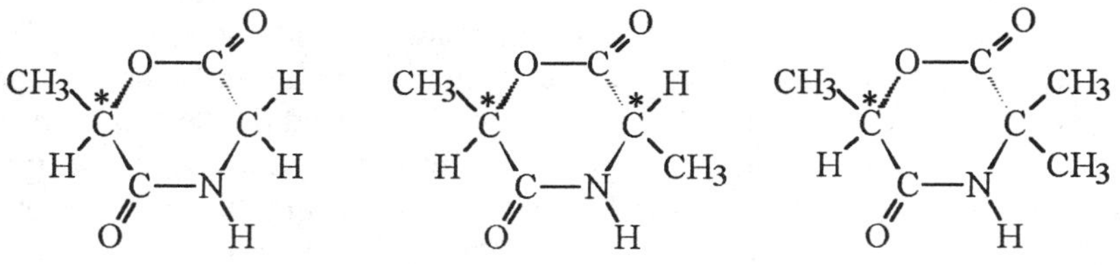

6-methyl-morpholine- 3,6-dimethyl-morpholine 3,3,6-trimethyl-morpholine
2,5-dione   2,5-dione   2,5-dione

Copolymerization of lactides with small amounts of 6-methyl morpholine-2,5-diones (<20%) gave random copolymers. On the other hand, cyclic monomers consisting of glycolic acid and α-amino acid residues were copolymerized with lactides to yield terpolymers (66).
Incorporation of trifunctional α-amino acids into PLA resulted in functionalized polydepsipeptides. L-aspartic acid, L-lysine and L-cysteine were used for the synthesis of polydepsipeptides with pendant carboxyl, amine and thiol groups,

respectively (66). Corresponding morpholine-2,5-diones were prepared using monoprotected α-amino acids, the protective groups being removed after polymerization :

The homopolymerization of monoprotected morpholine-2,5-diones was substituent-dependent. The larger the substituent, the lower the final molecular weights. In any case, low molecular weight alternating copolymers were obtained. In contrast, the copolymerization with lactides resulted in high molecular weights ($\overline{M}_w$ from 14.000 to 87.000 for 10 to 20% of incorporated 1,4-morpholine-2,5-dione).

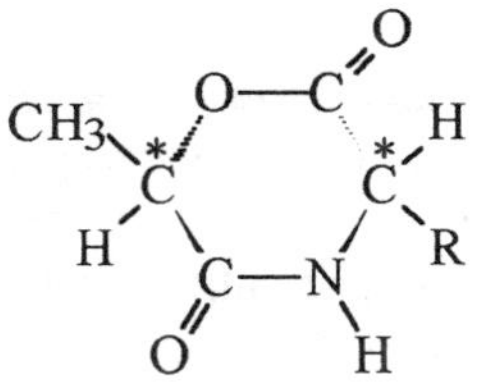

Incorporated amino acid	-R =
L-aspartic acid	-CH$_2$-COO-benzyl
L-lysine	-(CH$_2$)$_4$-NH-benzyloxycarbonyl
L-cysteine	-CH$_2$-S-p-methoxybenzyl

Alternating copolymers of DL-lactic acid with glycine, valine and alanine, respectively, were amorphous and their glass transition temperatures were relatively high (93 to 117°C) (67). Unfortunately, molecular weights did not exceed 30.000. In contrast, incorporation of small amounts of functionalized amino acids (L-aspartic acid, L-lysine and L-cysteine) into PLA50 decreased very slightly the glass transition temperature of the latter (66). Barrera (68) observed a decrease of both glass transition and melting temperatures as the ratio of L-lysine present in the copolymer increased. The copolymer with 10% of morpholine-dione units was amorphous.

Recently, John et al (69) introduced hydroxyl functions into PLA100 by copolymerization of 3-(O-benzoyl)-L-serinyl-morpholine-2,5-dione with L-lactide and deprotection of benzoyl groups by hydrogenolysis. High molecular weight polyesteramides could only be prepared when molar ratio of morpholine-2,5-dione derivative did not exceed 2.5%. Hydroxyl functions gave a possibility of chemical modification by introducing acrylate groups which can serve as crosslinkers.

Polydepsipeptides consisting of natural monomer residues appear as valuable addition to existing degradable polymers. Incorporation of amino acids into PLA backbone results in degradable polyesteramides with a wide range of chemical and physical properties. Functionalization of PLA offers a possibility of coupling small molecules (dyes, bioactive molecules, etc.).

Polyestercarbonates

Trimethylene carbonate or TMC, 2,2-dimethyltrimethylene carbonate or DTC and 2,2-[2-pentene-1,5-diyl]trimethylene carbonate or °HTC were used for copolymerization with lactide :

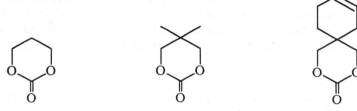

Trimethylenecarbonate   2,2-dimethyltrimethylene   2,2-[2-pentene-1,5-diyl]
        (TMC)   carbonate (DTC)   trimethylene carbonate (°HTC)

The copolymerization of lactides with TMC initiated by stannous octoate (70, 71) or with DTC initiated by diethyl zinc (72), respectively, gave random copolymers. The glass transition temperatures lied in all cases between those of the corresponding homopolymers ($T_{g(PolyTMC)} \approx -18°C$, $T_{g(PolyDTC)} \approx 27°C$). One can note that incorporation of about 10% of TMC or 50% of DTC into PLA backbone resulted in a decrease of T_g below the body temperature. These copolymers were proposed for certain medical applications, for example soft tissue implants, for which stiff and brittle materials are not suitable (70).

Block copolymers of L-lactide and DTC were prepared by successive addition of comonomers in the presence of metallic initiators (72). In this case, two melting temperatures were observed. For example, a copolymer with a molar ratio of L-lactide/DTC of 55/45 showed melting endotherms at 106 and 163°C, whereas the melting of pure homopolymers occured at 124°C (PolyDTC) and between 160 and 192°C (PLA100). The T_g values of the block copolymers have not been reported.

Recently, L-lactide was copolymerized with °HTC, a new cyclic monomer containing a C=C bond (73). In the presence of stannous octoate, copolymers with a blocky structure were formed, as a result of different comonomer reactivities. Increase in reaction time led to almost random distribution of repeat units. In all cases, only one T_g was observed which decreased with increased copolymer °HTC content ($T_{g(Poly°HTC)} \approx 33°C$). The vinyl substituents could be converted to epoxy functionalities which gave the possibility of further conversion to diols and other groups. The authors suggested the use of cyclohexene side groups as sites for free radical crosslinking.

Polyesteranhydrides

Poly(lactide-co-adipic anhydride).

Copolymers of DL-lactide and adipic anhydride were synthesized in bulk using stannous octoate as catalyst (74). The randomness of the comonomer units distribution has not been studied. Nevertheless, the copolymers exhibited unique T_g and/or T_m values which were lower than those of the homopolymers (Fig. 7). These new materials showed typical surface erosion behaviour leading to zero-order drug release and were proposed for elaboration of protein delivery systems.

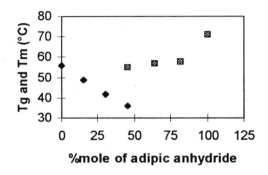

Figure 7 : Thermal properties of poly(DL-lactide-co-adipic anhydride), ■ T_m, ◆ T_g, *adapted from (74) with permission*

Polyestersiloxanes

Polylactide-co-polydimethylsiloxane.

As already mentioned, the stiffness of PLA appeared often as a disadvantage in contact with soft tissue. Polydimethylsiloxane or PDMS (Fig. 16) was chosen as a rubber component of novel flexible degradable materials for biomedical applications (75) :

$$\text{HO} \longrightarrow \left(\begin{array}{c} \text{CH}_3 \\ | \\ \text{Si} \longrightarrow \text{O} \\ | \\ \text{CH}_3 \end{array} \right)_{\!\! x} \!\!\! \text{H}$$

The presence of PDMS to PLA or PLAGA improved the surface smoothness and thermal stability of the polyesters.

The introduction of small amounts of PDMS (<4%) into PLA decreased slightly its T_g (from 63 to 58°C). The same effect was observed for PLAGA-PDMS copolymers.

Conclusion

Linear copolymers of lactic acid form a family of degradable materials that has enlarged very much during the last decade. Most of them exhibit attractive properties. However, the presence of comonomers in PLA chains provokes generally a decrease of glass transition temperature (with the exception of alternating polydepsipeptides with $\overline{M}_w$<30.000). This can be of interest for biomedical and pharmaceutical uses but can come short when confronted to the requirements of commodity applications.

So far, the attempts to functionalize PLA were scarce. The functionalizing comoOnomers could only be introduced in small amounts and the molecular weights of resulting copolymers were low. Anyhow, the family of lactic acid copolymers can still be enlarged because many comonomers deserve to be prospected, especially in attempts to improve thermal properties of PLA homopolymers aimed at applications in the packaging area.

References

1. Vert, M., Chabot, F., Leray, J., Christel, P. *Makromol. Chem. Suppl.* **1981**, *5*, 30.
2. Kulkarni, R. K., Pani, K. C., Neuman, C., Leonard, F. *Arch. Surgery* **1966,** *93*, 839.
3. Cutright, D. E., Hunsuck, E. E., Beasley, J. D. *J. Oral. Surg.* **1971**, *29*, 393.
4. Dittrich, D. W., Schulz, R. C. *Angew. Makromol. Chem.* **1971**, *15*, 109.
5. Kricheldorf, H. R., Kreiser-Saunders, I. *Makromol. Chem.* **1990**, *191*, 1057.
6. Nijenhuis, A. J., Grijpma, D. W., Pennings, A. J. *Macromolecules* **1992**, *25*, 6419.
7. Schwach, G., Coudane, J., Engel, R., Vert, M. *Polym. Bull.* **1996**, *37*, 771.
8. Vert, M., Chabot, F., Leray, J., Christel, P. *French Patent* **1978**, 78.29878.
9. Törmälä, P., Vainionpää, S., Kilpikari, J., Rokkanen, P. *Biomaterials* **1987**, *8*, 42.
10. Bos, R. R. M., Boering, G., Rozema, F. R., Leenslag J. W. *Oral Maxillofac. Surg.* **1987**, *45*, 751.
11. Suuronen, R. ,Wessman, L., Mero, M., Tormala, P., Vasenius, J., Partio, E., Vihtonen, K., Vainionpaa, S. *J. Mater. Sci., Mater. Med.* **1992**, *3*, 288.

218

12. Vert, M., Christel, P., Chabot, F., Leray J. *Macromolecular Biomaterials*, CRC Press Inc., Boca Raton, Florida, USA, 1984, pp. 119.
Beck, L., R. Cowsar, D. R., Lewis, D. H., Coxgrove, R. L., Riddle, C. T., Lowry, S. L., Epperly, T. *Fert. Ster.* **1979**, *31*, 545.
14. Bodmeier, R., Chen, H., *J. Pharm. Pharmacol.* **1988**, *40*, 754.
15. Spenlehauer, G., Vert, M., Benoit, J.-P., Bodaert, A. *Biomaterials* **1989**, *10*, 557.
16. Smith, K. L., Schimpf, M. E., Thompson, K. E. *Advanced Drug Delivery Reviews* **1990**, *4*, 343.
17. Woodland, J. H. R., Yolles, S., Blake, D. A., Helrich, M., Meyer, F. J. *J. Med. Chem.* **1973**, *16*, 897.
18. Juni, K., Ogata, J., Nakano, M., Ichihara, T., Mori, K., Akagi, M. *Chem. Pharm. Bull.* **1985**, *33*, 313.
19. Ikada, Y., Hyon, S. H., Jamshidi, K., Higashi, S., Yamamuro, T., Katutani, Y., Kitsugi, T. *J. Control. Rel.* **1985**, *2*, 179.
20. Benoit, J.-P., Puisieux, F. *Actual Chem. Therm.* **1985**, *12* 141.
21. Jamshidi, K., Hyon, S. H., Ikada Y. *Polymer* **1998**, *29*, 2229.
22. Chabot, F., Vert, M., Chapelle, S., Granger, P. *Polymer* **1983**, *24*, 53.
23. Gilding D. K., Reed, A. M. *Polymer* **1979**, *20*, 1459.
24. Augurt, T. A., Rosensaft, M. N., Perciaccante V. A. *US Patent* **1976**, 3,960,152.
25. Shen, Z., Zhu, J., Ma, Z. *Makromol. Chem, Rapid Commun.* **1993**, *14*, 457.
26. Rebert, N. W. *Macromolecules* **1994**, *27*, 5533.
27. Kricheldorf, H. R., Kreiser, I. *Makromol. Chem.* **1987**, *188*, 1861.
28. Kricheldorf, H. R., Jonté, J. M., Berl, M. *Makromol. Chem., Suppl.* **1985**, *12*, 25.
29. Grijpma, D. W., Nijenhuis, A. J., Pennings, A. J. *Polymer* **1990**, *31*, 2201.
30. Zhang, Y., Gross, R. A., Lenz R. *Macromolecules* **1990**, *23*, 3206.
31. Reeve, M.S., McCarthy, S. P., Gross, R. A. *Macromolecules* **1993**, *26*, 888.
32. Hori, Y., Takahashi, Y., Yamaguchi, A., Nishishita, T. *Macromolecules* **1993**, *26*, 4388.
33. Nakayama, A., Kawasaki, N., Arvanitoyannis, I., Aiba, S., Yamamoto, N. *J. Environ. Polym. Deg.* **1996**, *4*, 205.
34. Nakayama, A., Kawasaki, N., Aiba, S., Maeda, Y., Arvanitoyannis, I., Yamamoto, N. *Polymer* **1998**, *39*, 1213.
35. Kurcok, P., Penczek, J., Franek, J., Jedlinski, Z. *Macromolecules* **1992**, *25*, 2285.
36. Schindler, A., Jeffcoat, R., Kimmel, G. L., Pitt, C. G., Wall, M. E., Zweidinger, R. *Contamporary Topics in Polymer Science*, Plenum Press, New York, USA, 1977, Vol. 2.; pp 251.
37. Pitt, C. G., Schindler, A. *Biodegradables and Delivery Systems for Contraception*, MPT Press Ltd., Lancaster, England, 1980, Vol. 1.; pp 17.
38. Perego, G., Vercellio, T., Balbontin, G. *Makromol. Chem.* **1993**, *194*, 2463.

39. Zhang, X., Wyss, U. P., Pichora, D., Goosen, M. F. A *J.M.S.-Pure Appl. Chem.* **1993**, *A30*, 933.
40. Grijpma, D. W., Pennings, A. J. *Polymer Bulletin* **1991**, *25*, 335.
41. Vanhoorne, P., Dubois, P,. Jérôme, R., Teyssié, P. *Macromolecules* **1992**, *25*, 37.
42. Kasperczyk J., Bero, M. *Makromol. Chem.* **1993**, *194*, 913.
43. Bero, M., Kasperczyk, J., Adamus, G. *Makromol. Chem.* **1993**, *194*, 907.
44. Grijpma, D. W., Zondervan, G. J., Pennings, A. J. *Polymer Bulletin* **1991**, *25*, 327.
45. Vion, J. M., Jérôme, R., Teyssié, P., Aubin, M., Prud'homme, R. E. *Macromolecules* **1986**, *19*, 1828.
46. Jacobs, C., Dubois, P., Jérôme, R., Teyssié, P. *Macromolecules* **1991**, *24*, 3027.
47. Bero, M., Adamus, G., Kasperczyk, J., Janeczek, H. *Polymer Bulletin* **1993**, *31*, 9.
48. Penning, J. P., Dijkstra, H., Pennings, A. J. *Polymer* **1993**, *34*, 942.
Grijpma, D. W., Pennings, A. J. *Macromol. Chem. Phys.* **1994**, *195*, 1633.
50. Song, C. X., Sun, H. F., Feng, X. D. *Polym. J.* **1987**, *19*, 485.
51. Sinclair, R. G. *US Patent to Gulf Oil Corp.*, **1977**, US 4045-418
52. Kimura, Y., Shirotani, K., Yamane, H., Kitao, T. *Polymer* **1993**, *34*, 1741.
53. Shalaby, S. W. *US Patent,* 4 190 720, Ethicon, **1980**
54. Albertsson, A. C., Löfgren, A. *Makromol. Chem., Macromol. Symp.* **1992**, *53*, 221.
55. Cohn, D., Younes, H. *J. Biomed. Mater. Res.* **1988**, *22*, 993.
56. Deng, X. M., Xiong, C. D., Cheng, L. M., Xu, R. P. *J. Polym. Sci. : Part C* **1990**, *28*, 411.
57. Jedlinski, Z., Kurcok, P., Walach, W., Janeczek, H., Radecka, I. *Makromol. Chem.* **1993**, *194*, 1681.
58. Li, S. M., Rashkov, I., Espartero, J. L., Manolova, N., Vert, M. *Macromolecules* **1996**, *29*, 57.
59. Chen, X., McCarthy, S. P., Gross, R. A. *Macromolecules* **1997**, *30*, 4295.
60. Li, Y. X., Kissel, T. *J. Control. Rel.* **1993**, *27*, 247.
61. Lee, C. W., Kimura, Y. *Bull. Chem. Soc. Jpn.* **1996**, *69*, 1787.
62. Shakaby, S. W., Koelmel, D. F. *Eur. Pat. Appl.* 0086613, **1983**
63. Yonezawa, N., Toda, F., Hasegawa, M. *Makromol. Chem., Rapid Commun.* **1985**, *6*, 607.
64. Helder, J., Feijen, J., Lee, S. J., Kim, S. W. *Makromol. Chem., Rapid Commun.* **1986**, *7*, 193.
65. Samyn, C., Van Beylen, M. *Makromol. Chem., Macromol. Symp.* **1988**, *19*, 225.
66. in't Veld, P. J. A., Dijkstra, P. J., Feijen, J. *Makromol. Chem.* **1992**, *193*, 2713.
67. Dijkstra, P. J., Eenink, M. J. D., Feijen, J. *Degradable Materials: Perspectives, Issues and Opportunities*, CRC Press Inc., Boca Raton, Florida, USA, 1990, pp. 99.

68. Barrera, D. A., Zylstra, E., Lansbury, P. T., Langer, R. *J. Am. Chem. Soc.* **1993**, *115*, 11010.

69. John, G., Tsuda, S., Morita, M. *J. Polym. Sci. Part A-Polym. Chem.* **1997**, *35*, 1901.

70. Buchholz, B. *J. Mater. Sci. : Mater. Med.* **1993**, *4*, 381.

71. Grijpma, D. W., Pennings, A. *J. Macromol. Chem. Phys.* **1994**, *195*, 1633.

72. Schmidt, P., Keul, H., Höcker, H. *Macromolecules* **1996**, *29*, 3674.

73. Chen, X., McCarthy, S. P., Gross, R. A. *Macromolecules* **1998**, *31*, 662.

74. Castaldo, L., Corbo, P., Maglio, G., Palumbo, R. *Polym. Bull.* **1992**, *28*, 301.

75. Shinoda, H., Ohtaguro, M., Iimuro, S. *Eur. Pat. Appl.* 90305667.9, **1991**

Chapter 15

Solid-State Structure of Poly(lactide) Copolymers

James Runt[1], Jiang Huang[1], Melissa S. Lisowski[1], Eric S. Hall[2], Robert T. Kean[2], and J. S. Lin[3]

[1]Department of Materials Science and Engineering, The Pennsylvania State University, University Park, PA 16802
[2]Central Research, Cargill Inc. and Cargill-Dow, Minneapolis, MN 55440
[3]Solid State Division, Oak Ridge National Laboratory, Oak Ridge, TN 37831

Inferences from small-angle x-ray scattering, optical microscopy and DSC experiments were used to establish a model of the lamellar microstructure for random poly(L-lactide/meso-lactide) copolymers. From analysis of the SAXS data it was concluded that the copolymers contained significant interfibrillar regions whose concentration increased with comonomer content. Comparison of equilibrium melting points (derived using the Gibbs-Thomson approach) with those calculated from models for random copolymer melting leads to the proposal that R stereochemical defects are rejected from the crystalline regions.

Introduction

The family of aliphatic polyesters based on lactic acid has been the subject of widespread interest in recent years. In the biomedical field these materials have been used as therapeutic aids in surgery and in pharmacology in drug delivery systems (*1,2*). One particular advantage of high molecular weight poly(lactic acid) is its *in vivo* biodegradability.

There is continuing interest in replacing conventional petroleum-derived polymers with those from renewable resources. Lactic acid polymers are one prominent alternative, as the monomers can be derived from agricultural sources such as corn. Recent advances appear to have made it possible to convert corn by-products into monomer and polymer in a relatively economical manner (*3,4*). The huge waste disposal issue in the U.S. and other parts of the world is a longstanding and complex problem. An additional ecological benefit of poly(lactide)s is that they decompose rapidly and completely in a typical compost environment.

The lactide produced from an agricultural source like corn consists predominately of the S stereoisomer. As a result, polymers synthesized from such sources are often of high S content and consequently are capable of crystallization. In fact, many of the useful properties of this class of materials stem for their semi-crystalline nature. However, such poly(lactide)s are rarely composed of all S stereoisomer but are copolymers in which S units are interrupted in the chain by R stereoisomer defects. Defect arrangement and concentration have a profound effect on crystallizability and crystalline microstructure -- and hence ultimate properties.

In the present paper we summarize some of our initial research on the microstructure and crystallization of well-defined copolymers of L-lactide and meso-lactide (5).

Experimental

Materials

Lactic acid contains a chiral carbon atom, and two stereochemical isomers are possible. These are referred to as the S and R configurations. High molecular weight polymers containing lactic acid repeat units are synthesized via the ring-opening polymerization of one or more cyclic lactides; a lactide consists of two lactic acid units and hence has two chiral carbon atoms. The possible dimeric lactides will be referred to using the prefixes L-, D- and meso-. When polymerized, L-lactide results in the placement of two S units in the chain. Likewise, D-lactide results in two R units, while meso-lactide results in the introduction of one R and one S unit.

Poly(L-lactide) [PLLA] and three model L-lactide/meso-lactide copolymers were synthesized using a tin (II) octanoate catalyst at 180 °C (5). Copolymers prepared under these reaction conditions have been shown to be random, or nearly so (6). $\overline{M}_w$ and $\overline{M}_w$ of PLLA and the copolymers are the same within experimental uncertainty (see Table 1) and hence differences in crystallization behavior reflect differences in comonomer content only. The polymerization feed contained 0, 3, 6 and 12% meso-lactide, which resulted in 0.4, 2.1, 3.4 and 6.6% R stereochemical defects (determined by chromatographic analysis) in chains containing predominately the S isomer.

Table I. PLLA and poly(L-lactide/meso-lactide) copolymers

Copolymer	meso%	$\overline{M}_n$*	$\overline{M}_w$*	R%**	T_m (°C)	T_g (°C)
A	--	65,500	127,400	0.4	174	59
B	3	65,800	122,600	2.1	160	58
C	6	63,900	119,100	3.4	153	58
D	12	65,500	121,300	6.6	126	58

  * with reference to polystyrene standards
 ** from chiral liquid chromatography

Characterization

Specimens for small-angle x-ray scattering (SAXS) experiments were first melted in a Mettler FP-82 hot stage at a temperature ~40 °C above the nominal melting point (for 5 min.), then rapidly transferred to a second hot stage held at the desired crystallization temperature (T_c). SAXS experiments were carried out on the Oak Ridge National Laboratory 10 m pinhole-collimated SAXS camera using Cu Kα radiation ($\lambda = 0.154$ nm) and a 20×20 cm^2 position-sensitive area detector. Data were collected at sample to detector distances of 2 and 5 m, and suitably appended and superimposed. Data were converted to an absolute differential cross-section by means of precalibrated secondary standards. Scattering profiles were transformed into correlation functions (7) from which linear crystallinities (i.e., the average crystallinity associated with the lamellar stacks), lamellar thicknesses (1_c) and amorphous thicknesses in lamellar stacks (1_a) were determined. Additional experimental details and details of the analysis can be found in refs. 5 and 8.

Samples were prepared in a similar fashion for spherulite growth rate experiments. Optical microscopy was used to measure spherulitic growth rates as a function of crystallization temperature (T_c). A self-seeding technique was used at high T_cs to accelerate the nucleation process. Differential scanning calorimetry (DSC) was used to measure degrees of crystallinity, employing a value for the ultimate heat of fusion of 100 J/gm (which was derived from slowly polymerized, highly crystalline poly(L-lactide)). Melting temperatures were also determined from DSC experiments, employing small sample weights (~0.15 mg) to minimize thermal lag. Nominal melting points (T_m) and glass transitions temperatures (T_g) of PLLA and the L-lactide/meso-lactide copolymers are reported in Table 1.

Results and Discussion

Morphology

All of the polymers under consideration are semi-crystalline and exhibit spherulitic superstructures when crystallized from the melt (see Figure 1 for the morphology of PLLA partially crystallized at 140 °C). The influence of a small amount of R stereoisomer is dramatic: the bulk crystallinity (for materials crystallized under isothermal conditions) decreases from ~50% PLLA (material A in Table 1), to about 15% for the copolymer containing 6.6% of the R stereoisomer.

Lamellar microstructure

The lamellar microstructure of the copolymers, crystallized isothermally at a number of different temperatures, was characterized using small-angle x-ray scattering. Except for PLLA, SAXS long periods were found to increase with increasing R stereoisomer content, consistent with lower degrees of crystallinity. The bulk crystallinities of the polymers under investigation are too low to assume that the

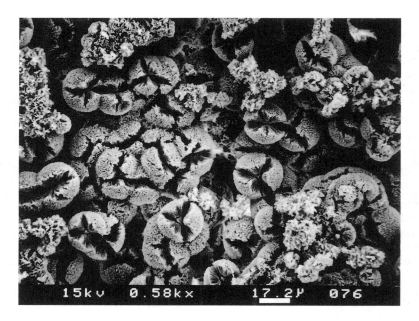

Figure 1: Spherulitic morphology of PLLA (material A), crystallized at 140°C (580x). This specimen was quenched before crystallization was complete and the amorphous material removed by etching in 50/50 DMSO/0.02N KOH at 80°C for 4 min. Note that portions of the spherulites facing the surface have also been etched away.

amorphous component is located only within the lamellar stacks. It is clear from optical microscopy experiments that there is no significant amount of amorphous material between spherulites, as all spherulites impinge upon the completion of primary crystallization. By comparing the linear crystallinity derived from the SAXS correlation function (after distinguishing l_c from l_a) and the bulk crystallinity from DSC, the fraction of the amorphous material residing in interfibrillar regions was estimated. From the definition of the linear crystallinity (w_c) it follows that w_c must be greater than or equal to the bulk crystallinity (ϕ_c). The case where $w_c = \phi_c$ indicates that effectively all of the amorphous material resides between lamellae in stacks, i.e., the material consists essentially of continuous lamellar stacks. The situation where $w_c > \phi_c$ indicates that a certain amount of the amorphous component resides in interfibrillar regions and the partitioning between the interfibrillar and interlamellar regions can be calculated from w_c and ϕ_c.

It was concluded from analysis of the SAXS data that PLLA and the L-lactide/meso-lactide copolymers contain a significant concentration of amorphous material between fibrils. This varies from about 35% for PLLA, to ~60% for the copolymer containing 6.6% R co-units. The remainder of the amorphous fraction resides within lamellar stacks. In other words, the solid state microstructure of these copolymers can be represented by a finite lamellar stack model [see, for example, *9* and *10* and Figure 2], as opposed to the 'infinite' stack model which is appropriate for more highly crystalline polymers like poly(ethylene oxide) (*8*).

Figure 2: Schematic view of 'finite lamellar stack' model for semi-crystalline polymers and blends (reprinted from ref. 9). The lines represent lamellar bundles (fibrils) in a spherulitic superstructure. The 'squiggley' lines represent pockets of amorphous material between the fibrils.

Defect location

One of the key questions in any crystalline random copolymer system is the location of the defects (i.e., the R stereoisomers in this case) within the lamellar microstructure: are they completely excluded from the crystals (i.e., like in the classic Flory model (*11*)) or are some included as defects within crystals? Insight, albeit indirect, can be gained by determining equilibrium (ultimate) melting points (T_m°) and comparing these with values predicted from models for crystallization and melting of random copolymers. The popular Hoffman-Weeks approach (*12*) for determining T_m°s is not appropriate here due to curvature in the T_m vs. T_c plots (see also ref. *13*). T_m°s determined using the classic Gibbs - Thomson approach (*12*) are the most reliable since the expression is general, not requiring assumptions about the nature of the crystal surface. End (fold) surface free energies (σ_e) were also extracted from the slope of the linear Gibbs-Thomson fits. The results are presented in Table 3, along with literature values for poly(L-lactide) (*14,15*). T_m° and σ_e for material A, which is essentially PLLA, are very similar to the PLLA literature values. Only data where the linear crystallinities are in the range of < 0.4 or > 0.6 were used in our analysis, so no reliable T_m° and σ_e could be determined for copolymer B using the Gibbs-Thomson approach. The T_m° value in brackets in Table 2 is an estimate arrived at using a data-fitting approach (*16*). The origin of the decrease in σ_e is unclear at present but it has been associated previously with the decreased formation of 'tight' folds and adjacent reentry folding with increasing comonomer content (*17*).

Table II. Equilibrium melting points and surface free energies derived from Gibbs-Thomson analysis (5)

Material	$T_m^{\,o}$ (°C)	σ_e (erg/cm^2)
PLLA (14,15)	207 ~ 212	61
A (0.4% R)	214	60
B (2.1%)	(200)	----
C (3.4%)	187	37
D (6.6%)	166	27

Equilibrium melting points were then compared with predictions from various models for melting of random copolymers: i.e., where the comonomer is completely excluded from the crystallites (11), uniformly included in the crystals (the model of Sanchez and Eby (18)) and a modification of the exclusion model that incorporates the average sequence length of the crystallizable comonomer (19). As seen in Table 3, the predicted values from the modified exclusion model (which is also equivalent to the expression in the exclusion limit of the recent model of Wendling and Suter (20)) were found to be in excellent agreement with the experimental T_m^os, implying *exclusion* of R stereochemical units from the crystal lattice, at least under the crystallization conditions used in our studies. This result is not surprising in itself as comonomer units are frequently found to be excluded from lamellar crystals. However, in an early paper on solution grown crystals of poly(L-lactide/D,L-lactide)s, Fischer, et al. argued that at least some R units are included in S crystals (21). However, they also found that rejection of R co-units increased significantly with increasing degree of supercooling ($\Delta T = T_m^{\,o} - T_c$), the opposite of what would be expected based on crystallization kinetics arguments. One of our current interests is the direct characterization of R stereochemical defect location under a variety of crystallization conditions, using selective hydrolysis experiments. One such measurement has been conducted recently by others on a hydrolyzed poly(lactide) copolymer, which initially contained ~4% R units (22). About 25% of the original R units (~1% of the total repeat units) remained in microcrystalline fragments after hydrolysis, in broad agreement with implications from equilibrium melting points. More work is needed to determine if the remaining R stereounits are actually incorporated in crystalline lamellae or are located on the crystal surfaces (i.e., are an 'artifact' of a reaction time that is too short).

Spherulite Growth Rates

Crystallization kinetics of PLA and the copolymers were followed by measuring spherulitic growth rates (G) as a function of crystallization temperature and meso-lactide content. As expected, the rate is a strong function of meso concentration: at T_c

Table III. Equilibrium melting point predictions from copolymer models vs. experimentally-derived values (5)

Copolymer	T_m^o ($^\circ$C)	Exclusion	Inclusion	Modified Exclusion
0.4% R	214	----	----	----
2.2%	(200)	209	211	198
3.4%	187	206	209	189
6.6%	166	197	204	168

= 117 °C, the growth rate of PLLA is about 60x faster than that of the copolymer containing 6.6% R units, whereas at 135 °C the dependence of growth rate on R concentration is much more significant (the ratio of the growth rates of these copolymers is more than 340) -- see Figure 3. Models of random copolymer crystallization (*18,23*) predict a linear dependence of ln G on the mole fraction of non-crystalline comonomer at a given T_c. The data for the meso-lactide copolymers also appear to follow this relationship.

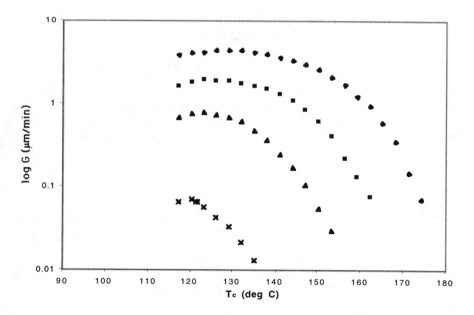

Figure 3. Spherulitic growth rates of PLLA and the L-lactide/meso-lactide copolymers as a function of crystallization temperature [5]. ◆ - PLLA; ■ - 3% meso-lactide copolymer; ▲ - 6% meso-lactide copolymer; ✕ - 12% meso-lactide copolymer.

Current research is focusing on the role of defect size on crystallization and microstructure of poly(lactides). A series of well-defined poly(L-lactide/D-lactide) copolymers have been synthesized using the same reaction conditions as used for the meso-lactide copolymers. In the D-lactide copolymers, two stereochemical R lactic acid units are necessarily in adjacent positions, resulting in a significantly larger interruption in crystallizable sequences compared to the meso-lactide materials.

Spherulite growth rates have been determined recently for the D-lactide copolymers and several general conclusions can be drawn (24). The growth rates for the meso-lactide copolymers are significantly higher (by ~30%) than L-lactide/D-lactide copolymers with the same comonomer content (and molecular weight). This is not surprising given the defect size resulting from incorporation of D-lactide compared with meso-lactide. However, the D-lactide copolymers exhibit significantly higher growth rates than the meso-lactide copolymers when compared at similar optical compositions. This ranges from somewhat less than a factor of two higher for the D-lactide copolymer containing 1.7% R stereochemical counits vs. the 3% meso-lactide copolymer (2.1% R), to a factor of more than six higher for a D-lactide copolymer containing 6.2% R counits vs. the 12% meso-lactide copolymer (6.6% R). This presumably arises from a reduction in the number of defect sites for the D-lactide copolymers at equivalent R stereoisomer content (that is, a larger average crystallizable sequence length for the same optical composition). Research is continuing on the D-lactide copolymers, particularly on lamellar organization and defect location.

Acknowledgments

We would like to express our appreciation to the Advanced Technology Program of NIST for partial support of this work (contract # 70NANB5H1059). We would also like to acknowledge the donors of the Petroleum Research Fund, administered by the American Chemical Society, for partial support. This research was also supported in part by the Division of Materials Sciences, U.S. Department of Energy under contract No. DE-AC05-96OR22464 with Lockheed Martin Energy Research Corporation.

References

1. Vert, M.; Christel, P.; Chabot, F.; Leray, J. In *Macromolecular Materials*, Hastings, G. W.; Ducheyne, P., Eds., CRC Press: Boca Raton, FL, 1984; p. 119.
2. Lewis, D. H. *Drugs Pharm. Sci.* **1990**, *45*, 1.
3. Gruber, P. R.; Hall, E. S.; Kolstad, J. J.; Iwen, M. L.; Benson, R. D.; Borchardt, R. L. *U. S. Patent 5,142,023* **1992**.
4. Gruber, P. R. *6th Annual Mtg. Bio/Enironmentally Degradable Polym. Soc.* San Diego, CA, **1997**.
5. Huang, J.; Lisowski, M. S.; Runt, J.; Hall, E. S.; Kean, R. T.; Buehler, N.; Lin, J. S. *Macromolecules* **1998**, *31*, 2593.

6. Thakur, K. A. M.; Kean, R. T. ; Hall, E. S.; Kolstad, J. J.; Lindgren, T. A.; Doscotch, M. A.; Ilja Siepmann, J.; Munson, E. J. *Macromolecules* **1996**, *30*, 2422.

7. G. R. Strobl, M. J. Schneider and I. G. Voigt-Martin, *J. Polym. Sci. Polym. Phys. Ed.* 18, 1361 (1980).

8. Talibuddin, S.; Wu, L.; Runt, J.; Lin, J.S. *Macromolecules* **1996**, *29*, 7527.

9. Hsiao, B. S. ; Sauer, B. B.; Verma, R. K.; Zachmann, H. G.; Seifert, S.; Chu, B.; Harney, P. *Macromolecules* **1995**, 28, 6931.

10. Hsiao, B. S.; Sauer, B. B. *J. Polym. Sci. Polym. Phys. Ed.* **1993**, *31*.

11. Flory, P. J. *Trans. Faraday Soc.* **1995**, *51*, 848.

12. Hoffman, J. D.; Davis, G. T.; Lauritzen, Jr., J. I. In *Treatise on Solid State Chemistry*, Hanny, N. B., Ed. , Plenum Press, New York, 1976; Vol. 3, Chap. 7.

13. Alamo, R. G.; Viers, B. D.; Mandelkern, L. *Macromolecules* **1995**, *28*, 3205.

14. Vasanthakumar, R.; Pennings, A. J. *Polymer* **1983**, *24*, 175.

15. Tsuji, H.; Ikada, Y. *Polymer* **1995**, *36*, 2709.

16. Huang, J. Prasad, A.; Marand, H. *Polymer* **1994**, *35*, 1896.

17. Lambert, W. S.; Phillips, P. J. *Macromolecules* **1994**, *27*, 3537.

18. Sanchez, I. C.; Eby, R. K. *Macromolecules* **1975**, *8*, 638.

19. Baur, H. *Makromol. Chem.* **1966**, *98*, 297.

20. Wendling, J.; Suter, U. W. *Macromolecules* **1998**, *31*, 2516.

21. Fischer, E. W.; Sterzel, H. J.; Wegner, G. *Kolloid Z. Z. Polym.* **1973**, *251*, 980.

22. Leatham, G. F. (Cargill Dow) Unpublished results.

23. Andrews, E. H.; Owen, P.J.; Singh, A. *Proc. Roy. Soc. Lond. A* **1971,** *324*, 79.

24. Baratian, S.; Hall, E. S.; Lin, J. S.; Runt, J. Unpublished results.

Chapter 16

Degradation Kinetics of Poly(hydroxy) Acids: PLA and PCL

Georgette L. Siparsky

TDA Research, Inc., Wheat Ridge, CO 80033

Polymers of α-hydroxy acids, like polylactic acid (PLA), are known to hydrolyze in the presence of water, and are widely used as degradable materials. Other hydroxy acid polymers, like poly-ε-caprolactone (PCL), are also degradable, but they require catalysis by an enzyme or an external acid such as HCl. The hydrolysis kinetics of the two systems were analyzed specifically for their mode of catalysis, and a kinetic equation was derived to describe the two processes. As an α-hydroxy acid, Polylactic acid was hydrolyzed by the polymer acid end-groups in a self-catalized process, and the reaction was found to be autocatalytic in nature. On the other hand, poly-ε-caprolactone hydrolysis obeyed pseudo-first order kinetics, with the rate of hydrolysis being dependent on the external acid concentration.

Aliphatic polyesters have been the subject of intense research in the development of environmentally degradable commodity plastics [1]. The interest in these materials was fueled by their successful application in the biomedical industry, where polyglycolic acid (PGA) and polylactic acid (PLA) have been used as degradable sutures, drug delivery systems and orthopedic implants [2-13]. Polycaprolactone (PCL), is another aliphatic polyester used in the preparation of degradable plastics; it is generally used to modify the material properties of other degradable plastics for specific needs [7]. Degradation of these polymers occurs by ester hydrolysis, with main chain cleavage giving oligomers and monomers that are biocompatible with the environment. Applications of these polymers are determined by their physical properties and their respective rates of degradation.

Introduction

The use of PLA and PCL as a commodity plastics has sparked a new interest in the hydrolysis mechanism, since degradation is anticipated to take place over a broad range of environments. Ideally, degradable commodity plastics are expected to be stable on the shelf, but quickly hydrolyze when they enter the waste stream. In order to design such a material, the hydrolysis process, and its mode of catalysis, must be well understood.

The degradation of PLA was shown to be a hydrolysis process, but the rate of degradation in the solid state was dependent on a number of physical characteristics like the degree of crystallinity and water vapor permeability [14, 19-23]. These variables are influenced by material processing, so it is not surprising that the reported measured rate of degradation has varied from 0.13 h^{-1} to 0.41 day^{-1} [24, 25]. For example, the presence of crystallinity in PLA film was shown to slow down the degradation rate, probably because water was not accessible to the crystalline domain. Similarly, thick samples degrade more quickly than thin films, presumably because the thick samples have a higher concentration of catalytic acid end-groups that do not easily leach out, as they do in thin films [14].

Since PLA was primarily used for biomedical applications, many degradation studies were conducted in a specific medium and temperature (pH 7.4 buffered system at 37°C) to simulate degradation in the human body. Under these buffered conditions, the hydrolysis process is controlled by that specific environment, and does not reflect the degradation process of a commodity packaging plastic.

High molecular weight aliphatic polyesters like PLA and PCL have been prepared by the ring opening polymerization of lactones, giving polymer chains that are terminated by hydroxyl groups and carboxylic acid end-groups. The degradation of such a polyester by ester hydrolysis produces smaller chains, also terminated by hydroxyl and acid groups, as shown in Equation 1, where R and R' are hydrogen or alkyl groups.

$$-\left[C(RR')COO-\right] + H_2O \rightarrow -C(RR')COOH + -C(RR')COH \qquad (1)$$

As the reaction continues, the acid end-group concentration, $[COOH]$, increases, since each hydrolyzed ester bond produce an acid end-group. Ester hydrolysis is generally catalyzed by acids and bases, so that the increasing concentration of acid end-groups is expected to lead to an accelerated rate of hydrolysis. Such a process is then said to be autocatalytic. It should be noted that in acid-catalyzed hydrolysis, the hydrogen ion, H_3O^+ is the true catalyst, so that the catalytic influence of the increasing carboxylic acid should be related to its degree of dissociation in the presence of water.

A unique feature of α-hydroxy polyesters (where ester groups are separated by only one carbon and written as $[-C(RR')COO-]$ is the high degree of dissociation of the acid-end group, and the expected contribution of that acidity to the overall hydrolysis process [15]. Polylactic acid has been known to hydrolyze more readily than other aliphatic polyesters [16]. The acid dissociation constant (pK$_a$) of oligomeric PLA is 3.1 (compared to many carboxylic acids with a pK$_a$ of 5) so that the dissociation of the acid end-group is expected to result in an acidic environment and a significant contribution to acid-catalyzed hydrolysis. The concept of "self catalysis", whereby the acid end-groups of the PLA chain catalyze the hydrolysis reaction, will be examined in this paper.

The early degradation studies conducted on PLA blocks and films examined the influence of acid end-group catalysis on the rate of hydrolysis. As in a typical ester hydrolysis, the rate was measured by the increasing concentration of the product, $[COOH]$, and was dependent on the concentration of the reactants: ester, $[E]$ and water $[H_2O]$, as shown in Equation 2. It was argued that if the reaction was acid catalyzed, then the reaction rate was also dependent on the acid concentration, $[COOH]$, as shown in Equation 3 [17, 18]. Since the rate equation accounted for the increasing acid concentration, it was referred to as the equation for autocatalysis [19-22].

$$\frac{d[COOH]}{dt} = k[H_2O][E] \qquad \text{for uncatalyzed hydrolysis} \qquad (2)$$

$$\frac{d[COOH]}{dt} = k[COOH][H_2O][E] \qquad \text{for acid-catalyzed hydrolysis} \qquad (3)$$

The concept of "self-catalysis" of PLA hydrolysis, whereby the catalyst concentration is quantitatively related to the concentration of one of the products, has not been adequately addressed. In this case, the acid end-group (catalyst) concentration increases at the same rate that the ester (reactant) concentration decreases. This is a direct result of the special case whereby the hydrolysis of a mole of ester results in a mole of carboxylic acid end-groups, which in turn, catalyze further hydrolysis.

In order to determine the relationship that best describes the hydrolysis kinetics, especially the contribution of acid catalysis to the overall rate of hydrolysis, we conducted our studies in solution, eliminating any solid state factors on the reaction rate. Thus diffusion of water *into* the PLA film, as well as the diffusion of high acid content oligomeric PLA *out* of the film, did not play a role in the kinetics. Similarly, we eliminated the influence of the degree of crystallinity of the film on the concentration of water and ester groups participating in the hydrolysis process [23-25]. Since PLA was insoluble in water, a mixed solvent system was required to carry out solution hydrolysis. Therefore, the hydrolysis of PLA in solution was carried out in a mixed, homogeneous system.

Accelerated Rate of Hydrolysis

Studies of PLA hydrolysis in solution have been carried out as early as 1936, in acetone/water using oligomers with an average degree of polymerization of 18 and HCl as the catalyst [16]. The measured rate was found to increase as hydrolysis proceeded. When the linear dimer, lactyl lactic acid, was isolated from the higher oligomers, it was found to have a higher hydrolysis rate constant than that of the remaining oligomers. It was thus determined that as the concentration of the dimer increased with hydrolysis, so did the overall rate constant. However, in water there is a complex equilibrium between lactic acid, its linear dimer, cyclic dimer and other oligomers, and there is some doubt as to whether these species were really isolated [26].

The effect of acid-end group catalysis was more recently examined by solution hydrolysis of PLA in acetone/water [25]. Since the water concentration did not change significantly, the reaction was treated as pseudo first order. There was no measurable hydrolysis without the addition of an external acid source, even after 60 days at 37° C. The addition of acetic acid had no effect until the acetic acid to ester ratio, $\dfrac{[CH_3COOH]}{[E]}$ was as high as 1:10. However, when HCl was used as the catalyst, hydrolysis did occur, with a pseudo first order rate constant of 0.41 day^{-1}. In this study, the rate of hydrolysis was found to be independent of the water concentration. Furthermore, hydrolysis did not take place in the absence of an external acid source. Despite the fact that hydrolysis of PLA is known to take place in the solid state (films and blocks) without an external acid like HCl, no measurable hydrolysis was observed in the acetone/water solution in this study.

Hydrolysis of PLA in solution was also conducted in the more polar solvent mixture of dioxane/water, this time to determine if hydrolysis was a random chain scission process, in which every ester in the chain has an equal probability of being hydrolyzed. Using NMR end-group analysis, two chain scission processes were observed during PLA hydrolysis at 60°C: a slow, random ester scission process ($k' =$ 0.01 h^{-1} where k' is the pseudo first order rate constant), and a faster chain end scission of the terminal ester ($k' = 0.13$ h^{-1}) [24]. The latter could dominate the process as hydrolysis proceeded and the number of terminal esters increased. The observed increase in the rate of degradation with time was then attributed to the increase in the concentration of terminal esters and the dominance of the chain-end scission process in the overall hydrolysis. A similar dual rate of hydrolysis (reaction of water with terminal ester groups and that of water with random internal chain ester groups) has been observed with polyethylene terephthalate films placed in boiling water [27].

Another explanation for the increasing rate of hydrolysis lies in the fact that as more esters are converted to acid and alcohol end-groups, the polymer medium becomes more hydrophilic, increasing the effective interaction with water molecules that is more likely to lead to hydrolysis. This phenomenon was observed in a study of the degradation of PLA films in water, whereby the water content of the films increased as hydrolysis proceeded [28]. The increase in water content of the films was attributed to the increase in the hydrophilicity of the polymer as the number of hydrophylic end-groups increased. The rate equation for ester hydrolysis included a water concentration term, $[H_2O]$, so that increasing $[H_2O]$ was expected to influence the rate of hydrolysis.

In summary, the three theories developed to explain the accelerating rate of hydrolysis as the reaction proceeded are summarized as follows:

1. *The increase in hydrophilicity of the polymer*: As hydrolysis proceeded, the number of OH and $COOH$ end groups increased, making the film more hydrophilic. This was followed by a corresponding increase in the water uptake of the films, and a faster rate of hydrolysis.

2. *The dual rate of hydrolysis*: Terminal esters hydrolyzed faster than mid-chain esters. As the reaction proceeded, the number of terminal esters increased, leading to an acceleration of the degradation process.

3. *The autocatalysis by the carboxylic acid end-groups*: Hydrolysis was catalyzed by the increasing number of carboxylic acid end-groups, and the increase in the rate of PLA degradation was due to a rise in the catalyst concentration.

The nature of the solvent, as well as the solvent to water ratio used in polyester solution hydrolysis have a significant effect on the reaction kinetics, so that results need to be interpreted with great care. Generally, rates of acid catalyzed ester hydrolysis in mixed solvents have been influenced by the solvent to water ratio [15]. For example, the rate of hydrolysis of the polyester derived from adipic acid and 1,6-hexanediol in a dioxane/water mixture has been found to increase linearly with water content, then drop sharply when the solution had more than 6% water [29, 30]. The influence of solvent ratio on the rate of hydrolysis was also examined with ethyl acetate in a water/DMSO mixture [31]. The rate increased with increasing DMSO content, reached a plateau at 20% water, then increased again once the water ratio was 60% or more. Therefore solution hydrolysis studies of PLA conducted here were interpreted with the understanding that limitations existed in comparisons between results obtained in the presence of a mixed solvent and those obtained in the presence of water only.

The two solvent systems used so far to examine PLA hydrolysis have been acetone and dioxane, with low dielectric constants, ε (acetone $\varepsilon = 21$ and dioxane $\varepsilon = 2$, compared to water $\varepsilon = 79$) [32]. Such non-polar environment are expected to retard carboxylic acid dissociation and misrepresent the hydrolysis process that takes place in the presence of water only. It was not surprising that self-catalyzed hydrolysis (catalysis by the PLA acid end-groups) did not take place in these solvent systems. Acetonitrile ($\varepsilon = 36$) was chosen for this study to better simulate the hydrolysis process of PLA.

Autocatalysis

Autocatalysis has been defined as the catalysis of a reaction by the product [33]. It has been observed in oxidation reactions (auto-oxidation) and some enzyme catalyzed reactions [33]. Autocatalysis kinetics have accounted for the increasing rates of reactions by including the product concentration in the rate equation. Such a computation has been used to examine the hydrolytic degradation of polyethylene terephthalate in solution [34]. Isothermal autocatalysis, where the increased rate of reaction was not due to an increase in reaction temperature, has been described by Equation 4

$$A = nB \rightarrow (n+1)B \qquad (4)$$

where A and B were the reactants, in this case ester, $[E]$, and carboxylic acid, $[COOH]$, respectively. The resulting increase in concentration of the product/reactant B led to a correspondingly increasing reaction rate, or autocatalysis. For n = 1, the reaction has been referred to as "quadratic isothermal autocatalysis" [17].

We have examined the hydrolysis kinetics of PLA in an acetonitrile solution and found that conventional first and second order kinetics, as well as quadratic isothermal autocatalysis kinetics, did not describe the process adequately. Our objective was:

1. To gain a better understanding of the influence of self-catalysis by the terminal carboxylic acid end-groups.

2. To quantitatively demonstrate that autocatalysis leads to the observed accelerated rate of hydrolysis of PLA.

An alternative autocatalysis expression was derived to accommodate the partial ionization of the terminal carboxylic acid catalyst. The derived equation fit the analytical data for hydrolysis of PLA at 50°C and 60°C in acetonitrile/water. On the other hand, PCL hydrolysis was not self-catalyzed and the reaction followed typical pseudo first order kinetics.

Experimental

Ring opening polymerization of lactide was used to prepare low molecular weight PLA (Mn 55,000, Mw 95,000 from Chronopol, Golden, CO). The residual monomer and low molecular weight oligomers were removed by dissolving the polymer in methylene chloride and precipitating the higher molecular weight fraction by the addition of methanol. Solutions of PLA for hydrolysis were made up in a 200 mL volumetric flask by dissolving a weighed amount of polymer in dry spectroscopic grade acetonitrile. Aliquots of that bulk solution were then added to 25 mL volumetric flasks, along with a specified amount of water. The concentration of water in most cases was greater than the ester concentration. This was done to ensure that the water concentration changed very little for the duration of the experiment, and thus treated as a constant value. The volume was made up to the 25 mL mark with acetonitrile. These working solutions were then aliquotted into 4 mL vials (leaving no dead space), and placed in an incubator/shaker at 60°C. The initial concentrations of ester, $[E]_0$, carboxylic acid end groups, $[COOH]_0$, water $[H_2O]_0$, and hydrochloric acid $[HCl]_0$, are listed in Table 1. At regular intervals, one of the vials was removed and refrigerated until the study was completed. To examine the influence of specific acid catalysis, dilute hydrochloric acid was added to a specified set of samples. The PLA solutions were titrated against standardized potassium hydroxide made up in acetonitrile/water. Phenol red was used as the indicator.

Table I. Concentration of reactants in solution hydrolysis of PLA.

Sample	$[COOH]_0$ $\times 10^3 M$	$[E]_0$ M	$[H_2O]_0$ M	$[HCl]_0$ $\times 10^2 M$
E-1	1.2	0.4	4.4	0
E-2	2.1	0.8	4.4	0
E-3	4.1	1.5	4.4	0
E-4	5.8	2.2	4.4	0
A-1	5.8	2.2	4.4	0
A-2	5.8	2.2	4.4	1.6
A-3	5.8	2.2	4.4	2.6
A-4	5.8	2.2	4.4	5.8
W-1	5.0	1.9	1.1	0
W-2	5.0	1.9	3.3	0
W-3	5.8	2.2	4.4	0
W-4	4.9	1.9	6.6	0

Polycaprolactone was not very soluble in acetonitrile. Low molecular weight (Mn 15,000) was used and titrations were carried out at 40°C in very dilute solutions, where the carboxylic acid concentration was in the order of 10^{-4} M.

The variables influencing the hydrolysis kinetics of PLA in solution (water, ester and acid concentrations) were studied independently. Three sets of samples were prepared with each set examining one of these variables. Table 1 shows the E series with varying ester concentration and constant water and acid, the A series with varying hydrochloric acid concentration and the W series with varying water concentration. Using $[E]_o$ and $[E]$ as the initial ester concentration and at time t respectively, hydrolysis was measured by the change in the ester group concentration, as shown in the plots of $\ln \dfrac{[E]_0}{[E]}$ versus time in Figure 1.

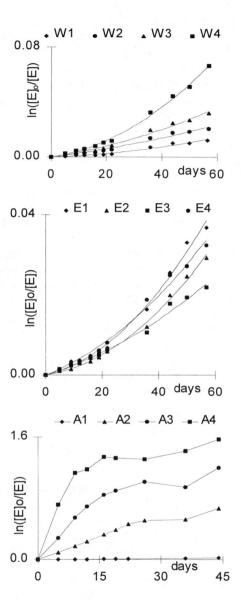

Results and Discussion

We initially examined the influence of the variables, the ester and water concentration, on the hydrolysis of PLA. The reactions were carried out by varying the initial concentration of one of these variables and applying pseudo first order kinetics to the results determined by titration. It was evident that hydrolysis kinetics of PLA were dependent on the water concentration (Figure 1, top), since increasing the initial water concentration resulted in a higher slope and reaction rate. Similarly,

Figure 1. Hydrolysis of PLA at 60 °C. (top) varying initial water concentration (middle) varying initial ester concentration (bottom) Varying HCl concentration.

hydrolysis rates were also dependent on the ester concentration (Figure 1, middle). It is interesting to note that in both cases the rate curves appeared linear in the early stages of the reaction, but as hydrolysis progressed, the slope of the curves increased, indicating an accelerated reaction. The ester hydrolysis consumed only 0.8% of ester groups after 22 days, and almost 5% after 60 days (sample E4). We also examined the influence of HCl acid catalysis on the reaction rate (Figure 1, bottom). The HCl catalyzed reaction was considerably faster than that of the other solutions, with complete hydrolysis by day 22 for the solution with an HCl concentration of 5.8×10^{-2} M.

Pseudo First Order and Second Order Kinetics

Pseudo first order kinetics were examined by using three levels of HCl catalyst. The reaction rate was determined by the plot of the rate constants obtained for HCl catalyzed hydrolysis of PLA in acetonitrile/water versus the HCl concentration as shown in Figure 2. The error in the rate measurements using duplicate solutions was less than 5%. The measured rate constant of the HCl catalyzed reaction, k', was related to the rate constant of the uncatalyzed reaction, k, by Equation 5 and was calculated from the slope in Figure 2 as 0.34 day^{-1}.

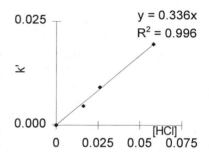

Figure 2. Reaction rates for HCl catalyzed PLA hydrolysis.

$$k' = k \ [HCl] \tag{5}$$

The linearity of this plot ($R^2 = .996$) indicated that the rate of reaction was directly proportional to [H$^+$] or catalyst concentration (assuming that HCl is fully ionized in the water/acetonitrile solution). In the presence of HCl, the reaction was dominated by the HCl concentration, and any contribution by the carboxylic acid end-groups to the catalysis process was undetectable. In order to understand the "self catalysis" process, we evaluated the titration results using the reactant concentrations, $[E]$ and $[H_2O]$, in the absence of HCl and applied second order kinetics, as shown in Figure 3.

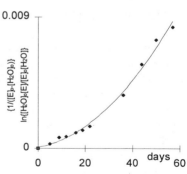

Figure 3. Second order kinetics of PLA hydrolysis.

Since the rate of hydrolysis increased with the water and ester concentration, a second order rate plot shown in Figure 3 provided a better evaluation of the reaction kinetics. The second order rate kinetics are described by Equation 6 and the solution in Equation 7 was used to find the rate constant k.

$$\frac{-d[E]}{dt} = k[H_2O][E] \tag{6}$$

$$\frac{1}{\left([E]_0 - [H_2O]_0\right)} \ln \frac{[H_2O]_0[E]}{[H_2O][E]_0} = kt \tag{7}$$

The rate plot in Figure 3 appeared to be linear for the first few days, but by day 22 the rate of hydrolysis appeared to increase with time. At this point, less than 1% of the ester groups initially present were hydrolyzed. It became obvious that a third reactant played a key role in the kinetics of the system and that its concentration was increasing with time.

Quadratic Isothermal Autocatalysis

The increasing carboxylic acid concentration resulting from ester hydrolysis was expected to lead to increased catalysis of the same reaction by the acid groups (autocatalysis). To examine this autocatalytic influence of the carboxylic acid end-groups on the reaction rate, the hydrolysis results were interpreted using quadratic isothermal autocatalysis kinetics. Appendix A shows the textbook derivation of the quadratic isothermal autocatalysis as it applies to polyester hydrolysis. To simplify the kinetics, the following assumptions were made:

1. In order to reduce the number of variables, the initial water concentration, $[H_2O]_0$, was treated as a constant term incorporated into the measured rate constant.

2. The term $x = ([COOH]K_a)^{1/2}$ was used to represent the acid catalyst concentration in the calculation, accounting for the dissociation of the acid end groups; the integration of [x] in the quadratic equation was then conducted with respect to time.

The hydrolysis rate is dependent on the concentration of the reactants, ester and water, and that of the dissociated acid catalyst, [x]. The rate of appearance of the product can be defined by Equation [8], and the solution in Equation [9] is used to find the rate constant.

$$\frac{d[x]}{dt} = k[E][H_2O][x] \tag{8}$$

$$\frac{1}{[H_2O]}\ln\frac{[COOH]^{1/2}}{[E]} = at + \frac{1}{[H_2O]}\ln b \tag{9}$$

where the constants $a = k\{[E]_0 + ([COOH]_0 K_a)^{1/2}\}$ and $b = \dfrac{[COOH]^{1/2}}{[E]}$

Equation [8] has described the catalyst concentration as $x = ([COOH]K_a)^{1/2}$ since the dissociated acid defines the level of catalytic activity. On the other hand, the concentration of the acid produced is best described as $[COOH]$. Therefore, autocatalysis in PLA hydrolysis is not truly quadratic in nature, as indicated by n=1 in Equation 4. A mole of acid $[COOH]$ is produced for every mole of ester ([E]) consumed. However, the influence of the acid catalyst concentration on the reaction rate depends on $([COOH]K_a)^{1/2}$ and that dependence was half order in nature. Indeed, values of the constants a, [E]0 and k listed in Table II indicate that the measured values of k ranged from 2.35 to 12.7x 10^{-3} M^{-2} day^{-1}. A true rate constant is expected to have the same value, regardless of the initial concentration of the reactants Therefore we do not have true "quadratic isothermal autocatalysis".

Table II. Rate constants for the hydrolysis of PLA using conventional quadratic isothermal autocatalysis.

Sample	"a"(10^3 $M^{-1}day^{-1}$)	[E]0	k =a/[E]0 ($10^3 M^{-2}day^{-1}$)
E-1	5.6	0.44	12.7
E-2	5.1	0.79	6.48
E-3	5.9	1.54	3.80
E-4	5.2	2.21	2.35

Autocatalysis: Distinguishing the Product from the Catalyst

A new equation for autocatalysis was derived in order to make the distinction between the acid catalyst concentration, $([COOH]K_a)^{1/2}$ and the acid product concentration, $[COOH]$, while taking their relationship into account. The computation was carried out by monitoring the ester concentration. The derivation of the new autocatalysis kinetics is shown in Appendix B, using Equation 10 as the starting point,

$$\frac{d[E]}{dt} = k'[E]([COOH]K_a)^{1/2} \tag{10}$$

where $k' = k[H_2O]$

The water concentration was initially assumed to be constant and taken to be a factor in the measured rate constant. This was a reasonable assumption because for most cases the water concentration was much greater than the ester concentration and therefore changed very little for the duration of the experiment. The solution for Equation [10] is

$$\frac{1}{a^{1/2}} \ln \frac{(a-[E])^{1/2} - a^{1/2}}{(a-[E])^{1/2} + a^{1/2}} + c = (k' K_a^{1/2}) \tag{11}$$

where the constants a and c are related to the initial ester and acid concentrations by

$$a = [E]_0 + [COOH]_0$$

$$c = \frac{1}{a^{1/2}} \ln \frac{(a-[E]_0)^{1/2} - a^{1/2}}{(a-[E]_0)^{1/2} + a^{1/2}}$$

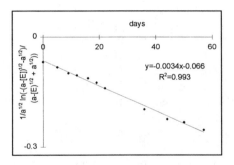

Plotting the measured values for the acid concentration in Equation 11 yielded a slope $(k'K_a^{1/2})$, with a typical plot shown in Figure 4 for hydrolysis of sample W-4 at 60°C. The slope was 0.0034 $M^{1/2}day^{-1}$ and the water concentration was 4.4 M. The constant $(kK_a^{1/2})$ was computed as follows

Figure 4. Autocatalytic Hydrolysis of PLA using the half order power of [COOH].

$$kK_a^{1/2} = \frac{(k' K_a^{1/2})}{[H_2O]}$$
$$= 0.0034 \, M^{1/2}day^{-1} / 4.4 \, M \tag{12}$$
$$= 7.7 \times 10^{-4} \, M^{3/2}day^{-1}$$

Results of solution hydrolysis using Equation 11 are shown in Table III for hydrolysis at 50°C and Table IV for hydrolysis at 60°C. Values of $kK_a^{1/2}$ that did not meet the 95% confidence limits are marked NC and are discussed below. Additional samples, labeled 1 through 11, were also analyzed. The autocatalysis equation derived in Appendix B using the half order power of the carboxylic acid concentration gave a linear plot with a constant value of $kK_a^{1/2}$ for several samples.

However, our results have demonstrated that the solution to the autocatalysis equation is valid within a set of limits defined by the initial water to ester ratio, $\dfrac{[H_2O]_0}{[E]_0}$. The derived autocatalysis equation was based on an acid dissociation constant that is not affected by the water content in the mixed solvent medium. It was also assumed that in the presence of a large excess of water, its concentration remained relatively unchanged as hydrolysis proceeded. The values of $k(K_a)^{1/2}$ marked as NC (non-conforming values) in Tables III and IV were obtained in hydrolysis conditions that did not fall within these assumptions.

Carboxylic acids are known to associate in acetonitrile by forming hydrogen-bonded dimers as shown in Figure 5, reducing their dissociation into free carboxylate and hydrogen ions. In the presence of water, the dimer association is reduced as water acts to form hydrogen bonds with the dissociated acid. In essence, the value of the dissociation constant K_a is expected to increase with increasing water content in the acetonitrile/water

Table III. Autocatalysis rate constants for the hydrolysis of PLA using the half order power of [COOH] at 50°C.

Sample	$[H_2O]_0/[E]_0$	$10^2\,kKa^{1/2}\,M^{-3/2}\,day^{-1}$	
E-1	11.1	0.104	NC
E-3	2.9	0.045	
E-4	2.0	0.038	
W-2	1.6	0.036	
W-4	3.0	0.025	
1	4.8	0.055	
2	4.5	0.043	
3	4.8	0.104	NC
4	3.7	0.051	
5	4.5	0.069	
6	3.0	0.023	
7	5.3	0.180	NC
8	2.0	0.045	
9	1.9	0.044	
10	3.3	0.061	
11	1.1	0.059	

NC: Non-conforming values.

Table IV. Autocatalysis rate constants for the hydrolysis of PLA using the half order power of [COOH] at 60°C.

Sample	$[H_2O]_0/[E]_0$	$10^2\,kKa^{1/2}\,M^{-3/2}\,day^{-1}$	
E-1	10.0	0.18	NC
E-2	5.6	0.12	NC
E-3	2.9	0.070	
E-4	2.0	0.077	
W-1	0.57	0.18	NC
W-2	1.8	0.084	
W-4	3.6	0.082	

NC: Non-conforming values.

mixture. Therefore, when the water concentration in our experiments was high, the measured $k(K_a)^{1/2}$ was also high. The dependence of K_a on the acetonitrile/water ratio set the experimental limitation that $k(K_a)^{1/2}$ was constant only within narrow ranges of that solvent to water ratio.

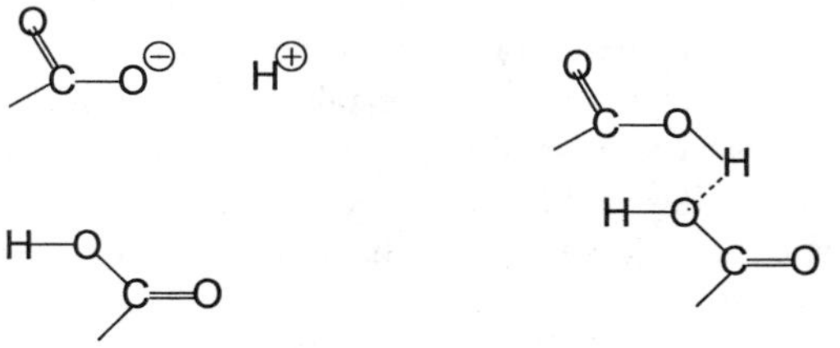

Carboxylic acid dissociation in a dilute solution with free H^+ ions

Carboxylic acid association in a concentrated solution reduces $[H^+]$

Figure 5. Association of carboxylic acids in acetonitrile.

The assumption that the water concentration is high enough so that it does not change significantly during the reaction becomes invalid when $[H_2O]_0 < [E]_0$. In this case, the ratio of the water concentration relative to its initial value, $\dfrac{[H_2O]}{[H_2O]_0}$ changes more rapidly that the same ratio for the ester concentration, $\dfrac{[E]}{[E]_0}$ during the course of the reaction. This discrepancy were observed when the $\dfrac{[H_2O]_0}{[E]_0}$ ratio was less than 1, as shown by sample W-1 in Table 4.

The kinetics of autocatalysis observed in PLA hydrolysis have been examined; we have demonstrated that the derived autocatalysis equation of the hydrolysis of PLA in acetonitrile/water is valid within the constraints of the experiment. These include working in a narrow range of acetonitrile/water ratios so that acid dissociation is not varied by the solvent/water ratio, and ensuring that the water concentration was high enough to be treated as a reactant whose concentration changed little during the hydrolysis experiment.

It must be noted that the absolute value of $k(K_a)^{1/2}$ has limited value, since it applies to a narrow range of experimental conditions. The goal of this research was to demonstrate that the accelerated rate of hydrolysis observed during the degradation of the polymer was directly related to the increasing acid end-group concentration.

Thus we have shown that the hydrolysis process is autocatalytic. Additionally, we have derived a relationship that can be further developed to predict degradation behavior in specific environments.

In summary, the concept of "self catalysis", whereby the acid end-group catalyzed the hydrolysis of PLA, has been established. Furthermore, the PLA hydrolysis process was shown to be autocatalytic and found to fit the derived equation relating the observed increase in hydrolysis rates to the acid end-group dissociation. Previous studies of PLA hydrolysis had not adequately addressed the issue of carboxylic acid dissociation, but we have demonstrated that it is a key factor in the autocatalysis process.

PCL Hydrolysis

The acid dissociation constant, K_a, of PCL is two orders of magnitude lower than that of PLA, so that acid catalyzed hydrolysis was expected to be very slow. The "self catalysis" concept was not observed in the hydrolysis of PCL, even at temperatures approaching the polymer melting point. The hydrolytic degradation of PCL required the use of an external acid catalyst. In a biological environment, PCL hydrolysis has been observed and measured in the presence of specific enzymes. We found that PCL did not hydrolyze in an acetonitrile/water solution, even after a period of 60 days at 60°C. The addition of hydrochloric acid was necessary to study PCL hydrolysis. The results of PCL hydrolysis at 60°C using pseudo first order kinetics (Equation 13) are shown in Figure 6 for four concentrations of HCl and rate constants listed in Table 5.

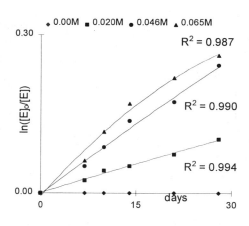

Figure 6. *Hydrolysis of PCL using pseudo first order kinetics.*

$$\ln \frac{[E]}{[E]_0} = kt \tag{13}$$

where $k = [HCl][H_2O]$

The acid catalyzed hydrolysis of PCL in acetonitrile/water (k_{H+} 0.039 $M^{-2}day^{-1}$) was much slower than previously measured in dioxane/water at 60°C [24]. The low solubility of PCL in acetonitrile may well be an indication that the polymer chains in solution are tightly clustered so as to minimize the energy of the system [34]. In this case, the ester groups are not accessible for protonation and addition of water, leading to a reduced rate of hydrolysis. In comparison, low molecular weight PLA (Mn 50,000) was soluble in acetonitrile and the HCl catalyzed rate constant was an order of magnitude higher than that of PCL (k_{H+} of 0.039 $M^{-2}day^{-1}$ for PCL and 0.34 $M^{-2}day^{-1}$ for PLA).

Table V. First order rate constants for HCl catalyzed hydrolysis of PCL.

Sample	[HCl] M	k'_{H+} $(M^{-1}day^{-1})$	k_{H+} $(M^{-2}day^{-1})$
PCL-1	0	0	0
PCL-2	0.020	0.0036	0.041
PCL-3	0.046	0.0086	0.042
PCL-4	0.065	0.0096	0.033

Unlike PLA, PCL was not observed to undergo self-catalyzed hydrolysis in acetonitrile/water solution, even at the temperatures approaching the polymer melting point of 65°C. This result is not surprising since PLA carboxylic acid terminal groups dissociate 100 times more than those of other carboxylic acids (pK_a of 3 compared to 5). The solvent influence on the rate of hydrolysis was also substantial as PCL was observed to hydrolyze much faster in dioxane than in acetonitrile [24].

Conclusion

The self-catalyzed hydrolysis of PLA was examined in a mixed solvent medium since PLA was not soluble in water. Acetonitrile was used as the organic solvent since it is a more polar solvent that acetone or dioxane. The hydrolysis of PLA was carried out in acetonitrile/water solution to determine the influence of terminal carboxylic acid end-group concentration on the hydrolysis kinetics. We found that the reaction was self catalyzed by the increasing carboxylic acid end-groups produced by ester hydrolysis. An equation was derived to describe the autocatalysis process, by differentiating the product concentration, [COOH], from the catalyst concentration, $([COOH] K_a)^{1/2}$. The value of a reaction constant, $k(K_a)^{1/2}$ was measured for our specific experimental conditions. However the kinetic study in acetonitrile could only be evaluated in a narrow range of PLA solution concentrations as the solvent/water ratio was found to influence the dissociation of the carboxylic acid end-groups. Thus, we have demonstrated that experimental conditions governing the dissociation of the acid end-groups can influence the rate of degradation.

The hydrolysis kinetics of PLA differed from those of PCL, where the addition of an external acid was necessary for hydrolysis to take place. Therefore "self-catalysis" was not observed in the hydrolysis of PCL in the acetonitrile/water medium. Pseudo first order kinetics describe the hydrolysis kinetics of PCL.

Appendix A: Derivation of Isothermal Quadratic Autocatalysis

The equation for self catalyzed ester hydrolysis can be expressed as follows:

$$A + B \rightarrow 2B$$

where A is the ester and B is the carboxylic acid in the ester hydrolysis reaction

$$E + H_2O + COOH \Leftrightarrow 2COOH + ROH$$

The forward reaction is that of hydrolysis (rate = $k\,[E][H_2O]$) and the reverse reaction is that of esterification (rate = $k_{est}\,[COOH][ROH]$). Since the product of the ester and water concentrations is much larger than the product of the carboxylic acid and alcohol concentrations, the esterification reaction can be neglected at the early stages of the reaction.

The true catalyst is H_3O^+ since ester hydrolysis of simple aliphatics is known to occur by specific acid catalysis. A dissociation equilibrium exists between the carboxylic acid group and water. The rate equation can be written as

$$\frac{d[COOH]}{dt} = k[COOH][H_2O][E]$$

If x is the carboxylic acid produced at time t, then the equation can be written as

$$\frac{d[x]}{dt} = k_1([E]_0 - x)([COOH]_0 + x) \text{ where } k_1 = k[H_2O]$$

Integrating,

$$\frac{dx}{([E]_0 - x)([COOH]_0 + x)} = \int k_1 dt$$

The denominator on the left hand side of the equation can be rearranged as follows:

$$\frac{1}{([E]_0 - x)([COOH]_0 + x)} = \frac{([E]_0 + [COOH]_0)}{([E]_0 - x)([COOH]_0 + x)([E]_0 + [COOH]_0)}$$

$$= \frac{([E]_0 - x) + ([COOH]_0 + x)}{([E]_0 - x)([COOH]_0 + x)([E]_0 + [COOH]_0)}$$

$$= \frac{1}{([COOH]_0 + x)([E]_0 + [COOH]_0)} + \frac{1}{([E]_0 - x)([E]_0 + [COOH]_0)}$$

$$= \frac{1}{([E]_0 + [COOH]_0)} \bullet \left\{ \frac{1}{([E]_0 - x)} + \frac{1}{([COOH]_0 + x)} \right\}$$

$$\int \frac{dx}{([E]_0 - x)([COOH]_0 + x)} = \int k_1 t$$

$$= \frac{1}{([E]_0 + [COOH]_0)} \cdot \int \left\{ \frac{1}{([E]_0 - x)} + \frac{1}{([COOH]_0 + x)} \right\} dx$$

The limits of $x = 0$ when $t = 0$ and $x = x$ when time $= t$ are applied, so that

$$k_1 t = \frac{1}{([E]_0 + [COOH]_0)} \cdot \left\{ \ln \frac{([E]_0)}{([E]_0 - x)} + \ln \frac{([COOH]_0 + x)}{([COOH]_0)} \right\}$$

$$k_1 t = \frac{1}{([E]_0 + [COOH]_0)} \cdot \ln \left\{ \frac{([COOH]_0 + x)[E]_0}{([E]_0 - x)[COOH]_0} \right\}$$

By Setting $[COOH]_0 + x) = [COOH]$ and $([E]_0 - x) = [E]$, then

$$k_1 t = \frac{1}{([E]_0 + [COOH]_0)} \cdot \ln \left\{ \frac{([COOH][E]_0}{[E][COOH]_0} \right\}$$

But if $k_1 = k[H_2O]$, or $k = k_1 / [H_2O]$, then

$$kt = \frac{1}{[H_2O]([E]_0 + [COOH]_0)} \cdot \ln \left\{ \frac{([COOH][E]_0}{[E][COOH]_0} \right\}$$

$$([E]_0 + [COOH]_0)kt = \frac{1}{[H_2O]} \cdot \ln\left\{\frac{([COOH]}{[E]}\right\} + \frac{1}{[H_2O]} \cdot \ln\left\{\frac{[E]_0}{[COOH]_0}\right\}$$

By setting $a = ([E]_0 + [COOH]_0 k$, and $b = \dfrac{[COOH]_0}{[E]_0}$ the equation converts to

$$\frac{1}{[H_2O]} \ln\left\{\frac{[COOH]}{[E]}\right\} = at + \frac{1}{[H_2O]} \ln b$$

a plot of $\left(\dfrac{1}{[H_2O]}\right) \ln\{[COOH]/[E]\}$ versus time, t, will result in a slope of

$a = k([E]_0 + [COOH]_0)$ and an intercept of $(1/[H_2O]) \ln b$. The concentration of water is assumed to be constant when the water-to-ester ratio is high.

The carboxylic acid dissociation providing the catalyst for the reaction
$$H_2O + COOH \Leftrightarrow H_3O^+ + COO^-,$$
so that the real acid catalyst is $[H_3O^+] = ([COOH]K_a)^{1/2}$
Then by substitution, the equation can be written as

$$\frac{1}{[H_2O]} \ln\left\{\frac{[COOH]^{1/2}}{[E]}\right\} = at + \frac{1}{[H_2O]} \ln b$$

where a $= k\{[E]_0 + (K_a[COOH]_0)^{1/2}\}$ and $b = \left(\{K_a[COOH]_0\}^{1/2} / [E]_0\right)$

The substitution of $[H_3O^+]$ for the acid catalyst is performed after the integration, introducing an error in the computation.

APPENDIX B: Derivation of the isothermal "half-order" autocatalysis equation

The autocatalysis equation is set up to distinguish the product concentration $[COOH]$ from the catalyst concentration as $([COOH]K_a)^{1/2}$ before integration.

$$Rate = k[E]([COOH]K_a)^{1/2}[H_2O]$$

By setting x as the concentration of reacted ester, or

$$x = \left([E]_0 - [E]\right) = \left([COOH] - [COOH]_0\right)$$

then

$$\frac{d[E]}{dt} = k'[E][COOH]K_a)^{1/2}, \quad \text{where } k' = k[H_2O]$$

$$\frac{d[E]}{dt} = k'[E]([COOH]_0 + x)^{1/2} K_a^{1/2}$$

$$= k'[E]([COOH]_0 + [E]_0 - [E])^{1/2} K_a^{1/2}$$

Let $a = [E]_0 + [COOH]_0$

$$\frac{d[E]}{dt} = k'[E]([a\text{-}[E])^{1/2} K_a^{1/2}$$

Integrating with $[E]_0 = [E]$ at time $t = 0$, and $[E] = [E]$ at time $t = t$

$$\int \frac{d[E]}{[E](a - [E])^{1/2}} = \int k' K_a^{1/2} dt = (k' K_a^{1/2})t$$

The general solution, when $a > 0$ is

$$\int \frac{dy}{y(a + by)^{1/2}} = \frac{1}{a^{1/2}} \ln \left\{ \frac{(a + by)^{1/2} - a^{1/2}}{(a + by)^{1/2} + a^{1/2}} \right\}$$

for $b = -1$ and substituting $y = [E]$,

$$\int \frac{d[E]}{[E](a - [E])^{1/2}} = \frac{1}{a^{1/2}} \ln \left\{ \frac{(a - [E])^{1/2} - a^{1/2}}{(a - [E])^{1/2} + a^{1/2}} \right\} - \frac{1}{a^{1/2}} \ln \left\{ \frac{(a - [E]_0)^{1/2} - a^{1/2}}{(a - [E]_0)^{1/2} + a^{1/2}} \right\}$$

Let $c = \dfrac{1}{a^{1/2}} \ln \left\{ \dfrac{(a - [E]_0)^{1/2} - a^{1/2}}{(a - [E]_0)^{1/2} + a^{1/2}} \right\}$

then $\dfrac{1}{a^{1/2}} \ln\left\{ \dfrac{(a-[E])^{1/2} - a^{1/2}}{(a-[E])^{1/2} + a^{1/2}} \right\} + c = (k' K_a^{1/2})$

A plot of $\dfrac{1}{a^{1/2}} \ln\left\{ \dfrac{(a-[E])^{1/2} - a^{1/2}}{(a-[E])^{1/2} + a^{1/2}} \right\}$ versus time yields a slope

of $(k' K_a^{1/2})$ and an intercept $c = \dfrac{1}{a^{1/2}} \ln\left\{ \dfrac{(a-[E]_0)^{1/2} - a^{1/2}}{(a-[E]_0)^{1/2} + a^{1/2}} \right\}$

References

1. E. Lipinsky and R.G. Sinclair (1986) *Chem. Eng. Progress*. Aug 26.
2. A. C. Albertsson and S. Karlsson (1994)*Chem. Techn. Biodeg. Polym.* **23**, 7.
3. K. P. Andriano, T. Pohjonen and P. Tormala (1994) *J.Appl. Biomat.* **5**, 133.
4. R. Bodmeier, K. and H. Chen (1989) *Interntl. J. Pharmaceutics* **51**, 1.
5. P. Cerrai, L. Tricoli, G. D. Guerra, R. Sbrabati Del Guerra, M. G. Cascone and P. Guisti (1994) *J. Mater.Sci: Mater. Med.* **5**, 308.
6. E. Chiellini and R. Solaro (1993) *Chemtech.* July, 29.
7. M. Coffin and G. Mc Ginity (1993) *J. Pharm. Res.* **9** (2), 200.
8. L. Fambri, A. Pegiretti, M. Mazzurana and C. Migliaresi (1994) *J. Mat. Sci: Mat. in Med.* **5**, 679.
9. R. A. Kenley, M. O. Lee, T. R. Mahoney and L. M. Sanders (1987) *Macromolecules* **20**, 2398.
10. S. Li, H. Garreau and M. Vert (1990) *J. of Materials Sc. :Mater. in Med.* **1**, 123.
11. S. Li, H. Garreau and M. Vert (1990) *J. Mater. Sci. Mater. Med.* **1**, 198.
12. G. Rafler and M. Jobmann (1994) *Drugs Made in Germany.* **37** (3), 83.
13. M. Therin, P. Christel, S. Li, H. Garreau and M. Vert (1992) *Biomaterial*13 (9), 594.
14. M. Vert, S. Li and H. Garreau (1994) *J. Biomater. Sci. Polym. Edn.* **6**, 639.
15. S. Patai (1969) *The Chemistry of Carboxylic Acids and Esters.* Part 2. Chap 11. Interscience. N.Y..
16. S. Bezzi, L. Riccoboni and C. Sullam (1936) Dehydration Products of Lactic Acid: Typifying the Transformation of Cyclic Esters into Linear Esters. *Italian Academy of Science.*
17. Gray P. and Scott S. (1990) *Chemical Oscillations and Instabilities: Non-linear Chemical Kinetics.* Clarendon Press. Oxford.
18. P. W. Atkins (1990) *Physical Chemistry.* Fourth Edition. Oxford University Press, N.Y.
19. Y. Doi and K. Fukuda (1994) *Biodegradable Plastics and Polymers. Proceedings of the Third International Scientific Workshop on Biodegradable Plastics and Polymers.* Osaka, Japan.

20. X. Feng, C. Song and W. Chen (1983) *J of Polym Sc: Polym Let. Ed.* **21**, 593.
21. I. Grizzi, S. Garreau and M. Vert (1995) *Biomaterials.* **16** (4), 305.
22. C. Pitt and Z. Gu (1987) *J. of Controlled Release.* **4**, 283.
23. H. Cai, V. Dave, R. A. Gross and S. P. McCarthy (1995) *Polym Prepr.* **36**(1), 422.
24. C. Shih (1995) *J. Control. Release* **34**, 9.
25. X. Zhang, U. Wyss, D. Pichora and M. Goosen (1994). *J. Bioactive and Compatible Polymers* 9: 80.
26. C. H. Holten (1971) *Lactic Acid.* Verlag Chemie.
27. A. Ballara and J. Verdu (1989) *Polym. Degrad. and Stability* **26**, 361
28. E. A. Schmitt, D. R. Flanagan and R. J. Lindhardt (1994) *Macromolecules* **27**, 743.
29. M. Maniar, D. Kalonia and A. Simoneli (1991) *J. Pharm. Sc.* **80**(8), 778.
30. E. Szabo-Rethy and I. Vancso-Szmercsanyi (1972) *Chem Zvesti.* **26**, 390.
31. E. Tommila and M. Murto (1963) *Acta Chem. Scand* **17**(7), 1957.
32. *CRC Handbook of Chemistry and Physics.* 55th Edition. (1974-1975). CRC Press, Cleveland, OH.
33. P. W. Atkins (1990). *Physical Chemistry.* Fourth Edition. Oxford University Press, New York.
34. H. Zimmerman and N. Kim (1980) *Polym. Eng. & Sci.* **20**(10), 680.
35. F. W. Billmeyer (1984) *Textbook of Polymer Science*, Third Edition. Wiley Interscience. New York.

Biodegradability and Recycling

Chapter 17

Broad-Based Screening of Polymer Biodegradability

Thomas M. Scherer[1], Mary M. Rothermich[2], Robin Quinteros[2], Matthew T. Poch[2], Robert W. Lenz[1], and Steve Goodwin[2]

Departments of [1]Polymer Science and Engineering and [2]Microbiology, University of Massachusetts, Amherst, MA 01003

Biodegradable polymers have evolved significantly over the past fifteen years. The earliest products were modifications of olefinic polymers, such as the addition of varying amounts of corn starch to polyethylene. Significant improvements in biodegradability have been made through compounding natural and synthetic polymers in a variety of combinations. Our growing understanding of the biodegradability of naturally-occurring polymers has led to the development of a new generation of synthetic materials whose structures are based on natural synthesis pathways. While these developments are quite promising, they have greatly increased the need for rapid and efficient screening for biodegradability of polymers. It is also important to remember that biodegradability is not a single property, but is dependent on the biological, physical, and chemical conditions prevailing in a given environment. Screening-level testing of biodegradability needs to take all of these factors into account.

The current work presents several approaches to screening-level testing of biodegradability. Examples are drawn from our work with aliphatic polyesters, both the naturally occurring poly(ß-hydroxyalaknoates) and the group of synthetic aliphatic polyesters commercially available under the trade name "Bionolle". The approaches examined include visual observations of zones of clearing in polymer containing overlays, weight loss studies in both laboratory and field conditions, isolation of microorganisms capable of depolymerization and growth on the materials of interest, and enzymatic assays.

Introduction

Biodegradable polymers are still undergoing rapid evolution. First generation products included minor changes to the lowest cost, highest volume olefinic polymers, such as the addition of 5-15% corn starch to polyethylene. These products were claimed to be biodegradable but failed to meet the required specifications. Second generation products are based on more sophisticated compounding of synthetic and natural polymers which appears to offer combinations of processability, economics, performance, and true biodegradability. The next generation materials are the result of new chemistries and scale-up of production which make them economically acceptable.

Polymer biodegradation by microorganisms is mediated by extracellular degradative enzymes that produce water-soluble products from macromolecular substrates. The role of biocatalytic enzyme systems in the degradation of natural polymeric materials, such as cellulose, lignin, amylose, chitin, and polypeptides has been studied in great detail. The environmental degradation of polyesters was recognized to proceed through similar biological processes as early as 1963[1]. The enzymatic degradation of naturally occurring poly-B-hydroxyalkanoates (PHAs) and other aliphatic polyesters has received increasing attention in the last decade [2].

Cell-free bacterial poly-3-hydroxybutyrate (PHB) and poly-3-hydroxybutyrate-co-3-hydroxyvalerate (PHB/V) are readily degraded by microbial populations in marine, activated sewage sludge, soil, and compost ecosystems. Bacteria which have been isolated from these environments and are known to secrete extracellular PHB depolymerases include *Pseudomonas lemoignei, Comamonas testosteroni, Alcaligenes faecalis, Pseudomonas stuzteri, Streptomyces sp.*, and *Pseudomonas picketti* [3-8]. Their enzymes have been purified and characterized. These enzymes catalyze the hydrolysis of short side chain length PHAs (SCL-PHAs), including PHB/V and poly(3-hydroxybutyrate-co-4-hydroxybutyrate) (P(3HB/4HB)) copolymers [9,10]. In addition, a poly(ß-hydroxyoctanoate) (PHO) depolymerase from *Pseudomonas fluorescens* has been studied in detail [11].

Synthetic substrates such as poly-3-propriolactone (PPL) and poly(ethylene adipate) (PEA) are also susceptible to enzymatic hydrolysis by PHB depolymerases, although other synthetic aliphatic polyesters such as poly-4-butyrolactone (P4BL), poly-5-valerolactone (PVL), and poly-6-caprolactone (PCL) are not hydrolyzed. Synthetic polymers consisting of high molecular weight hydrocarbons such as polyethylene have been shown to be unsusceptible to enzymatic attack [12]. A number of synthetic polymers have been designed so as to contain "weak links" which may permit the controlled biodegradation of the polymer [13]. One approach involves the insertion of ester groups into the main chain of the polymer. Examples of synthetic polyesters include polycaprolactone (PCL) and the BIONOLLE products. The biodegradability of PCL has been studied by Oda *et al.* [14], while that of the BIONOLLE products has been studied by Nishioka *et al.* [15] and Scherer *et al.* [16].

The role of fungi during aliphatic polyester biodegradation was recognized early in the study of biodegradable synthetic polymers. The environmental degradation of polyester materials occurs very rapidly in compost [8,17,18], sewage [8], and soil [19-23]. Such environments also comprise the primary means of managed solid waste disposal for biodegradable polymers. The important role of fungal degradative activity in such environments is well documented with respect to other natural polymeric materials (i.e. cellulose, amylose, lignin), and a growing number of studies have recognized the ability of fungi to degrade bacterial PHB and its PHB/V copolymers [22,24]. Matavulj and Molitoris examined the ability of fungi to degrade PHB by systematic affiliation, and found that members of basidiomycetes, deuteromycetes, ascomycetes, zygomycetes, chytridiomycetes and myxomycetes are able to degrade PHB [25]. PHB depolymerase from the fungus *Penicillium funiculosum* [26] has also been shown to play an important role in PHA biodegradation. Although eukaryotic organisms have been isolated as participants in the environmental degradation of synthetic and biosynthetic polyesters [14,15,27,28], investigations of the enzyme systems responsible for fungal polyester degradation have not been as detailed as the investigations of bacterial polyester degradation.

Lipases (triacylglyceride hydrolases), and closely related cutinases, also play a role in polyester biodegradation [29,30]. These esterases hydrolyze unbranched synthetic aliphatic polyesters such as PCL and PEA, but are unable to catalyze PHA degradation [31]. The examination of lipase substrate specificity *in vitro* has demonstrated a broad hydrolytic activity towards aliphatic polyester [32-36].

Fungal PHB depolymerase substrate specificity also was evaluated in the hydrolysis of a series of aliphatic condensation polyesters. Polymers of this type are poised to enter the market as biodegradable packaging and containers. Synthetic aliphatic polyester, based on diacid-diol repeating units that differ fundamentally from the B-hydroxyalkanoate repeating unit, were previously found to be biodegradable in various ecosystems [15]. However, little is known about the enzymatic processes involved in the biodegradation of synthetic aliphatic polyesters, especially which types of hydrolases are active in their environmental degradation.

PHA depolymerases from procaryotic and eucaryotic organisms appear to be biochemically similar, in general. The isolated enzymes have been determined to have catalytic features common to the serine esterase family [3,37]. In addition to the catalytic features of the enzymes responsible for polyester hydrolysis, intrinsic substrate properties of the polyesters determine the rate of biodegradation. The degradation of PHAs has been the most thoroughly investigated in this regard. Substrate features which were found to influence the enzymatic degradation rate of PHB applied to the degradation behavior of other polyesters as well. Copolymer composition[10], stereochemical structures[38-40], sequence distribution[41], surface area [42], and morphology[43] all were found to influence the enzymatic degradation of aliphatic polyesters.

A variety of approaches have been developed for the testing of biodegradability [18,44-50]. The present study evaluates methods that are applicable for the screening of biodegradability in diverse environments.

Materials and Methods

Unless otherwise indicated, all materials were used as obtained from Sigma Biochemical Co. (St. Louis, MO) and Aldrich Chemical Co. (Milwaukee, WI).

Polymers. Bacterial PHB and PHB/V were obtained from Aldrich Chemical Co., Milwaukee, WI. Bacterial PHO was produced from *Pseudomonas oleovorans* as described previously [51,52]. The Bionolle resins are from the family of synthetic aliphatic polyesters being produced by the Showa High Polymer Co., LTD. of Japan. These aliphatic polyesters of dicarboxylic acids and glycols are low molecular weight condensation polyesters which have been further polymerized using isocyanate chain extension reactions. Four grades of Bionolle were tested. Grade 1000 is a poly(butylene succinate), grade 3000 is a poly(butylene succinate-co-adipate), grade 6000 is a poly(ethylene succinate), and grade 7000 is poly(ethylene succinate-co-adipate). The properties of the Bionolle family of polymer has been recently reviewed[53]. Bionolle polymers were obtained as extrusion blown films, which were rinsed with ethanol and dried before use.

The synthetic polyesters used as growth substrates were processed to obtain a high surface area to volume ratio. The Bionolle materials (Showa High Polymer Co. Ltd., Tokyo, Japan) obtained as pellets (7.0 g) were dissolved in 250 ml chloroform at 60 °C and reprecipitated in menthanol in a commercial glass blender (Osterizer) with vigorous mixing. The precipitated polymer particles were retained by filtration and dried under vacuum for 3 days. These processed polymers were sterilized by suspending the particles in 70% ethanol/ water and evaporating the solutions from covered vials *in vacuo*.

PHO Latex . PHO can form a latex in water suspension [54]. One gram of PHO is dissolved in 300 ml acetone. The mixture is poured into a separatory funnel and dripped into 125 ml of cold deionized water, at a rate of four drops a second. The mixture is placed into a rotovap apparatus to remove the acetone.

Preparation of ^{14}C-PHB. The two step PHB production method employed here was adapted from Doi *et al.* [55]. In the first step a culture of *Alcaligenes eutrophus* is grown to high density. *A. eutrophus* cells from a plate were used to inoculate a baffled 500-mL Erlenmeyer flask containing 100 mL of "rich" medium (5 g trypticase peptone, 5 g peptone, 10 g yeast extract, 5 g meat extract, and 5 g $(NH_4)_2SO_4$ per liter, pH to 7.0). The culture was incubated, with shaking, at 30°C overnight. Seven mL of the turbid overnight culture were used to inoculate 700 mL of rich medium in a baffled 2800 mL Fernbach flask. The culture was incubated with shaking at 30°C for 21 hours.

In the second step the cells are grown in a nitrogen-free butyric acid medium in the presence of sodium ^{14}C butyrate. At the end of the 21-hour incubation the cells were harvested by centrifugation at 5000 rpm for 15 min. The supernatant was discarded and each cell pellet was resuspended by vortexing with 5 mL of sterile saline. The re-suspended cells were removed from the centrifuge bottles by pipette and pooled in a sterile mixing cylinder. The total volume of cell suspension was 25 mL. The 25 mL of cell suspension was transferred to 625 mL of nitrogen-free 53 mM butyric acid medium (5 mL n-Butyric acid, 3.8 g Na_2HPO_4, 2.65 g KH_2PO_4, 0.2 g

$MgSO_4$, 1.0 mL micronutrient solution [9.7g $FeCl_3$, 7.8g $CaCl_2$, 0.156g $CuSO_4 \cdot 5H_2O$, 0.2181g $CoCl_2 \cdot 6H_2O$, 0.1344g $CrCl_3 \cdot 6H_2O$, and 0.118g $NiCl_2 \cdot 6H_2O$ per L O.1 N HCl]) in a 2800 mL Fernbach flask.

At the same time as cell transfer, 50μCi (specific activity 20 mCi/mmol) of sodium ^{14}C-butyrate (New England Nuclear, Boston) was added to the butyric-acid culture flask. The label was on the carbonyl carbon-$CH_3(CH_2)_2^{\bullet}COONa$. The Fernbach flask was then tightly closed with a rubber stopper. The rubber stopper was fitted with glass inlet and outlet tubing, and air was pumped into the system with an aquarium air pump. The outlet gasses were bubbled into 100 mL, 1N NaOH traps at 2-3 bubbles /sec. The culture was stirred with a magnetic stirrer and maintained at 30°C in a waterbath.

The extraction of the ^{14}C-PHB was adapted from Williams and Wilkinson [56].

After 10.5 hours, the cells were distributed to three 500-mL bottles and harvested by centrifugation at 5000 rpm for 15 minutes. A subsample of the supernatants was saved for counting and the rest discarded. Each cell pellet was resuspended in 4.5 mL saline; cell suspensions were pooled in a graduated mixing cylinder and saline was added to make a total volume of 36 mL. Twenty-four 25-mL screw-top Corex tubes were prepared with enough 5% sodium hypochlorite solution (Chlorox laundry bleach) in each to balance with all other tubes (about 19-20 mL ea); 1.5 mL of the cell suspension were added to each tube and each was shaken by hand; samples were placed in 37°C room for 1 hour (no shaking). After the digestion period, samples were centrifuged at 7000 rpm for 20 min. A subsample of the bleach supernatant was saved for counting and the rest discarded. The pellets from the hypochlorite digestion were washed twice with water. Each time the suspension was centrifuged at 7500 rpm for 20 minutes, a subsample of the wash water was saved for counting, and the rest discarded. The washed residue (PHB) was air-dried and stored as a powder. PHB films were made by adding chloroform to the powder, heating at 100 °C for 15 minutes, casting the solution into a glass Petri plate, evaporating the chloroform, and finally peeling the film from the surface of the plate.

Microorganisms. A strain of *Aspergillus fumigatus* M2A was isolated from the municipal compost pile in Springfield Massachusetts for its ability to hydrolyze PHB[17]. *Commomonas* strain P37 was isolated from a residential compost pile in Northampton Massachusetts based on its ability to hydrolyze PHO. The marine, aerobic, heterotrophic bacterium TM007 was isolated from Great Sippewisset Marsh, West Falmouth, Massachusetts. This unidentified strain was also isolated based on its ability to hydrolyze PHB.

Growth Media. Enrichment and isolation assays were performed using a mineral salts medium (mineral salts medium A) which was prepared by combining 1.15 g K_2HPO_4, 0.625 g KH_2PO_4, 0.02 g $MgSO_4.7H_2O$, 1.00 g NH_4NO_3, 10 mL trace mineral solution, and 0.19 g yeast extract, with enough distilled water to make 1 L.

The trace mineral solution contained 12.8 g nitriloacetic acid (adding the nitriloaceic acid to 200 mL H_2O and adjusting the pH to 6.5 with KOH), 0.1 g $FeSO_4.7H_2O$, 0.1 g $MnCl_2.4H_2O$, 0.17 g $CoCl_2.6H_2O$, 0.1 g $CaCl_2.2H_2O$, 0.1 g $ZnCl_2$, 0.02 g $CuCl_2.2H_2O$, 0.01 g H_3BO_3, 0.01 g $Na_2MoO_4 \cdot 2 H_2O$, 1.0 g NaCl, 0.017 g Na_2SeO_3, and 0.026 g $NiSO_4.6H_2O$ combined with enough distilled water to make 1 L.

All other assays were performed using the following mineral salts medium (mineral salts medium B). This was prepared by combining 3.57 g Na_2HPO_4, 1.50 g KH_2PO_4, 1.00 g NH_4Cl, 0.20 g $MgSO_4.7H_2O$, 0.50 g $NaHCO_3$, 0.10 mL 0.2% $FeNH_4$-Citrate solution, and 2.00 mL Ho-Le solution, with enough distilled water to make 1 L. One liter of Ho-Le solution contained 2.85 g H_3BO_3, 1.80 g $MnCl_2.4H_2O$, 1.36 g $FeSO_4$, 1.77 g Na-tartrate, 26.9 mg $CuCl_2.2H_2O$, 20.8 mg $ZnCl_2$, 40.4 mg $CoCl_2.6H_2O$, 25.2 mg $Na_2MoO_4.2H_2O$, and enough distilled water to make 1 L.

The PHA-degrading isolate, strain P37C, was enriched from residential compost. The enrichment culture was prepared using PHO as the sole carbon source. Serial dilutions of the final enrichment culture were spread on PHO overlay plates [17]. Colonies, which showed good growth or clearing of the PHO, were selected and isolated. The isolate, strain P37C, was chosen for its ability to form clear zones in PHO and PHB/agar tubes. *Aspergillus fumigatus* and the marine isolate TM007 were isolated in a similar manner except the sole carbon source was PHB.

To measure growth on insoluble substrates, protein analysis was performed using a Bradford protein assay kit purchased from Bio-rad Laboratories Inc. (Hercules, CA) with bovine serum albumin as the standard. Growth on soluble substrates was measured as the increase in optical density at 660 nm.

^{14}C-PHB degradation studies. Mineralization of PHB was monitored by measurement of evolved $^{14}CO_2$. Reactor vessels were tall 300 mL straight-walled, spoutless beakers (figure 1). Four different reaction mixtures were set up as follows: 1) 43 g (wet weight, corresponding to about 30 g dry weight) of microbial mat from the Great Sippewissett Salt Marsh of Cape Cod near Woods Hole, Massachusetts, 65 mL filtered seawater, and 30 mg of shredded [^{14}C]PHB film of 337,000 DPM / mg specific activity (10^7 DPM total). The components were mixed to produce a slurry. 2) 43 g (wet weight) of microbial mat, 65 mL filtered seawater, and 30 mg of powdered [^{14}C]PHB of 337,000 DPM / mg specific activity. 3) 35 g (wet weight, corresponding to 30 g dry weight) of home compost, 65 mL of carbon-free minimal salts medium described above, and 30 mg of shredded [^{14}C]PHB film. Components were mixed to a slurry. 4) 35 g (wet weight) of home compost, 65 mL of carbon-free minimal salts medium, and 30 mg of powdered [^{14}C]PHB. The amount of PHB added to the systems was 10,000 μg per gram of dry microbial mat or compost. The reaction beakers were tightly closed with rubber stoppers. The rubber stoppers were fitted with glass inlet and outlet tubing, and air was pumped into the system with an aquarium air pump. The air was bubbled through water before being passed into the reactor flasks to avoid excessive drying of the reaction materials. The outlet gasses were bubbled into 100 mL, 1N NaOH traps at 2-3 bubbles /sec. Replicate 1 mL samples were removed from the traps at the time intervals and the traps were refilled whenever their volumes had become reduced to 88 mL. The 1.0 mL samples were diluted with 3 mL of dH_2O and combined with 12 mL of Scintiverse II (Fischer Scientific) in glass 20 mL capacity scintillation vials; the mixture was held undisturbed in the dark for 24 hours before counting with a Beckmann Model LS5000TD scintillation counter. The evolution of $^{14}CO_2$ was monitored for 86 days.

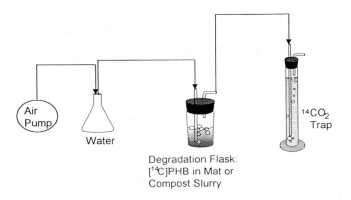

Figure 1. Schematic diagram of the apparatus used to measure ^{14}C-PHB degradation

Field Weight Loss Studies. Two field sites were selected for weight loss studies, The Great Sippewissett Salt Marsh in West Falmouth, Massachusetts and the municipal compost facility in Springfield Massachusetts. For the compost studies, rectangles of each film (approximately 2 cm x 9 cm) were placed on wooden stakes under a nylon mesh. Stakes were driven 50 cm into the compost pile. Each stake contained duplicate films and all stakes were placed within a one m^2 area. During the normal mechanical turning of the compost piles the stakes were removed and replaced. At each sampling period, the stakes were returned to the laboratory for a 24 hour period. The films were removed, carefully cleaned and then weighed. The films were then replaced on the stakes and returned to the compost pile.

Films of Bionolle 6000 and 7000 were placed into the microbial mat in the Great Sippewisset Salt Marsh. The 2 x 2 cm films were placed in 1mm nylon mesh bags that were sewn closed with nylon thread. The bags were placed at the interface between the microbial mat and the underlying sediments. Films were placed in triplicate and polyethylene films were used as negative controls.

Clear zone assays. A molten mixture containing PHA latex or PHA suspension, agar, and minimal growth medium, was prepared so that the concentration of polyester repeating units was approximately 7 mM. A 3-mL volume of this mixture was added to screw cap tubes. This represents 5.3×10^{-4} mmol of repeating unit for every millimeter of tube length. This factor is used in converting the length of clear zones into rates of polymer hydrolysis. The mixture was allowed to gel and the surface of the agar mixture in each tube was inoculated with either 50 µl cell culture or with enzyme preparation. Control tubes were inoculated with 50 µl sterile growth medium. The tubes were observed daily for clear zone formation. All clear zone tubes were prepared in triplicate, and clear zone lengths were measured using a ruler with millimeter gradations. For each set of tubes any differences in clear zone length between the three tubes were so small that they were not detected using this method of measurement. Therefore, figures showing clear zone tube results do not have error bars.

Respirometry. Complete mineralization was measured by the conversion to carbon dioxide. The design of the respirometry system was based on the ASTM standard [57] and on Bartha and Pramer [58]. Air, which had been passed through a CO_2-purging system, was passed through growth flasks containing the PHAs, and CO_2 produced in the flasks as a result of mineralization of the PHAs was collected in a 0.5 M KOH solution. A schematic of the CO_2-purging apparatus, the growth flasks, and the CO_2-collection apparatus is shown in figure 2. The respirometry studies described here were performed with a pure culture of *Commomonas* P37C.

***In vitro* enzymatic degradation of film samples.** Triplicate samples of 2.0 cm^2 surface area (1.0 cm x 1.0 cm dimensions) were placed in 2.0 mL of 50 mM Tricine-NaOH/ 0.01 mM $CaCl_2$, pH 8 buffer with 1.0 mM cycloheximide fungicide. 20 µL (@42 µg/mL protein) of purified *A. fumigatus* PHB depolymerase preparation was added to each sample in buffer to catalyze degradation. Control samples were placed in identical buffer solutions which lacked only the enzyme. All samples were 9.3 L/hr.

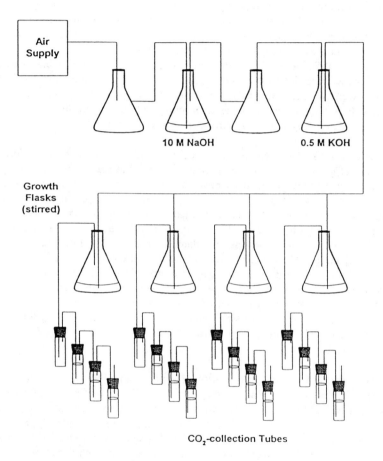

Figure 2. Schematic diagram of the respirometry apparatus

The reservoir was filled with 1.0 L of 30 mmol NaHCO$_3$ to act as a buffer and periodic pH adjustments were made with KOH in order to keep the system neutral. A Masterflex pump was used to circulate the leachate. The reactor was filled with a 30/70 mix of seed to bulk refuse. The seed was obtained from an active landfill at a depth of about 40 feet. The bulk refuse was obtained from a plant in Hartford, Connecticut that processes refuse as a fuel source. Films of PHBV and Bionelle 6000 were placed in a mesh nylon bag and centered in the reactor. Gas samples were taken twice a day and analyzed by gas chromatography as described above.

Results

^{14}C radiolabeling of PHB. The data in Table I show that only a very small fraction of the labeled butyric acid was metabolized all the way to CO$_2$, with only 0.3% of the original radioactivity recovered in the CO$_2$ trap. In this particular procedure, the cells apparently took up only about 59% of the sodium ^{14}C-butyrate added as label, while 41% of the original radioactivity remained in supernatant of the cell harvest. It should be possible to increase the net incorporation of the sodium ^{14}C-butyrate by carefully monitoring the decrease of butyric acid from the stage 2 medium and timing the harvest of the cells to correspond with complete disappearance of the substrate. After the cells were harvested, they were subjected to an oxidizing digestion by a 5% sodium hypochlorite solution. PHB is not degraded by short-term exposure to 5% sodium hypochlorite solution, but most other cellular components are. Therefore, non-PHB cell material will be solubilized and will remain in the "bleach supernatant" after centrifugation of the digest. In this case, 4.3% of the counts were recovered in this supernatant. The resultant pellet of PHB is pure white and somewhat powdery. Care must be taken when removing the supernatant at this point in the procedure, and from the water washes that follow. The PHB pellet is very easily dislodged and disrupted and in some cases, it is necessary to remove the supernatant with a pipette rather than by decanting. The water washes in the labeling procedure discussed here had only residual amounts of radioactivity. Finally, 37% of the 50 μCi label used for this procedure ended up in the PHB product. The ^{14}C-PHB produced in this manner can be used as either a powder, or can be dissolved in hot chloroform and cast as a plastic film. Using this procedure, we have produced ^{14}C-PHB with specific activity as high as 337,400 DPM / mg.

^{14}C-PHB polymer degradation studies. After 7 days 40% of the radioactivity in the ^{14}C-PHB powder, in both the microbial mat and compost, was recovered as ^{14}CO$_2$ (figure 3). By day 16 in the compost and 28 in the microbial mat, 50% of the radioactivity was recovered, thereafter a slow, steady increase continued through the end of the period for which evolution of ^{14}CO$_2$ was monitored. By day 86, the recovery was about 55% in both systems. In contrast, ^{14}C-PHB films degraded more slowly with a distinct difference in rates between microbial mat and compost. Forty percent of the radioactivity in the film was recovered from the microbial mat system after 22 days, but that level of recovery was not reached in the compost until 63 days. By day 43 of the experiment, the ^{14}C-PHB film in microbial mat was degraded to

Table I. Incorporation of ^{14}C into bacterial PHB

Fraction	Total activity of fraction (DPM)	Percentage of original label
sodium ^{14}C-butyrate added at beginning of labeling period	1.0×10^8	
Whole-cell suspension at end of labeling procedure	1.0×10^8	100%
^{14}CO$_2$ trapped during labeling procedure	3.5×10^5	0.3%
Supernatant from whole cell harvest	4.1×10^7	41%
Supernatant from bleach digestion of cell mass	4.3×10^6	4.3%
1st water wash of residue	5.4×10^5	0.5%
2nd water wash of residue	3.3×10^5	0.33%
film cast from dried PHB	3.7×10^7	37.5 %
total all fractions	8.4×10^7	83.9%

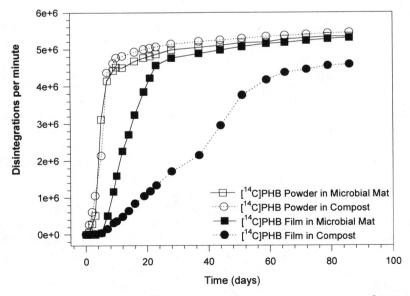

Figure 3. Biodegradation of ^{14}C-PHB in microbial mat and compost (10^7 DPM total)

incubated at 45°C. Sample buffers were exchanged every 2-3 days with the addition of 20 µL new enzyme solution to continue degradation. After exposure to the enzyme solutions, followed by drying under vacuum for 6 hours, the samples were weighed on an analytical balance. The more rapidly degrading films of bacterial PHB/V and P3HB/4HB were weighed after 4, 6, 12, and 24 hours incubation with enzyme or buffer alone. Chemical hydrolysis of polyester samples in abiological controls was substracted from the total hydrolysis rate to determine the rate of enzymatic degradation.

Films (1-2.5 cm^2) were placed in enzyme suspensions and weight loss of the films was monitored over time. The bacterial enzyme suspension was a crude preparation obtained by growing TM007 on PHB and then removing the enzyme containing supernatant. Enzyme activity towards PHB was demonstrated before testing against the other films. The bacterial enzymatic studies were performed at 30°C. A commercial preparation of lipase from the fungus *Rhizopus arrhizus* was obtained from Sigma Biochemical Co. (St. Louis, MO). The lipase film degradation studies were run at 37°C.

Anaerobic degradation. The anaerobic enrichment medium consisted of KH_2PO_4, 0.30 g/L; NH_4Cl, 1.00 g/L; Na_2HPO_4 ($7H_2O$), 2.10 g/L; $MgCl_2$ (6 H_2O), 0.20 g/L; $FeSO_4$ (2.5%), 25.00 µL/L; Yeast Extract, 0.30 g/L; a trace mineral solution, 10.00 mL/L; and a vitamin solution, 10.00 mL/L; resazurin, 1.00 mL/L[59,60]. The medium is heated, gassed with nitrogen to remove oxygen, and autoclaved. After cooling, the vitamin solution is added and the medium is brought to pH 7.

Inocula for enrichments were obtained from three sources: a commercial anaerobic digester in Pittsfield Massachusetts, a manure slurry pit in South Deerfield Massachusetts, and an anaerobic vat maintained in our laboratories at the University of Massachusetts. The latter is a 30 gallon vat which has received anaerobic samples from both contaminated and uncontaminated samples over a period of five years. The vat, maintained at 34°C, is referred to as the universal inoculum. The inocula were prepared within an anaerobic glove bag containing a nitrogen atmosphere by straining through cheesecloth to remove the larger debris. The strained liquid was poured into a 160 mL serum bottle, and then gassed with nitrogen to remove any other gas. One mL of the inoculum was injected into the pressure tubes containing 10 mL of medium.

Gas Analysis. Samples of gas were removed from the headspace of the culture tubes and the concentration of methane was determined. Glass syringes were used with a 22 gauge needle to remove 0.40 mL of gas, and locked with a gas tight minert valve. The sample was injected into a Shimadzu GC-8A TCD Gas Chromatagraph, with a column of Silica Gel 60/80 (Alltech). The carrier gas was helium and the column temperature was 40°C.

Landfill Reactor. A landfill-simulating reactor was developed to examine anaerobic degradation under more realistic conditions. The reactor was attached to a gas collector that measured the volume of gas produced as the displacement of water. The reactor was run at a temperature of 34°C, with the leachate recycling at a rate of about the same extent as the powders. However $^{14}CO_2$ evolution from ^{14}C-PHB film in the compost leveled off after day 70 at about 45% of the initial radioactivity.

Field weight loss studies. The results of the biodegradation tests on the "Bionolle" films in compost are presented in Tables II and III. Bionolle 3000 degraded the fastest with an average weight loss of 41% after 18 days. Bionolle 7000 showed an average weight loss of 22% after 37 days, followed by Bionolle 6000 12% after 37 days and Bionolle 1000 which lost an average of only 6% after 37 days. With the exception of Bionolle 3000, the Bionolle films showed very little visual evidence of disintegration. Disintegration is used here to distinguish the break down of films into small pieces as opposed to actual loss of weight from the film.

On the other hand Bionolle films placed into the microbial mat showed little weight loss. The Bionolle 6000 films were retrieved after 26 days and had only 0.2% weight loss. The bionolle 7000 films were left in place for one year and demonstrated no weight loss.

Clear zone studies. The clear zone tube method was tested using a suspension of PHO with *Commomonas* P37C. The initial rates of clear zone formation were proportional to the inoculum volumes applied to the tubes (figure 4), and inversely proportional to the PHO concentrations in the tubes (figure 5). In tubes with a constant concentration of PHO, initial rates of hydrolysis measured after 24 hours were 2.8 x 10^{-4} mmoles of repeating units per day (mmoles RU/day), 5.5 x 10^{-4} mmoles RU/day, 8.2 x 10^{-4} mmoles RU/day, 11.0 x 10^{-4} mmoles RU/day, and 13.7 x 10^{-4} mmoles RU/day for tubes inoculated with 10, 20, 30, 40, and 50 μL cell-free supernatant respectively.

The fungus *Aspergillus fumigatus* was compared to *Commomonas* strain P37 for the ability to hydrolyze PHB and PHB/V. Two sets of clear zone tubes were prepared and inoculated in duplicate, with either $5x10^6$ fungal spores or with $5x10^6$ bacterial cells. The results indicate that under these conditions, degradation of PHBV is more rapid than that of PHB for each organism (figure 6). Initial rates of hydrolysis of the PHAs were measured after 24 hours. Fungal strain hydrolyzed PHB at a rate of 23.4 x 10^{-4} mmoles RU/day, while strain P37C hydrolyzed PHB at a rate of 12.4 x 10^{-4} mmoles RU/day. The initial rates of hydrolysis of PHB/V were 33.0 x 10^{-4} mmoles RU/day for the fungal isolate, and 22.0 x 10^{-4} mmoles RU/day for the bacterial isolate.

Microbial growth studies. The fungus *Aspergillus fumigatus* was found to grow actively on PHB, poly(caprolactone), Bionolle 6000, and Bionolle 3000. Other aliphatic polyesters such as bacterial PHO, poly(l-lactide), poly(d-lactide), low molecular weight poly(ethylene succinate), and Bionolle 7000 did not support the growth of the fungus. Fungal growth on low molecular weight PES and polylactide which appeared after long periods of incubation (2 weeks) may have been stimulated by the products of partial chemical hydrolysis.

The PHA-degrading isolate, *Comamonas* strain P37C, was isolated for its ability to degrade the PHO. It was subsequently found that strain P37C is capable of growing on a wide variety of PHAs. The PHAs tested for biodegradability by strain

Table II: "Bionolle" Film Compost Degradation (Bionolle 3000)

Polymer		Film Weight (g)			
		Day 0	Day 6	Day 13	Day 18
"Bionolle" 3000 (A)		0.0592	0.0491	0.0449	0.0398
"Bionolle" 3000 (B)		0.0525	0.0382	0.0327	0.0265

Table III: Bionolle Film Compost Degradation (Bionolle 1000, 6000, 7000)

Polymer		Film Weight (g)			
		Day 0	Day 25	Day 32	Day 37
"Bionolle" 1000 (A)		0.0848	0.0714	0.0712	0.0741
"Bionolle" 1000 (B)		0.0822	0.0826	0.0829	0.0826
"Bionolle" 6000 (A)		0.0712	0.0682	0.0681	0.0675
"Bionolle" 6000 (B)		0.0730	0.0634	0.0628	0.0591
"Bionolle" 7000 (A)		0.0845	0.0787	0.0748	0.0724
"Bionolle" 7000 (B)		0.0880	0.0656	0.0647	0.0612

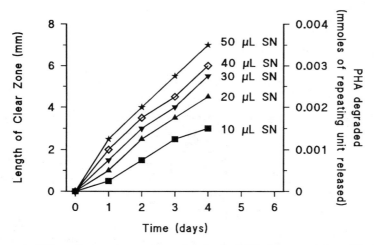

Figure 4. Effect of enzyme amount on clearing in PHO clear-zone tubes. The SN label indicates the volume of enzyme-containing supernatant added.

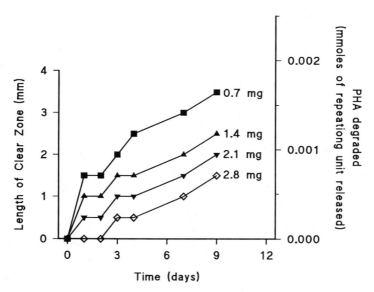

Figure 5. Effect of PHO concentration on clearing in PHO clear-zone tubes. Concentrations are given in mg of PHO per mL.

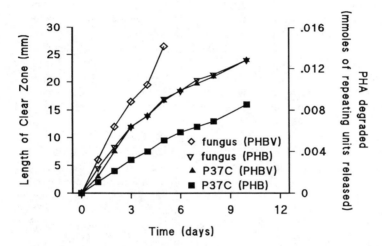

Figure 6. Comparison of the rate of clearing in PHB and PHB/V clear-zone tubes for the fungus *Aspergillus fumigatus* and the bacterium *Commonmonas* P37C.

P37C included PHB, PHBV, and PHO (figure 7). P37C also grows on R-β-hydroxybutyrate, R-hydroxyoctanoate and sodium octanoate (figure 8). Strain P37C does not ferment sugars.

Marine strain TM007, a Gram-negative non-motile rod, was originally isolated from marine microbial mat on poly-β-hydroxybutyrate (PHB) agar overlay plates. It was selected as a PHB-degrader because a clear zone formed around the colony in the opaque PHB agar medium. TM007 grows on PHB (figure 9) and PHB/V, but not on PHO. It also grows on glucose, succinate, and R-β-hydroxybutyrate, the racemic mixture of R - and S-, but not on S-β-hydroxybutyrate. The organism appears to grow more rapidly and to a greater density when cultured on 0.1%PHB as the sole carbon source in comparison to 10mM β-hydroxybutyrate as the sole carbon source (figure 9).

Enzymatic film degradation. The enzymatic degradation of the films using the bacterial depolymerase is presented in Table IV. The greatest rates of biodegradation were shown for Bionolle 6000 and 7000. In contrast to the rapid degradation of Bionolle 3000 in compost, attack by the bacterial depolymerase occurred rather slowly. There was also little degradation of Bionolle 1000. The enzymatic degradation of films using the fungal enzyme is presented in Table V. Bionolle 6000 was degraded most rapidly by the fungal enzyme. Again Bionolle 3000 degraded slowly. The three Bionolle films tested were not susceptible to attack by the commercially prepared lipase from *R. arrhizus* (Table VI).

Respirometry. An additional assay for assessing the biodegradability of PHAs, involves the measurement of carbon dioxide produced as a result of the microbial mineralization of the polyesters. Results from a trial run in which the sole source of carbon was PHBV are shown in figure 10. Here the rate of carbon dioxide production over the initial 19 hours, was 20×10^{-4} g/hour carbon dioxide, and after a period of 90 hours, 56% of the carbon in the PHBV was converted to carbon dioxide. On the other hand , none of the Bionolle films tested showed significant mineralization versus an uninoculated control (figure 10).

Anaerobic biodegradation. Anaerobic enrichments for both the synthetic polyesters (Bionolle 1000, 3000, 6000, 7000) and the bacterial polyesters (PHB/V and PHO) demonstrated methane production (Tables VII and VIII). Because the polymers were the sole source of carbon and energy in these enrichments, they demonstrate the potential for all of these polymers to be degraded anaerobically. The most readily degraded film was PHB/V which produced up to 6.441 μmol CH_4 / 10 mL culture / day. Films of PHB/V and Bionolle 6000 were also placed in the landfill reactor for forty days. Rates of methane production peaked at 2.7 mL CH_4 / g refuse / day and averaged 2.2 mL CH_4 / g refuse / day. This compares favorably with previous landfill reactor studies [61] where the rates of methane production peaked at 2.5 mL CH_4 / g refuse / day. After forty days the PHB/V film had lost 60 % of its weight while the Bionolle 6000 had only lost 4 % of it initial weight.

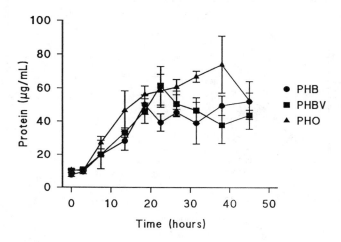

Figure 7. Growth of *Commonmonas* P37C on PHAs

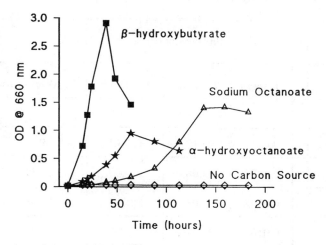

Figure 8. Growth of *Commonmonas* P37C on soluble substrates

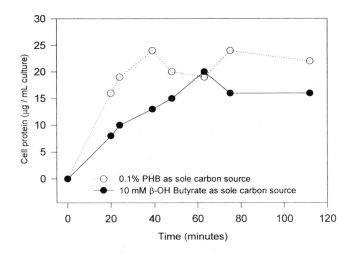

Figure 9. Growth of isolate TM007

Table IV: Enzymatic Degradation with the Bacterial PHB Depolymerase

Polymer	Film Weight (g)			
	0 Hours	90 Hours	185 Hours	% Weight Loss
"Bionolle" 1000	0.0111	0.0107	0.0101	9%
"Bionolle" 3000	0.0089	0.0083	0.0077	13.5%
"Bionolle" 6000	0.0102	0.0082	0.0031	69.6%
"Bionolle" 7000	0.0108	0.0076	0.0052	51.9%

Table V: Enzymatic Degradation with the Fungal PHB Depolymerase

Polymer		Film Weight (g)			
		0 Hours	91 Hours	204 Hours	% Weight Loss
"Bionolle" 1000		0.0117	0.0118	0.0117	0%
"Bionolle" 3000		0.0075	0.0073	0.0068	9.3%
"Bionolle" 6000		0.0061	0.0044	0.0017	72.1%

Table VI: Enzymatic Degradation with the Fungal Lipase

Polymer		Film Weight (g)			
		0 Hours	95 Hours	191 Hours	% Weight Loss
"Bionolle" 1000		0.0079	0.0078	0.0078	0%
"Bionolle" 3000		0.0051	0.0050	0.0050	0%
"Bionolle" 6000		0.0038	0.0037	0.0037	0%

Table VII. Methane Production Rates from Anaerobic Enrichments on Bionolle Films

	7000	6000	3000	1000
	μmol CH$_4$ / 10mL /day			
Anaerobic digestor, Pittsfield MA	2.823	3.056	2.408	2.705
Universal inoculum, Amherst MA	0.363	NT[1]	NT	NT
Anaerobic manure slurry pit, South Deerfield MA	0.020	1.090	0.803	0.844

[1] NT Not tested

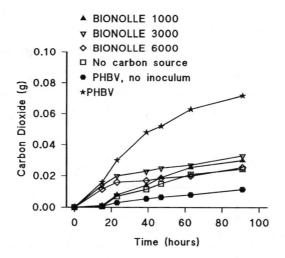

Figure 10.   Comparative respirometry of Bionolle and PHB/V using *Commonmonas* P37C

Table VIII. Methane Production Rates from Anaerobic Enrichments on Poly(β-hydroxyalkanoate) Films

	PHB/V	PHO
	μmol CH$_4$ / 10mL /day	
Anaerobic digestor, Pittsfield MA	6.441	0.963
Universal inoculum, Amherst MA	6.017	3.156
Anaerobic manure slurry pit, South Deerfield MA	5.630	1.343

Table IX. Weight Loss of Aliphatic Polyesters in Anaerobic Landfill Simulator

	PHB/V	Bionelle 6000
	(grams)	
initial weight	0.5367	0.6605
final weight	0.2121	0.6320
weight loss	0.3246	0.0285
% degradation	60	4

Discussion

A broad-based approach to the screening of polymer biodegradability begins with field studies in a range of environments. Field environments would include soils, sediments from marine and freshwater ecosystems, open water systems, compost, anaerobic digester, and activated sewage. It is important to screen potential biodegradable polymers in their target environment under realistic conditions. Polymers that readily degrade in a target environment under optimized conditions may perform poorly under more realistic conditions. It is also important to note that a product's final, degradation environment may not always be the target environment. For instance, some portion of those products targeted for compost degradation may ultimately be degraded in landfills under anaerobic conditions. This emphasizes the utility of screening in a range of environments.

We have drawn examples from two environments, a municipal compost system and a marine microbial mat. The compost system was actively managed; the windrows were monitored for heat production and were turned for aeration as necessary. The microbial mat is a very variable environment with salinity ranging from higher than seawater to nearly fresh. Field screening studies can be based on simple weight and do not require a large number of time points. The goal is to identify those environments that hold the greatest potential for degradation of the polymer and from which microorganisms and enzymes capable of depolymerization can be isolated.

A radiolabelled sample of the polymer can be put to several uses at the screening level. The first use is again to screen environments for biodegradation potential. In most cases this will involve bringing a small portion of the environment back into the laboratory. The advantages are that the tests can be run on very small quantities of polymer and that it is possible to identify intermediates of the degradation process.

Most polymer-degrading bacteria and fungi in pure culture today were initially isolated on clear zone plates. The basic approach is to add a fine suspension of polymer to a thin layer of agar to produce an opaque surface on which putative degrading organism can be spread. Those organisms producing extracellular depolymerases can be identified by the zone of clearing that develops around the colony as the polymer is hydrolyzed. It is very difficult to obtain quantitative data because of the ability of many microorganisms to move across the surface of the plate. This is especially true for fungi. However, clear zone plates are very useful for isolating polymer-degrading organisms that can be used in further testing. All three of the microorganisms discussed in this study were isolated on clear zone plates. One caution is that organisms producing extracellular depolymerases that remain associated with the surface of the cell may not be detected by this method.

Clear zone tubes can be used to show whether or not diffusable enzymes are secreted, as well as the rate at which the PHA is hydrolyzed by these enzymes under a given set of conditions. This assay, when the inoculum consists of bacterial cells or fungal spores, depends upon the rate of growth of the organism and the quantity of enzyme secreted. As the microorganisms grow, they secrete PHA depolymerases which diffuse through the agar, hydrolyzing the PHA. This forms a clear zone which moves down the tube. In order to determine whether the clear zone tube results were related to the applied concentration of enzyme preparation, or to the concentration of PHA in the tube, the assay was evaluated by varying either the concentration of enzyme preparation or the PHA substrate. For these two assays the tubes were inoculated with concentrated cell-free supernatant with PHO depolymerase activity. In this case, a given amount of enzyme is available for PHA hydrolysis. The enzyme diffuses through the agar, degrading the PHA as it goes. Over time the enzymes begin to lose activity, and the hydrolysis process and clear zone development stop when the enzymes cease to function.

When the inoculum consists of whole cells, the microorganisms secrete extracellular PHA-degrading enzymes, the enzymes degrade the PHA, and the hydrolysis products diffuse up to the bacteria. This allows the bacteria to grow and produce more enzyme which diffuses down through the agar to the PHA. The cycle continues until either all the PHA has been hydrolyzed or the bacteria lose their viability. The system in which clear zone tubes are inoculated with whole cells is clearly a much more dynamic and less predictable one. Comparisons between different organisms under these circumstances are therefore purely qualitative. Different organisms may grow and secrete enzymes at completely different rates, and different types of extracellular depolymerases may diffuse through the agar at different rates. This should be kept in mind with assays such as that shown in Figure 6, where PHA clear zone tubes were inoculated with either strain P37C or the fungus, *A. fumigatus*.

The ability of a microorganism to grow on a polymer as sole source of carbon or energy is a clear indication of its ability to degrade that polymer. In addition to organisms that have been isolated from clear zone plates, the films that remain after field weight loss studies (if any) are also a good source of potential degradative organisms. New polymers can fairly rapidly be screened for the ability to support the growth of microorganisms known to degrade related polymers (e.g. the screening of the Bionolle polymers for the ability to support the growth of known PHA degrading bacteria and fungi). Because of the heterogeneous nature of the systems (a solid growth substrate in a liquid culture often with cells attached to the polymer surface) growth can not always be measured by standard microbiological techniques such as increases in optical density. In these cases increases in total protein content can be used as a good measure of microbial growth in pure culture.

Of the four grades of Bionolle, the 3000 degraded most readily in the compost, losing 41% of its weight in 18 days. The order of biodegradability of the Bionolles when tested against a fungal depolymerase was 6000 > 3000 >> 1000. The order of degradation of the Bionolles when tested against a depolymerase from a bacterium was 6000 > 7000 >> 3000 >1000. The degradation rates of these aliphatic polyesters by *A. fumigatus* PHB depolymerase have also been examined in a separate study [16]. The hydrolysis rates for *A. fumigatus* PHB depolymerase with these synthetic polyesters demonstrate a broad enzyme substrate specificity. The rates of polyester degradation were normalized to units of surface area and eliminated contributions by chemical hydrolysis, as observed in control samples. These enzyme catalyzed "rates of degradation" should not be interpreted as a constant value, or as environmental rates of degradation, because the *in vitro* experiments bear little resemblance to the complex conditions of real ecosystems. The rates of *in vitro* enzymatic polyester degradation are indeed relevant only by comparison but do give a sense of the relative susceptibility of particular polymers to enzymes from specific organisms.

A previous study has demonstrated varying levels of Bionolle biodegradability in various environments [15]. The different screening approaches used also provided varying information on the biodegradability of the aliphatic polyesters tested in this study. Weight loss from the Bionolle films in compost field tests differed from the weight loss in films exposed to depolymerase from *A. fumigatus*, in spite of the fact that this organism had been isolated from the same compost. In addition, two of the Bionolle films showed no weight loss in field tests from the marine system but were degraded by the depolymerase from a bacterium isolated from that system. The Bionolle films also were shown to have the potential to be biodegraded anaerobically. However, Bionolle 6000 film showed no biodegradation from an anaerobic landfill reactor that was shown to actively degrade PHB/V. Testing for anaerobic biodegradation should be an important part of screening even polymers which are targeted for aerobic environments such as compost. Not all polymers will end up in their target environments and even environments such as compost can have a tendency to become anaerobic under certain conditions [62]. The latter fact may help to explain why some polymers, which degrade readily under optimized conditions in the laboratory, may not perform as well in full-scale tests. In the final analysis, biodegradability can not be thought of a as single property of a polymer. Biodegradability is dependent on the degradation environment and the physical and chemical conditions that prevail.

A broad-based approach to the screening of polymer biodegradability can incorporate field studies, enrichment and isolation of microorganisms, pure culture growth studies, and even screening with crude supernatants and partially purified enzymes. When undertaken early in the polymer development process both positive and negative results can help to focus further development of the polymer. Keeping the initial screening process as broad as possible can help to avoid unwanted surprises when polymer products face the final test, the marketplace.

Acknowledgments

This study was funded in part by grants from the Massachusetts Department of Environmental Protection, the National Science Foundation (MCB-92022419 to S.G. and R. W. L.) and the Biodegradable Polymer Research Center at the University of Massachusetts, Lowell. Technical contributions by Balint P. DeCheke and Aldric Tourres are also gratefully acknowledged.

References

1)Chowdhury, A. A. *Archives of Mikrobiology* **1963**, *47*, 167.
2)Jendrossek, D. *Polymer Degradation and Stability* **1998**, *59*, 317-325.
3)Schirmer, A.; Matz, C.; Jendrossek, D. *Canadian Journal of Microbiology* **1995**, *41*, 170-179.
4)Mukai, K.; Yamada, K.; Doi, Y. *Polymer Degradation and Stability* **1993**, *41*, 85-91.

5)Mukai, K.; Yamada, K.; Doi, Y. *Polymer Degradation and Stability* **1994**, *43*, 319-327.
6)Yamada, K.; Mukai, K.; Doi, Y. *International Journal of Biological Macromolecules* **1993**, *15*, 215-220.
7)Brandl, H.; Bachofen, R.; Mayer, J.; Wintermantel, E. *Canadian Journal of Microbiology* **1995**, *41*, 143-153.
8)Briese, B. H.; Jendrossek, D.; Schlegel, H. G. *FEMS Microbiology Letters* **1994**, *117*, 107-111.
9)Mukai, K.; Doi, Y.; Sema, Y.; Tomita, K. *Biotechnology Letters* **1993**, *15*, 601-604.
10)Doi, Y.; Kanesawa, Y.; Kunioka, M. *Macromolecules* **1990**, *23*, 26-31.
11)Schirmer, A.; Jendrossek, D.; Schlegel, H. G. *Applied and Environmental Microbiology* **1993**, *59*, 1220-1227.
12)Cook, T. F. *Journal of Polymer Engineering* **1990**, *9*, 171-211.
13)Lenz, R. W. *Biodegradable Polymers*; Lenz, R. W., Ed.; Springer-Verlag: New York, 1993, pp 1-40.
14)Oda, Y.; Asari, H.; Urakami, T.; Tonomura, K. *Journal of Fermentation and Bioengineering* **1995**, *80*, 265-269.
15)Nishioka, M.; Tuzuki, T.; Wanajyo, Y.; Oonami, H.; Horiuchi, T. *Biodegradable Plastics and Polymers* **1994**, 584-590.
16)Scherer, T. M.; Fuller, R. C.; Lenz, R. W.; Goodwin, S. *Polymer Degradation and Stability* **1998**, *(submitted)*.
17)Gilmore, D. F.; Antount, S.; Lenz, R. W.; Goodwin, S.; Austin, R.; Fuller, R. C. *Journal of Industrial Microbiology* **1992**, *10*, 199-206.
18)Pagga, U.; Beimborn, D. B.; Boelens, J.; de Wilde, B. *Chemosphere* **1995**, *31*, 4475-4487.
19)Nishida, H.; Tokiwa, Y. *Kobunshi Ronbunshu* **1993**, *50*, 739-746.
20)Lepidi, A. A.; Nuti, M. P.; de Bertoldi, M. *Proceedings of the workshop on "Plant Microbe Interaction" Ital Soc Soil* **1972**.
21)Kasuya, K.; Doi, Y.; Yao, T. *Polymer Degradation and Stability* **1994**, *45*, 379-386.
22)Mergaert, J.; Webb, A.; Anderson, C.; Wouters, A.; Swings, J. *Applied and Environmental Microbiology* **1993**, *59*, 3233-3238.
23)Lopez-Llorca, L. V.; Valiente, M. F. C. *Micron* **1994**, *25*, 457-463.
24)Mergaert, J.; Anderson, C.; Wouters, A.; Swings, J.; Kersters, K. *Fems Microbiological Reviews* **1992**, *103*, 317-321.
25)Matavulj, M.; Molitoris, H. P. *Fems Microbiological Reviews* **1992**, *103*, 323-331.
26)Brucato, C. L.; Wong, S. S. *Archives Biochemistry Biophysics* **1991**, *290*, 497-502.
27)McLellan, D. W.; Halling, P. J. *FEMS Microbiology Letters* **1988**, *52*, 215-218.
28)Pranamuda, H.; Tokiwa, Y.; Tanaka, H. *Applied and Environmental Microbiology* **1995**, *61*, 1828-1832.
29)Tokiwa, Y.; Suzuki, T. *Agric Biol Chem* **1977**, *41*, 265-274.
30)Murphy, C. A.; Cameron, J. A.; Huang, S. J.; Vinopal, R. T. *Applied and Environmental Microbiology* **1996**, *62*, 456-460.
31)Tokiwa, Y.; Suzuki, T. *Nature (London)* **1977**, *270*, 76-78.
32)Tokiwa, Y.; Ando, T.; Suzuki, T.; Takeda, K. *American Chemical Society* **1990**, *Chapter 12*, 136-148.
33)Walter, T.; Augusta, J.; RJ, M.; Widdecke, H.; Klein, J. *Enzyme and Microbial Technology* **1995**, *17*, 218-224.
34)Mukai, K.; Doi, Y.; Sema, Y.; Tomita, K. *Biotechnology Letters* **1993**, *15*, 601-604.
35)Jaeger, K. E.; Steinb_chel, A.; Jendrossek, D. *Applied and Environmental Microbiology* **1995**, *61*, 3113-3118.
36)Jaeger, K. E.; Ransac, S.; Dijkstra, B. W.; Colson, C.; van Heuvel, M.; Misset, O. *FEMS Microbiological Reviews* **1994**, *15*, 29-63.
37)Jendrossek, D.; Backhaus, M.; Andermann, M. *Canadian Journal of Microbiology* **1995**, *41*, 160-169.
38)Matsumura, S.; Shimura, Y.; Terayama, K.; Kiyohara, T. *Biotechnology Letters* **1994**, *16*, 1205-1210.

280

39)Kemnitzer, J. E.; McCarthy, S. P.; Gross, R. A. *Macromolecules* **1992**, *25*, 5927-5934.

40)Jesudason, J. J.; Marchessault, R. H.; Saito, T. *Journal of Envrironmental Polymer Degradation* **1993**, *1*, 89-98.

41)Hocking, P. J.; Timmins, M. R.; Scherer, T. M.; Fuller, R. C.; Lenz, R. W.; Marchessault, R. H. *JMS Pure Appl Chem* **1995**, *A32*, 889.

42)Timmins, M. R.; Gilmore, D. F.; Fuller, R. C.; Lenz, R. W. *19??* , 119-130.

43)Benedict, C. V.; Cook, W. J.; Jarrett, P.; Cameron, J. A.; Huang, S. J.; Bell, J. P. *Journal of Applied Polymer Science* **1983**, *28*, 327-334.

44)Müller, R. J.; Augusta, J.; Pantke, M. *Material und Organismens* **1992**, *27*, 179-189.

45)Augusta, J.; R.-J, M.; Widdecke, H. *Applied Microbiology Biotechnology* **1993**, *39*, 673-678.

46)Müller, R.-J.; Augusta, J.; Walter, T.; Widdecke, H. *The development and modification of some special test methods and the progress in standardisation of test methods in Germany*; Müller, R.-J.; Augusta, J.; Walter, T.; Widdecke, H., Ed.; Elsevier: New York, 1992, pp 237-249.

47)Urstadt, S.; Augusta, J.; Müller, R.-J.; Deckwer, W. D. *Journal of Environmental Polymer Degradation* **1995**, *3*, 121-131.

48)Sharabi, N.-D.; Bartha, R. *Applied and Environmental Microbiology* **1993**, *59*, 1201-1205.

49)Jendrossek, D.; Schirmer, A.; Schlegel, H. G. *Applied Microbial Biotechnology* **1996**, *46*, 451-463.

50)Pagga, U.; Beimborn, D. B.; Yamamoto, M. *Journal of Environmental Polymer Degradation* **1996**, *4*, 173-178.

51)Gagnon, K. D.; Fuller, R. C.; Lenz, R. W.; Farris, R. J. *Rubber World* **1992**, 32-38.

52)Fritzsche, K.; Lenz, R. W.; Fuller, R. C. *International Journal of Biological Macromolecules* **1990**, *12*, 92-101.

53)Fujimaki, T. *Polymer Degradation and Stability* **1998**, *59*, 209-214.

54)Ramsay, B. A.; Saracovan, I.; Ramsay, J. A.; Marchessault, R. H. *Journal of Environmental Polymer Degradation* **1994**, *2*, 1-7.

55)Doi, Y.; Kawaguchi, Y.; Nakamura, Y.; Kunioka, M. *Applied and Environmental Microbiology* **1989**, *55*, 2932-2938.

56)Williamson, D. H.; Wilkinson, J. F. *Journal of General Microbiology* **1958**, *19*, 198-209.

57)ASTM *D 5338-92* **1992**.

58)Bartha, R.; Pramer, D. *Soil* **1965**, *100*, 68-70.

59)Kenealy, W.; Zeikus, J. G. *Journal of Bacteriology* **1981**, *146*, 133-140.

60)Wolin, E. A.; Wolin, M. J.; Wolin, R. S. *Journal of Biological Chemistry* **1963**, *238*, 2882-2886.

61)Barlaz, M. A.; Schaefer, D. M.; Ham, R. K. *Applied and Environmental Microbiology* **1989**, *55*, 55-65.

62)Atkinson, C. F.; Jones, D. D.; Gauthier, J. J. *Journal of Industrial Microbiology* **1996**, *16J*, 182-188.

Chapter 18

Methacrylic Group Functionalized Poly(lactic acid) Macromonomers from Chemical Recycling of Poly(lactic acid)

Joshua A. Wallach and Samuel J. Huang

Polymers and Biosystems Group, Institute of Materials Science, University of Connecticut, Storrs, CT 06269–3136

A chemical recycling process was used to synthesize methacrylate functionalized poly (lactic acid) (PLA) macromonomers. PLA was equilibrated with caprolactone ethyl methacrylate (CLEMA) and stannous 2-ethyl hexanoate (SnOct) for thirty to sixty minutes depending on the feed ratios. The polymers formed were compared with macromonomers synthesized by the ring opening polymerization of L-lactide with CLEMA and SnOct, which requires at least twenty-four hours. Based on this comparison it was concluded that the recycling process is more effective in producing the methacrylate functionalized macromonomers as the ring opening process.

Introduction

Biodegradable polymers from renewable resources have received interest for many years due to the ever increasing amount of plastic waste produced (1). In spite of this interest, few commercial applications have been presented except in the medical field, preventing these materials from becoming a commodity plastic (2). Poly (lactide) (PLA) is perhaps the most widely used polymer of this type and is now commercially available in non-medical fields.

PLA in general has good mechanical properties comparable to that of polyolefins. It's limitations include brittleness for PLA produced from monomers with high enantiomeric purity. Modification either through reinforcing materials (blends and composites) (3-5) or through chemical modification (co-polymerization) (6-14) is often required to obtain a useful material. Copolymers containing lactide or PLA macromonomers have received the most attention due to the favorable range of properties that can be achieved. Several processes have been used for co-polymerization including; ring opening polymerization of lactide with other cyclic esters (10-13), ring opening polymerization of lactide with dihydroxy terminated oligomers of poly (caprolactone) (14) or poly (ethylene glycol) (6,7), or through polymerization of a methacrylate-terminated PLA macromonomers with other unsaturated monomers (8,9). This final process to produce copolymers is a promising approach to the use of PLA due to the large number of vinyl monomers, which may be used to produce fully or partially biodegradable materials with the PLA.

Synthesis of methacrylate terminated macromonomers of PLA has been investigated by several groups. Huang and Onyari (9) studied the ring opening polymerization of lactide initiated by hydroxyethyl methacrylate (HEMA) and catalyzed by stannous 2-ethyl hexanoate (SnOct). In similar studies Eguiburu et al (8,15) and Burakat et al (16,17) catalyzed the ring opening of lactide with an aluminum alkoxide/HEMA initiator system. End capping a hydroxy terminated PLA with acryloyl chloride (6,7) also produces the acrylate functionalized macromonomer. All of these procedures require a minimum of twenty-four hours, which adds to the already high cost of the polymers.

The current investigation presents a process of chemically recycling PLA to produce a methacrylate terminated PLA chain. This process proceeds through an ester interchange between a high molecular weight PLLA and a hydroxy terminated methacrylate (caprolactone ethyl methacrylate (CLEMA)). Equilibrium can be achieved within one hour as compared to the twenty-four hours required for the ring opening polymerization.

Experimental

Methacrylate terminated PLA

Poly (L-lactide) (PLLA) (Cargill Eco-PLA), CLEMA (Aldrich), and SnOct (Sigma) (2 % w/w PLLA) were added to a Brabender Prep Center set at 180 °C in varying ratios of the of the PLLA to CLEMA to provide a range of molecular weights. The combinations were mixed at 40 rpm for the times specified in Table (I). The reaction products were dissolved in chloroform and precipitated with hexanes for purification.

Table I. Reaction times for the recycling of PLA

Feed Ratio of Lactic Acid Units to CLEMA	Reaction Time
100:1	30 min
75:1	30 min
50:1	1 h
25:1	1 h
20:1	1 h

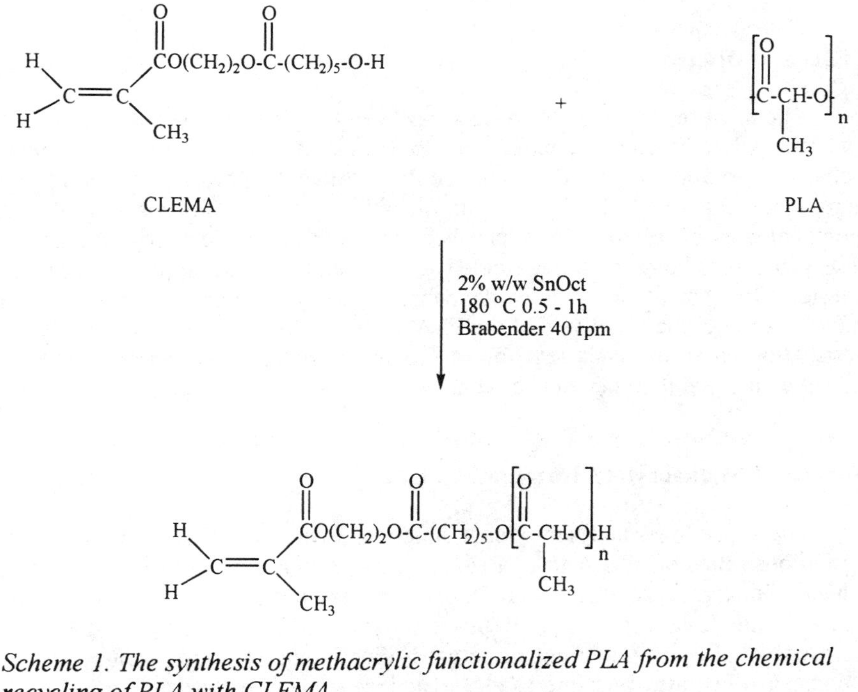

Scheme 1. The synthesis of methacrylic functionalized PLA from the chemical recycling of PLA with CLEMA.

For comparison the equivalent macromonomers were synthesized from the ring opening polymerization of lactide with CLEMA catalyzed by SnOct. The procedure follows that previously determined using HEMA and SnOct (9). The resultant polymer was purified by precipitation from chloroform with hexanes.

Characterization

Molecular weights of the macromonomers were determined using size exclusion chromatography (SEC) (Waters model 150C ALC) in THF using polystyrene standards. Chemical structure was determined with FT-IR (Nicolet 60SX) and NMR (Brucker 500 MHz CDCl$_3$ solvent) analysis. Thermal properties were analyzed by differential scanning calorimetry (Perkin Elmer DSC 7) at a heating rate of 20 °C per minute from -65 to 200 °C. To remove any previous thermal history, an initial scan to 200 °C followed by quenching to -65 °C was performed. Specific optical rotations were measured on a Jasco DIP-1000 digital polarimeter at 25 °C in chloroform.

Results and Discussion

Chemical recycling of polyesters has been extensively investigated with the focus on poly(ethylene terephthalate) (PET) (18,19). The process generally involves a solvolysis of the ester bonds ultimately forming monomers or macromonomers. Hydrolysis, alcolysis, acidolysis, and aminolysis fall into this general category of solvolysis and have been used in the chemical recycling of PET. These processes though are not limited to PET and may be applied toward any polyester. In light of this it is evident that a variety of macromonomers can be produced through the solvolysis of the PLA ester bond. Here we have focused our investigation on an alcolysis reaction to produce a methacrylate terminated PLLA due to the potential that they may pose as described in the Introduction.

Synthesis of Methacrylate Terminated PLLA

The breakdown and subsequent functionalization of PLLA was attempted with various ratios of PLLA to CLEMA and for various lengths of time. Previous methacrylate terminated PLA macromonomers were initiated by HEMA. Though HEMA could have been used in this study, CLEMA was chosen due to its decreased volatility at the temperature used, thus providing more accurate results. Scheme (1) illustrates this interchange reaction. The hydroxy end group of the methacrylate interchanges with the ester linkage of the PLA chain and subsequently produces a methacrylate terminated PLLA. Depending on the weight ratio of PLLA to CLEMA used, various molecular weights can be obtained. Knowledge of the original molecular weight is unnecessary as the final molecular weights will depend on the ratio of CLEMA to lactic acid units rather than the molar ratio of PLLA to CLEMA.

Transesterification is generally a slow process, however with the use of stannous octoate, a common transesterification catalyst used in the ring opening polymerization of lactide, equilibration could be achieved within one hour. An optimum amount of 2 % SnOct by weight of PLLA was determined based on the molecular weights and polydispersities of the resultant macromonomers from trials with 0, 1, and 2% catalyst [Table (II)]. Reaction times of 20, 30, 45, and 60 minutes were investigated. The times specified in the experimental were found to be the optimum for the specific formulation based on the polydispersities of the samples. It was observed that the dispersities of the samples achieved a minimum within these times, indicating equilibrium had been reached. The lowest time required for equilibration was given as the reaction time in the experimental section.

In order to analyze and compare the physical and structural properties of the macromonomers produced from the exchange reaction a series of macromonomers was synthesized from the ring opening polymerization of lactide. This reaction proceeds by an analogous ester exchange reaction between the hydroxy end group of the initiator or the growing polymer chain and the ester linkage of lactide. For this study CLEMA was used to initiate the ring opening in the presence of SnOct to produce a comparable polymer to the one formed through the exchange process. The ratio of SnOct to lactide was equivalent to that used in the exchange reaction (2% by weight) and is the same as that used in the previous studies.

Molecular weight

To determine the efficiency and control of the reactions, SEC was used. The number average molecular weight of the polymers indicates that transesterification is taking place whereas the polydispersities indicate equilibrium within the sample is being reached. Molecular weight results of the samples after purification are given in Table (III). With the exception of the 75:1 sample good control over the molecular weight is observed by varying the ratio of CLEMA to PLA. Comparing these weights to the molecular weights from the standard polymerization [Table (IV)] it appears that the recycling may have a better control over the molecular weight than the ring opening polymerization.

Due to the instability of PLLA at the temperatures and expose time used controls were run in order to determine the degree of unwarranted degradation. PLLA was mixed with SnOct for thirty and sixty minutes without CLEMA. After thirty minutes the PLLA was reduced from 83,000 to 24,000, and after an additional thirty minutes the molecular weight was reduced to 18,000. This reduction may be an effect of backbiting of the polymer a typical PLLA thermal degradation mechanism. It may also be a result of the SnOct participating in the interchange process in the same manner it can initiate lactide polymerization. This mechanism is not fully understood but believed to occur from impurities in the catalyst. In spite of this degradation it will be shown with NMR end group analysis that breakdown

Table II. Molecular weight and PDI dependence of recycled macromonomers on catalyst concentration

SnOct: % by weight of PLA	M_n[*] (Theory ~7200)	PDI[*]
0	62,100	2.1
1	5,600	2.4
2	5,800	2.0

[*]From SEC in THF using a PS calibration

Table III. Molecular weight results of the macromonomers from the recycling process after purification

Lactic Acid : CLEMA	M_n (Theory)	M_n (Exp.)[*]	PDI[*]
100:1	7476	7700	1.7
75:1	5586	8000	1.5
50:1	4062	6700	1.5
25:1	2044	4900	1.9
20:1	1684	3100	1.9

[*]From SEC in THF using a PS calibration

Table IV. Molecular weight results from the ring opening polymerization of lactide with CLEMA

Lactic Acid : CLEMA	M_n (Theory)	M_n (Exp.)[*]	PDI[*]
100:1	4776	13,000	1.8
75:1	5586	11,400	1.8
50:1	4062	10,400	1.9
35:1	2764	4,600	1.7
25:1	2044	4,400	1.4
20:1	1684	4,700	1.9

[*]From SEC in THF using a PS calibration

in the presence of CLEMA is due to an exchange reaction of the PLLA with the CLEMA.

Structural Properties

In order for a chemical recycling process to be useful it must be effective in breaking down the polymer as well as producing a species that can be reused. Though it is apparent that this process is effective in reducing the molecular weight of PLLA, the structure of the polymer and the presence of the functional group must be verified. To verify the presence of the methacrylate terminus and the PLLA composition NMR and IR studies were performed.

NMR spectra [Figure (1)] show the chemical structures from the two processes are nearly identical. The vinylic hydrogens corresponding to the methacrylate end group are observed at 5.6 and 6.1 ppm as singlets, while the methyl group appears at 1.8 ppm as a triplet. The characteristic PLA peaks are observed at 5.2 (CH) and 1.6 ppm (CH_3). IR analysis confirm the results obtained from NMR showing identical spectra for the two processes.

It is known that initiation of lactide in the ring opening synthesis can occur from both the SnOct and the CLEMA. Accordingly in the exchange reaction we have shown that the SnOct may play a similar role in the initiation of degradation. These processes will lead to polymers that do not have a methacrylate terminus. By correlating the functionality of the resultant macromonomers from NMR end-group analysis to their SEC molecular weight the relative efficiency of functionalization can be observed. Figure (2) shows a plot of M_n from SEC versus M_n from NMR for the two reactions. From the corresponding linear fit, the two reactions evidently are equally effective in producing the methacrylate functionalized PLLA rather than unfunctionalized products.

Thermal and Optical Properties

Beyond the specific chemical composition of these polymers thermal analysis can be used to determine and compare their physical characteristics. From DSC analysis the glass, crystallization, and melting transitions of the macromonomers were measured. Plots of these transition temperatures versus inverse molecular weight [Figures (3-5)] show a linear dependence on all three transitions for the polymers from both reaction mechanisms, which is expected for polymeric materials. The T_gs from both mechanisms have overlapping linear fits, while the T_m and T_c linear fits show some differences between the two reactions' products. The T_cs from the exchange reaction are greater than those from the ring opening polymerization for corresponding molecular weights. The T_ms show a different effect. At higher molecular weights the products from the ring opening polymerization have higher transition temperatures than the recycled polymers, while at lower molecular weights they agree better.

These differences in the T_cs and T_ms may be due to a decreased optical purity

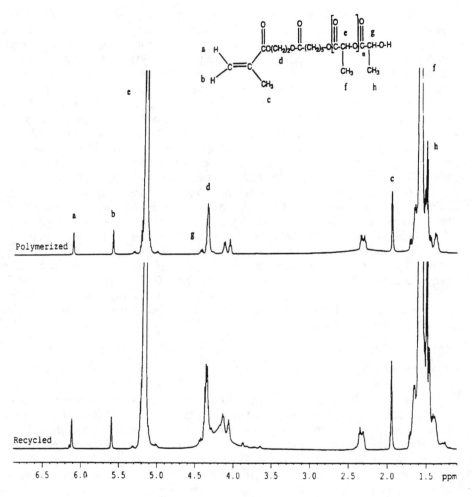

Figure 1. Typical NMR spectra of the macromonomers from both reactions.

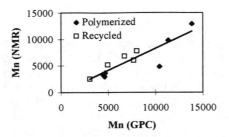

Figure 2. Comparison of M_n from SEC analysis and M_n from NMR end group analysis.

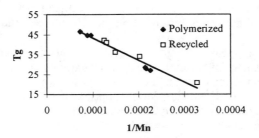

Figure 3. Dependance of T_g on M_n for PLA macromonomers.

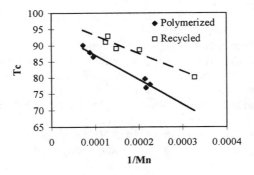

Figure 4. Dependance of T_c on M_n for PLA macromonomers.

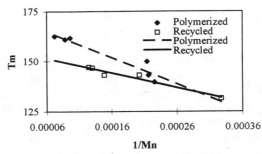

Figure 5. Dependance of T_m on M_n for PLA macromonomers.

of the polymers produced from the exchange reactions. For the ring opening polymerization 95% l-lactide was used, which under these polymerization conditions has been shown to have a small racemization effect (9). On the other hand, the PLLA used for the equilibration had already been subject to racemization in its polymerization. Additionally, racemization will occur under the conditions of the interchange reaction. Analyzing the degree of crystallinity of the samples reflects this theory [Table (V)]. The ring opening polymerization produced percent crystallinities in the range of 48 to 62% while the interchange reaction gave crystallinities between 33 and 52%. An increase in racemization has been shown (9) to lead to a decrease in the degree of crystallinity, a decrease in the melting temperature, and an increase in the crystallization temperature of PLLA samples. Furthermore, with an increase in racemization there would be little effect on T_g, as is observed.

Optical rotation measurements were made to confirm or deny this effect by determining the degree of racemization, which occurs under these reaction conditions. The specific rotations and the optical purity of the macromonomers are presented in Table (VI). From this data it appears that there is little difference between the two methods in the degree of racemization with the majority of the polymers having an optical purity around 75%. These results suggest that the exchange process has an equal or better retention of stereoregularity than the ring opening process. In the ring opening process the majority of the macromonomers lost 20% of their purity while those from the exchange reaction were reduced by only 10%. Though overall favorable these results do not account for the thermal irregularities between the two processes.

The irregularities may be due to the nature of the starting polymer used which leads to the irregularities observed. Another possibility is that there is some unfunctionalized macromonomers in the recycled polymer, which affect the crystallization process. For every initial polymer chain subject to recycling there will be one unfunctionalized chain. For example, if the 85,000 molecular weight chain was broken down into five 17,000 chains one in five (20%) chains would be unfunctionalized. When broken down to 8,500, one in ten (10%) would be unfunctionalized. Thus, at higher molecular weights the effects would be significantly more prominent as is seen in the melting transitions. This is, however, unlikely because from Figure (2) it was concluded that the recycling and ring-opening procedures produce the same amount of functionalized products.

Conclusions

Chemical recycling of PLLA has been shown to be an effective process of producing a methacrylate functionalized macromonomer. Within one hour a high molecular weight PLLA can be effectively broken down into a lower molecular weight reactive species. The macromonomers so produced have similar structural and stereochemical characteristics as those produced from the traditional ring opening process. Thermal properties do show some minor discrepancies between the two processes, however they do not appear to be drawbacks to this process.

Table V. Percent crystallinities of the PLA macromonomers

Sample	% Crystallinity[*]
Polymerized	
100:1	61.8
75:1	59.5
50:1	58.9
35:1	58.7
25:1	55.4
20:1	48.0
Recycled	
100:1	48.6
75:1	51.5
50:1	46.8
25:1	46.2
20:1	33.5

[*]From DSC analysis using 100% crystalline PLLA ΔH_m = 78.6 J/g (20)

Table VI. Specific optical rotation and percent optical retention of the macromonomers

Sample	[α][*]	% Optical Retention[**]
Cargill PLA	-149	86
L-Lactide	-270	95
Polymerized		
100:1	-145	83
75:1	-147	85
50:1	-132	76
35:1	-135	78
25:1	-129	74
20:1	-133	77
Recycled		
100:1	-131	75
75:1	-132	76
50:1	-130	75
25:1	-133	77
20:1	-118	68

[*]Measurements performed at 25 °C in chloroform
[**]100% L-lactide [α] = 285° (21), 100% PLLA [α] = 173° (22)

Acknowledgement

Financial support from the National Corn Growers Association and PLLA samples from Cargill are gratefully acknowledged.

References

1.   Huang, S.J. in *The Encyclopedia of Advanced Materials* Blour, D.; Brook, R.J.; Flemings, M.C.; Mahajam, S., Eds.; Pergamon Press: **1994**; pp 238-244.
2.   Sinclair, R.G. *J. Macromol. Sci. –Pure Appl. Chem.* **1996,** *A33,* 585.
3.   Davies, M.C.; Shakesheff, K.M.; Shard, A.G.; Domb, A.; Roberts, C.J.; Tendler, S.J.B.; Williams, P.M. *Macromolecules* **1996,** *29,* 2205.
4.   Gajria, A.M.; Dave, V.; Gross, R.A.; McCarthy, S.P. *Polymer* **1996,** *37,* 437.
5.   Zhang, L.L.; Xiong, C.D.; Deng, X.M. *Polymer* **1996,** *37,* 235.
6.   Hill-West, J.L.; Chowdhury, S.M.; Slepian, M.J.; Hubbell, J.A. *Proc. Natl. Acad. Sci. USA* **1994,** *91,* 5967.
7.   Sawhney, A.S.; Pathak, C.P.; Hubbell, J.A. *Macromolecules* **1993,** *26,* 581.
8.   Eguiburu, J.L.; Fernandez-Berridi, M.J.; San Roman, J. *Polymer* **1995,** *36,* 173.
9.   Huang, S.J.; Onyari, J.M. *J. Macromol. Sci. –Pure Appl. Chem.* **1996,** *A33,* 571.
10. Bero, M.; Kasperczyk, J.; Adamus, G. *Macromol. Chem. Phys.* **1993,** *194,* 907.
11. Kurcok, P.; Penczek, J.; Franek, J.; Jedlinski, Z. *Macromolecules* **1992,** *25,* 2285.
12. Nakayama, A.; Kawasaki, N.; Arvanitoyannis, I.; Iyoda, J.; Yamamoto N. *Polymer* **1995,** *36,* 1295.
13. Reeve, M.S.; McCarthy, S.P.; Gross, R.A. *Macromolecules* **1993,** *26,* 888.
14. Lostocco, M.R.; Huang, S.J. in *Polymer Modification* Swift, G. Ed.; Plenum Press, New York, **1997,** 45.
15. Eguiburu, J.L.; Fernandez-Berridi, M.J.; San Roman, J. *Polymer* **1996,** *37,* 3615.
16. Burakat, I.; Bubois, Ph.; Jerome, R.; Teyssie, Ph.; Goethals, E. *J. Polym. Sci. Polym. Chem. Ed.* **1994,** *32,* 2099.
17. Burakat, I.; Bubois, Ph.; Grandfils, Ch.; Jerome, R. *J. Polym. Sci. Polym. Chem. Ed.* **1996,** *34,* 497.
18. Paszun, D.; Spychaj, T. *Ind. Eng. Chem. Res.* **1997,** *36,* 1373.
19. Yoon, K.H.; DiBenedetto, A.T.; Huang, S.J. *Polymer* **1997,** *38,* 2281.
20. Iannace, S.; Ambrosio, L.; Huang, S.J.; Nicolais L. *J. Macromol. Sci. –Pure Appl. Chem.* **1995,** *A32,* 881.
21. *Aldrich Catalog Handbook of Fine Chemicals* **1996**
22. Bartus, J.; Wong, D.; Vogl, O. *Polym. Int.* **1994,** *34,* 433.

Biomedical Application of Polyesters

Chapter 19

In Vitro Cellular Adhesion and Proliferation on Novel Bioresorbable Matrices for Use in Bone Regeneration Applications

Cato T. Laurencin[1-4], Archel M. A. Ambrosio[1], Mohamed A. Attawia[1], Frank K. Ko[2,3], and Mark D. Borden[3]

Departments of [1]Chemical Engineering and [2]Materials Engineering, and [3]School of Biomedical Engineering, Science, and Health Systems, Drexel University, Philadelphia, PA 19104
[4]Department of Orthopaedic Surgery, MCP-Hahnemann School of Medicine, Philadelphia, PA 19104

The need for a synthetic alternative to conventional bone grafts stems from donor-site morbidity and limited supply. Using a tissue engineering approach, these replacements can be designed to provide the defect site with a temporary scaffold for bone regeneration while mechanically supporting surrounding tissue. This can be accomplished by fabricating porous matrices from bioresorbable materials. Our laboratory has developed a bioresorbable blend of a polyphosphazene bearing amino acid ester side groups and poly(lactide-co-glycolide), polymers which hydrolyze to metabolically benign products. A study of the pH of the medium in which the blend was degrading, revealed minimal acidity over a period of 6 weeks. In addition, *in vitro* experiments have demonstrated the ability of this material to support the adhesion and proliferation of osteoblast-like cells. Using the copolymer poly(lactide-co-glycolide), we have also developed several methods for fabricating porous matrices with mechanical properties similar to trabecular bone. In this study, matrices formed from the sintered microsphere method were evaluated *in vitro* as a scaffold for osteoblast attachment and proliferation. Scanning electron microscopy studies demonstrated cellular proliferation occurring on microsphere surfaces throughout the entire matrix.

Introduction

In the US alone, over 1 million operations are performed involving bone repair and as the population ages more and more of these procedures will be carried out (1). Traditionally, autografts and allografts have been used to repair or replace most bone defects. Each type of material has its own advantages and disadvantages. Autografts, which are bone grafts often obtained from the patient's iliac crest, do not elicit an adverse immunogenic response upon placement at the graft site (2). However, the removal of the graft from this area causes donor-site morbidity and is limited by inadequate supply. On the other hand, allografts, which are bone grafts usually obtained from cadavers, overcome the problem of supply, but their use could precipitate an unfavorable immunogenic response from the host (2). In addition, there is a significant risk of disease transfer involved with this procedure. Reinforcing prostheses such as metal plates and rods provide good support for load-bearing applications but their long-term viability leaves much to be desired. Since bone is a dynamic tissue that responds to changes in stress, implantation of a prosthesis with significantly higher mechanical properties causes a phenomenon called stress shielding (3). With a metallic implant providing the majority of the support, the surrounding tissue is shielded from normal physiological stresses. This leads to resorption of the bone surrounding the implant and a loss in bone density leaving the implant site prone to fracture. These limitations have led researchers to search for an alternative to conventional reconstructive treatment.

Advances in Biomaterials

Due to the limitations of current orthopaedic materials, an alternative is clearly needed. With advances in polymer chemistry and materials science, a variety of polymeric biomaterials have been developed. Polymers represent the most diverse class of biomaterials. Their physical, chemical and mechanical properties can be tailored to fit an intended application. Bioresorbable polymers, in particular, have generated strong interest in medical implant development primarily because their use obviates the need for a second surgical procedure required to remove an otherwise non-resorbable implanted device.

Thus far, the most commonly studied and widely accepted bioresorbable materials are the polymers of lactic acid and glycolic acid and the copolymer poly(lactide-co-glycolide) PLAGA (4-6). A major reason for this intense focus on these polymers is the fact that they have been approved by the FDA for certain clinical applications. However, these polymers exhibit some undesirable properties for bone regeneration. Both PLA and PGA degrade by bulk hydrolysis of the polymer backbone (7,8). The nature of this type of degradation results in a release of a significant amount of acidic monomers and oligomers at the end of degradation (9,10). This can lead to the formation of a sterile sinus when implanted in bone. This phenomenon has mainly been seen with the more quickly degrading poly(glycolic acid) than with the more slowly degrading poly(lactic acid), PLA. This has prompted many researchers and manufacturers to concentrate on the use of PLA and PLAGA containing high loading of the lactide monomer. However, these materials degrade slowly, over periods of 3 to 4 years. In addition, these bioresorbable polyesters lose their mechanical strength at rates much higher than loss of physical size (*11,12*). This early onset of mechanical strength deterioration is another major drawback that limits these materials' use in orthopaedic applications.

In an attempt to avoid these shortcomings, researchers have developed a number of new bioresorbable polymers. These include polyanhydrides (*13*), polyphosphazenes (*14,15*), polycarbonates (*16*) and poly(orthoesters) (*17*). Our laboratory has been involved in the development of new classes of polymers -- polyphosphazenes and poly(anhydride-co-imides) -- as well as blends and composites as part of our materials synthesis program (*14,15,18-20*). Ultimately, our goal is to develop a tissue engineered matrix for the repair and/or replacement of damaged or diseased bone.

Tissue Engineering of Bone

In recent years, tissue engineering has emerged as a potentially effective approach to the repair and/or replacement of damaged or diseased tissue (*1*). Tissue engineering has been defined by Laurencin as the application of biological, chemical, and engineering principles toward the repair, restoration, or regeneration of living tissues using biomaterials, cells, and factors alone or in combination. This ideology can be applied towards the development of synthetic alternatives to current bone grafts. In order to function as a synthetic alternative to bone, tissue engineered matrices must possess certain characteristics. The objective is to design a matrix that provides the newly regenerating tissue with a temporary site for cell attachment and proliferation. In essence, the matrix serves as a scaffold for tissue regeneration. Several researchers have shown that a 3-dimensional porous scaffolds can be created using bioresorbable copolymers such as PLAGA (*21-23*).

One of the main requirements for such a matrix is bioresorbability. The interest in bioresorbable materials stems from the fact that they do not elicit a permanent foreign body reaction since they are eventually resorbed and replaced by natural tissue. In addition to bioresorbability, porosity also plays an important role in the function of the implant. Porous networks extending through an implant increases the surface area allowing a greater number of osteoblasts to attach to the matrix. When using osteoconductive polymers such as PLAGA, the pore system provides a favorable surface for cell adhesion and growth. This enhances the regenerative properties of the implant by allowing the growth of bone tissue directly on the interior surface of the matrix.

The combination of bioresorbability and porosity are the basis for tissue engineered scaffolds. In an optimum situation, osteoblasts will permeate the implant and begin to deposit a mineralized layer of bone as they proliferate over the surface of the pore network. Since osteoblast mineralization is dynamic and changes with the onset of anatomical stress, degradation of the matrix and loss of mechanical strength will result in gradual loading of the newly formed bone. This will further stimulate additional mineralization through the pore system. The addition of bone mineral in the network will reinforce the matrix allowing it to fully withstand the anatomical stresses found at the implantation site. Upon complete resorption of the matrix, the regenerated bone will have completely replaced the initial structure.

Current Research

Our research in the area of matrix development follows a two-stage strategy. The first stage is the design and development of polymers, blends and composites. The second stage involves the design and development of processes to create 3-dimensional porous structures.

In the area of materials design and development, we have created a blend of PLAGA and a new bioresorbable polymer. As mentioned earlier, one of the main concerns about the use of bioresorbable polyesters is the release of a high concentration of acid during their degradation which can lead to tissue necrosis and other undesirable side effects. This acidity can be minimized by combining PLAGA with a polymer that produces basic degradation products. Following this approach, we have designed a polymer whose degradation products include ammonia and phosphate. The polymer is an amino acid-containing polyphosphazene, PPHOS-EG50 (18). Polyphosphazenes belong to a relatively new class of high molecular weight polymers with a backbone of alternating phosphorus and nitrogen atoms with two organic side groups on each phosphorus atom. In the polymer PPHOS-EG50, the ethyl glycinato side group imparts hydrolytic instability while the p-methylphenoxy side group imparts hydrophobicity. The loading ratio of these two side groups determines the rate of degradation of the polymer. Using a unicortical bone defect model in rabbits, we have demonstrated that this polymer is biocompatible (24). Our goal was to combine this polymer with PLAGA to produce a blend with minimal acidity, if any.

In the area of matrix processing, several researchers have used PLAGA and its homopolymers PLA and PGA to fabricate matrices for use in bone replacement applications (21-23). To date, the major drawback of many of these structures has been the relatively low mechanical properties of the matrices which may lead to problems with implant failure and stress overloading. In our lab, we have developed several methods for the fabrication of high modulus, porous structures (19, Laurencin et al., Cells and Materials, in press). With the goal of creating a porous matrix for use as a trabecular bone replacement, our lab has used the co-polymer poly(lactide-co-glycolide) [PLAGA] and synthetic hydroxyapatite [HA] in the fabrication of novel matrices. Previous work has introduced pores into a 50:50 PLAGA matrix using a novel salt leaching/microsphere technique (19). The basis of this method was the creation of an interconnected porous network by the random packing of polymer microspheres. The porous matrix was composed of PLAGA microspheres with particulate NaCl and HA. The particulate NaCl was used to widen the channels between the polymer microspheres and hydroxyapatite (HA) was used to provide added support to the matrix and to allow for osteointegration. This resulted in a structure with a compressive modulus in the upper range of trabecular bone (1.5 GPa) and ranges in pore diameters from 18-20 μm and 110-150 μm for micropores and macropores, respectively. Additional *in vitro* studies showed that osteoblasts attached and proliferated throughout the three-dimensional (3D), porous network while maintaining their characteristic phenotype (*25,26*).

The results of these studies have shown that the PLAGA/HA composite system and microsphere structure fulfills the osteoconductive and osteointegrative requirements of a bone graft substitute. All components of the matrix were biocompatible, and the presence of osteointegrative HA particles allows for direct bonding to bone. The only limitation of this method is that during degradation *in vitro*, the mechanical strength decreased to the lower limits of trabecular bone (*19*). During *in vivo* implantation this might cause the mechanical failure of the implant or stress overloading of the newly regenerated osteoblasts. In order to function as a synthetic bone replacement, matrices should be biomechanically similar to the type of bone they are replacing and have an internal pore structure to allow for bone ingrowth. Keaveny and Hayes (*27*) found compressive moduli in the range of 0.01 to 2.0 GPa for trabecular bone, and 14 to 18 GPa for cortical bone. A pivotal study by Hulbert *et al.* (*28*) showed that optimal pore size for bone ingrowth was in the range of 100-250 μm. By creating a matrix with these properties, bone ingrowth will occur in the pore system without the consequence of stress overloading of the implant and the cells.

In order to increase control of matrix structure, a solvent evaporation technique was used to fabricate the microspheres. This technique has been previously used to create microspheres for drug delivery applications (*29-31*). Incorporation of this technique into matrix processing has led to the development of a novel method for producing a 3-dimensional porous matrix, the sintered microsphere method (Laurencin *et al., Cells and Materials*, in press). In this method, PLAGA microspheres were placed in a mold and heated past the glass transition temperature of the copolymer (T_g=57.4°C). Once past this thermal transition, adjacent microspheres were thermally fused forming a 3-D structure. The random packing of the microspheres resulted in the formation of a highly interconnected pore system. Structural control of the matrix was obtained by using microspheres of a specific size range which allowed for specific control over internal porosity.

The object of our most current work was the *in vitro* evaluation of cellular attachment and proliferation on the surface of 1) a polyphosphazene/PLAGA blend, and 2) through a novel sintered PLAGA microsphere matrix.

Materials and Methods

Polymer Synthesis and Blend Preparation.

Poly[(50% ethyl glycinato)(50% p-methylphenoxy)phosphazene] (PPHOS-EG50) was synthesized according to a procedure reported previously (*14*). The polymer 50:50 poly(lactide-co-glycolide) (PLAGA) was obtained from American Cyanamid. A 50:50 blend of PPHOS-EG50 and PLAGA was prepared as described before (*18*). Circular matrices 14mm in diameter were cut out from the prepared matrix using a cork borer. The matrices were then lyophilized prior to use. PLAGA and PPHOS-EG50 circular matrices were also fabricated for comparison.

pH Study

The matrices were placed in scintillation vials containing 10 mL of distilled water that had been adjusted to pH 7.4. As a control, vials containing water only were prepared. The vials were placed in a water bath/shaker at 37°C and at various time points the media was collected. The pH of the collected media was measured using a pH meter.

Cell Culture Study

In order to assess the biocompatibility of the PPHOS-EG50/PLAGA blend, a preliminary *in vitro* study was performed. Osteoblast-like MC3T3-E1 cells were grown on the blends and cell adhesion and proliferation were examined. Prior to the study, the 14mm matrices were exposed to UV light to minimize bacterial contamination. MC3T3-E1 cells were plated onto the discs in 24-well plates at a seeding density of 1.0×10^5 cells/well. The cells were grown following standard procedures and conditions. For the adhesion study, the cells were examined over a period of 24 hours. Cell adhesion and spread were evaluated using a scanning electron microscope. As a control, MC3T3-E1 cells were also grown on tissue culture polystyrene (TCPS).

Thymidine Uptake

Cell proliferation over a period of 7 days was determined based on the uptake of ^{3}H-thymidine using standard protocol. Briefly, 6 hours before the designated time point, the media was replaced with new media containing ^{3}H-thymidine. After incubation in the thymidine-containing media, the cells were washed with phosphate buffered saline (PBS) followed by 10% trichloroacetic acid. The matrices with the cells were then incubated in 2N perchloric acid for 24 hours at 70°C to extract thymidine. The radioactivity of the extracts was then measured using a liquid scintillation counter.

In Vitro Microsphere Study

Microspheres prepared using a solvent evaporation technique were sintered to form 3-dimensional porous matrices as described previously (*31*, Laurencin *et al.*, *Cells and Materials*, in press). Prior to the tissue culture experiments, matrices (13 mm diameter by 2 mm thickness) were sterilized by UV radiation for 24 hours. Neonatal rat calvaria were isolated and grown to confluence in Ham's F-12 medium (GIBCO), supplemented with 12% fetal bovine serum (Sigma) (*32*). Cells were seeded onto UV sterilized microsphere matrices at a density of 1×10^5 cells/cm^2. Osteoblasts were cultured on the microsphere matrices for 7 and 14 days, fixed in glutaraldehyde, and dehydrated through a series of ethanol dilutions. Cells were prepared for scanning electron microscopy [SEM] by placing the samples in a series of Freon 113 dilutions (25%, 50%, 75% and 100%). Samples were sputter coated with gold (Denton Desk-1 Sputter Coater) and visualized by scanning electron microscopy [SEM] (Amray 3000) at an accelerating voltage of 20 kV.

Results and Discussion

The use of bioresorbable materials for the tissue engineered repair or replacement of bone provides a new and exciting alternative to current procedures. While current methods such as the use of autografts, allografts or metal alloys suffer from a number of drawbacks, the tissue engineering approach may become the most successful method because it utilizes principles from various disciplines such as engineering, biology and chemistry. In this approach, bioresorbable materials will serve as temporary scaffolds on which new cells will grow as the biomaterial degrades. Ultimately, the scaffold will be completely replaced by new tissue.

Bioresorbable Blend

PLAGA is a relatively hydrophilic polymer with hydrolytically unstable ester linkages. At the end of the degradation process, large amounts of acid are produced as monomers and oligomers dissolve. This has raised some concerns regarding its use in bone repair or replacement and in other applications in which a high localized concentration of acid is undesirable. By blending PLAGA with PPHOS-EG50, we have combined the desirable properties of each polymer while creating a material that does not release a significant amount of acid during degradation. Figure 1 shows the change in pH over a 6-week period. As shown in the graph, the media in which

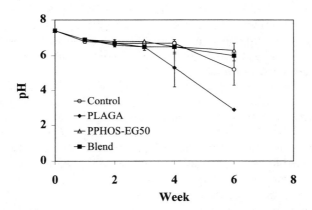

Figure 1. pH of media in which the various matrices were placed for a period of 6 weeks.

PLAGA was placed showed a significantly lower pH (2.9) after 6 weeks than those in which PPHOS-EG50 (pH 6.3) or the blend (pH 6.0) was immersed ($p < 0.05$). The control media, which had no polymer, had a pH of 5.2 which was not significantly different from those with the blend or PPHOS-EG50. This, however, was significantly higher than the media containing PLAGA indicating that the acidity of the PLAGA media was mainly due to the degradation products of the polymer. This validates the concerns raised about the use of medical devices made from PLAGA. Blending this polymer with a bioresorbable polyphosphazene has resulted in a material that does not generate a strongly acidic environment. While this experiment provides a semi-quantitative evaluation of the acidity of these biomaterials, a longer term study that more accurately measures the amount of acid that is produced during degradation is needed and is currently underway.

For a biomaterial to be used as an implant, it has to be biocompatible. Examining the adhesion and proliferation of cells on the surface of biomaterials is a facile and rapid *in vitro* method of initially screening materials for biocompatibility. Figure 2 shows low and high magnification SEM micrographs of MC3T3-E1 cells on the surface of the blend 24 hours after plating the cells. The cells adhered well to the matrix and are well spread. At higher magnification, nodules and a polygonal cell morphology characteristic of osteoblast cells are evident.

Cell proliferation can be quantified using different methods, the most common of which is cell counting. In this method the cells are lifted from the polymeric matrix using trypsin and then counted using a hemacytometer. To insure that all the cells have been lifted, the matrix is examined under the microscope. Clearly, this could only work if the matrix is transparent. Because our matrices were opaque, we decided to use ^{3}H-thymidine uptake as an indication of cell proliferation. Figure 3 shows a graph of the thymidine uptake by osteoblast-like cells over a period of 7 days. The results indicate that the proliferation of MC3T3-E1 cells on the surface of the blend is comparable to that on tissue culture polystyrene (TCPS). This demonstrates that the PPHOS-EG50/PLAGA blend is an excellent substrate for osteoblast growth and may be a good candidate as a scaffold for skeletal tissue regeneration. The next step would be to further characterize the material's *in vitro* properties and then perform an animal study to ascertain its biocompatibility with bone and surrounding tissue.

Sintered Microsphere Matrix.

The goal of creating a 3-dimensional porous matrix was accomplished using the sintered microsphere method. The random packing of the polymer microspheres resulted in a completely interconnected pore structure. By using microspheres of a specific diameter (178-297 μm), the size and structure of the pore network was able to be controlled. The electron micrograph of matrix indicated a structure consisting of a network of microspheres thermally fused together [Figure 4]. The *in vitro* osteoblast study showed promising attachment and proliferation throughout the entire microsphere matrix. At the 7 day time point, SEM images were taken of the top

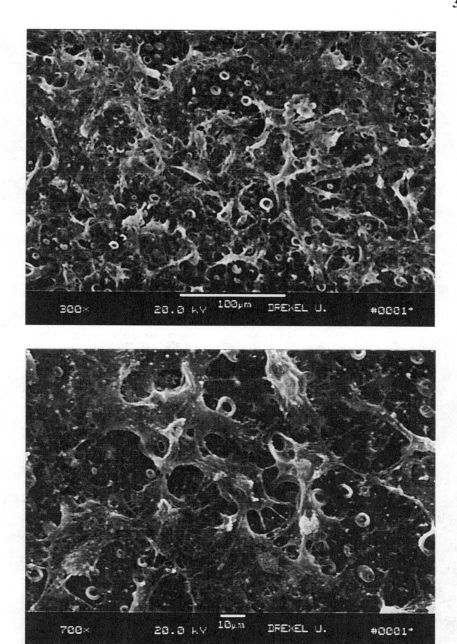

Figure 2. Electron micrographs of MC3T3-E1 cells on the surface of PPHOS-EG50/PLAGA blend at 24 hours, a) 300x magnification, and b) 700x magnification.

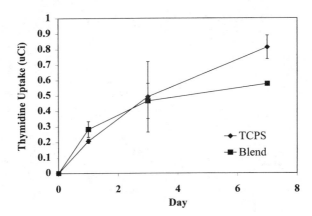

Figure 3. ^{3}H-Thymidine uptake of MC3T3-E1 cells cultured on the PPHOS-EG50/PLAGA blend surface and on TCPS.

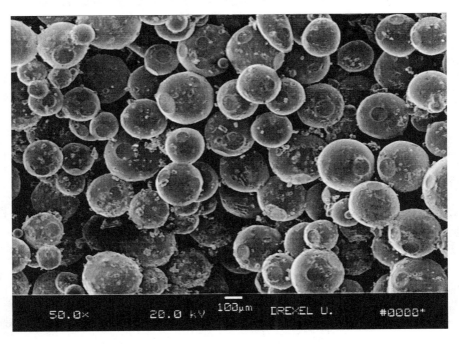

Figure 4. Electron micrograph of the sintered microsphere matrix.

(seeded surface), bottom and cross sections of the matrix. Figure 5 shows the top surface of the matrix at 7-days post seeding. This image shows significant osteoblast proliferation over the entire matrix surface. Cells have proliferated over the surface of individual microspheres as well the connections to adjacent spheres. Microsphere structure was clearly evident with cellular nodules localized at the apex of the spheres.

A close-up of the microsphere surface is shown in Figure 6. This image shows four PLAGA microspheres arranged circumferentially around a pore. Cellular nodules are seen on the surface of the two microspheres at the top of the image. Interestingly, this high magnification image reveals that the osteoblasts have uncharacteristically organized into concentric rings around the circumference of the pore. Typical morphology of osteoblasts proliferating 2-D PLAGA surfaces has shown no such organization (*33*). The consequence of such organization may be advantageous to eventual remodeling of the implant site. The natural structure of trabecular bone consists of mineralized sheets of collagen organized in concentric rings about a central haversian canal. By promoting cells to arrange in concentric layers through the pore system of the matrix, the regenerating tissue may organize in a manner similar to trabecular bone.

The SEM of the bottom of the matrix (Figure 7) indicates that cells have migrated through the thickness of the implant (approx. 2.5 mm). Individual cells are apparent on the surface of the microspheres surrounding a large cellular nodule. This image also shows the early stages of cellular proliferation between the adjacent microspheres. Cytoplasmic extensions are seen linking the cells on adjacent microspheres. The cross sectional image in Figure 8 also shows evidence of cellular migration through the implant (cross section was achieved by fracturing the matrix in half). By day 7 of cell culture, the proliferation front has progresses approx. 1 mm through the thickness of the matrix. Cells have migrated and proliferated along the pore network of the matrix. Cellular proliferation through the pore system is indicated by the spaces left from microspheres contained in the opposing fractured half. These spaces show that cell grew around the surface of the microsphere through the pore network. By 21 days of cell culture, osteoblasts have formed a cellular layer over the matrix closing of most of the pores. Figure 9 shows the surface of the matrix with very few pores remaining. In comparison to Figure 5 (7 days culture), microsphere structure is obscured by the layer of proliferating osteoblasts. The cells appear to cover the entire surface with penetration into the pores. The SEM image of the 21-day cross-section (not shown) was similar to Figure 8. There was only minor advancement of the proliferation front through the thickness of the matrix.

In summary, scanning electron microscopy has shown cellular proliferation on both the blend and sintered microsphere matrix. Cellular attachment on the blend surface was extensive. In addition, the cells retained a polygonal morphology characteristic of osteoblast cells indicating osteoconductivity of the blend material. The 3-dimensional surface of the sintered matrix provided an excellent medium for osteoblast attachment and proliferation. Based on the structure of the matrix, cells organized around pores in concentric rings. This initial study provided insight into the functional capability of the sintered matrix as a tissue engineered cellular scaffold.

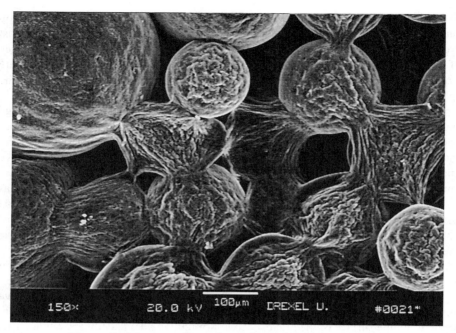

Figure 5. Matrix surface after 7 days of osteoblast culture.

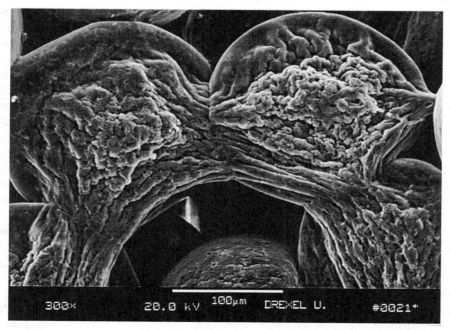

Figure 6. High magnification image of osteoblast organization around circumference of pore (7 days cell culture).

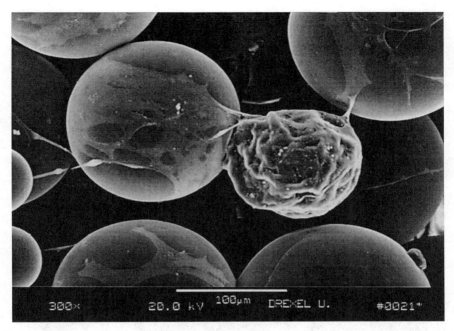

Figure 7. Electron micrograph of the bottom of matrix indicating cellular migration through the structure (7 days cell culture).

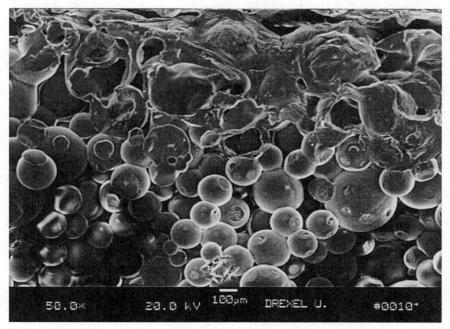

Figure 8. Fracture surface of matrix cross section showing proliferation of cells through the pore network (7 days cell culture)

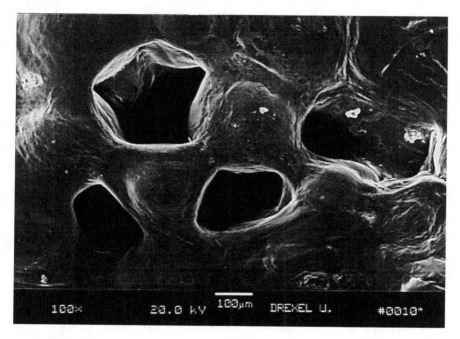

Figure 9. Matrix surface at 21 days of osteoblast culture.

Conclusion

As the population ages, there will be an increasing demand for bone grafts and orthopaedic devices for the treatment of injured or defective bone. With recent advances in biomaterials design and synthesis and the advent of tissue engineering, the use of autografts, allografts and possibly even metal implants will be lessened. We have demonstrated that by selecting the appropriate polymeric biomaterial and structural design, constructs supporting cellular growth and differentiation can be created. We foresee in the near future the use of tissue engineered constructs as a method of choice for the repair and/or replacement of damaged or diseased bone.

Literature Cited

1. Langer, R.; Vacanti, J. P. *Science* **1993,** *260*, 920.
2. Gadzag, A. R.; Lane J. M.; Glaser, D.; Forster, R. A . *J. Amer. Acad. Orthop. Surg.* **1995** *3*, 1.
3. Park, J. B. *Biomaterials Science and Engineering.* Plenum Press: New York, NY, 1984, pp. 357-378.
4. Kulkarni R. K., Pani K.C., Neuman C., Leonard F. *Arch Surg* **1966** 93, 839.
5. Miller, R.A., Brady, J.M., and Cutright, D.E. *J. Biomed. Mat. Res.* **1977,** 11, 711.
6. Reed
7. Chu, C. C. *J. Biomed. Mat. Res.* **1981,** *15*, 7895
8. Chu, C. C. *Polym. Sci. Technol.* **1980,** *14*, 279.
9. Hollinger, J. O.; Battistone, G. C. *Clin. Orthop.* **1986,** *207*, 290.
10. Bostman, O.; Hirvensalo, E.; Makinen, J.; Rokkanen, P. *J. Bone Joint Surg.* **1990,** *72B*, 592.
11. Pitt, C. G.; Gratzl, M. M.; Kimmel, G. L.; Surles, J.; Schindler, A. *Biomaterials* **1981,** *2*, 215.
12. Chu, C. C. *J. Appl. Polym. Sci.* **1981,** *26*, 1727.
13. Langer, R. *Annals Biomed. Engin.* **1995,** *23*, 101.
14. Laurencin, C. T.; Norman, M. E.; Elgendy, H. M.; El-Amin, S. F.; Allcock, H. R.; Pucher, S. R.; Ambrosio, A. A. *J. Biomed. Mater. Res.* **1993,** *27*, 963.
15. Laurencin, C.T.; El-Amin, S. F.; Ibim, S. E.; Willoughby, D. A.; Attawia, M.; Allcock, H. R.; Ambrosio, A. A.; *J. Biomed. Mater. Res.* **1996,** *30*, 133.
16. Ertel, S. I.; Kohn, J. *J. Biomed. Mater. Res.* **1994,** *28*, 919.
17. Heller, J.; Sparer, R. V.; Zentner, G. M. In *Biodegradable Polymers as Drug Delivery Systems*; Chasin, M.; Langer, R., Ed.; Marcel Dekker: New York, 1990, 121-162.
18. Ibim, S. E. M.; Ambrosio, A. M. A.; Kwon, M. S.; El-Amin, S. F.; Allcock, H. R.; Laurencin, C. T. *Biomaterials* **1997,** *18*, 1565.
19. Devin ,J.E.; Attawia, M.A.; Laurencin, C.T. *J. Biomater. Sci. Polymer Edn.* **1996,** *7*, 661.

20. Attawia, M.A.; Uhrich, K.E; Botchwey, E.; Langer, R.; Laurencin, C.T. *J. Orthop. Res.* **1996,** *14,* 445.
21. Coombes, A. D.; Heckman, J. D. *Biomaterials.* **1992,** *13,* 217.
22. Mikos, A. G.; Thorsen, A. J.; Czerwonka, L. A.; Bao, Y.; Langer, R. **1994,** *35,* 1068.
23. Thomson, R. C.; Yaszemski, M. J.; Powers, J. M.; Mikos, A. G. *J. Biomater. Sci. Polymer Edn.* **1995,** *7,* 23.
24. Laurencin, C. T.; Ambrosio, A. M. A.; Bauer, T. W.; Allcock, H. R.; Attawia, M. A.; Borden, M. D.; Gorum, W. J.; Frank, D. *Trans. Soc. Biomat.* **1998,** *21,* 436.
25. Attawia, M. A.; Devin, J. E.; Laurencin, C. T. *J. Biomed. Mater. Res.* **1995,** *29,* 843.
26. Attawia, M. A.; Herbert, K. M.; Laurencin, C. T. *Biochem. and Biophys. Res. Comm.* **1995,** *213,* 639.
27. Keaveney, T. M; Hayes W. C. *Bone Volume 7: Bone Growth-B.* Hall, B. K., Ed., CRC Press: Boca Raton, FL, 1992, pp. 285-344.
28. Hulbert, S. F.; Young, F. A.; Matthews, R. S.; *et al. J. Biomed. Mat. Res.* **1970,** *4,* 443.
29. Jalil, R.; Nixon, J. R. *J. Microencapsulation.* **1990,** *7,* 297.
30. Cohen S.; Yoshioka T.; Lucarelli M.; Hwang L.; Langer R. *Pharm. Res.* **1991,** *8,* 713.
31. Zhifang Z.; Mingxing A.; Shenghao W.; Fang L.; Wenzhao S.. *Biomat., Art. Cells & Immob. Biotech.* **1993,** *21,* 71.
32. Schartz, E. R. *Culture of Animal Cells,* Freshney, I., Ed., Liss Inc.: New York, NY, 1987, pp. 273-276.
33. Laurencin, C. T; Morris, C. D; Pierre-Jacques H.; Schwartz E.R.; Keaton A. R.; Zou L. *Polymers for Advanced Technologies.* **1992,** *3,* 359.

Chapter 20

Novel Approaches for the Construction of Functionalized PEG Layer on Surfaces Using Heterobifunctional PEG–PLA Block Copolymers and Their Micelles

Hidenori Otsuka[1], Yukio Nagasaki[2], and Kazunori Kataok[1]

[1]**Department of Materials Science, Graduate School of Engineering, The University of Tokyo, Hongo 7–3–1, Tokyo 113–8656, Japan**
[2]**Department of Materials Science and Technology, Science University of Tokyo, Noda, Chiba 278–8510, Japan**

This paper reviews novel approaches established by our group for the construction of a functionalized poly(ethylene glycol) (PEG) layer, PEG-brushed layer possessing a reactive group at the free end of tethered PEG chain, on various substrates. Two different approaches were involved in our strategy to prepare PEGylated surfaces. An AB-type block copolymer composed of α-acetoxy-poly(ethylene glycol) (PEG) as the hydrophilic segment and polylactide (PLA) as the hydrophobic segment was synthesized, and utilized to construct the functionalized PEG layer on the biodegradable polylactide surface by simple coating. In this way, a PEG-brushed layer with a terminal aldehyde group was readily prepared which may have both non-fouling and ligand-binding properties. Further, core-polymerized reactive micelles were prepared by the association and subsequent polymerization of α-acetal-ω-methacryloyl-PEG/PLA as a surface modifier. The micelle with the polymerized core maintained its spherical structure even after the anchoring on the amintaed surface through azomethine formation, allowing the surface to have extremely high non-fouling character. Based on the characterization of these PEGylated surfaces from a physicochemical (contact angle, ζ potential) as well as biological (protein adsorption) point of view, our strategy to construct a functionalized PEG layer was confirmed. Active functional groups were present at the tethered PEG-chain end, these materials will have a high utility in the biomedical field.

Surface modification with polymers is the most common way to control the surface properties in biological and biomedical applications[1, 2]. Poly(ethylene glycol) (PEG) coating has been used to minimize non-specific fouling of materials surface with biocomponents, particularly plasma proteins. For example, a PEGylated surface, which means the surface is covered with tethered chains of poly(ethylene glycol) using the functionality of PEG end groups, extremely reduces protein adsorption[3, 4] resulting in a high blood compatibility[5-8]. This is considered to be due to unique properties of tethered PEG, that is, steric stabilization related to high flexibility[9-11] and the large exclusion volume[12-14] of PEG strands in water. PEG-coating can be performed using various methods such as covalent grafting of PEG having reactive chain-ends to the surface[15, 16], graft copolymerization of PEG macromonomer with the surface[17, 18], and direct adsorption of PEG onto surfaces in the form of a surfactant or block copolymer in which one of the blocks is a PEG[19, 20]. However, most of the PEG-coated surfaces possess no reactive group on the PEG chain end. To provide additional functionality on the PEG-coated surface, we designed a block copolymer having an end-functionalized PEG segment. Here, PEG/PLA block copolymers having different functionalities at each of the two chain ends can be utilized as surface modifiers in two ways, to provide the reactive sites on a PEGylated surface (Figure 1). The first approach is the construction of a functionalized PEG layer on a biodegradable PLA surface through a simple coating end-functionalized PEG/PLA copolymers. Another approach is the surface modification with an assembly, i.e., polymer micelles prepared from their functionalized PEG/PLA block copolymers. Apparently, in both approaches, a reactive group located on the PEG chain end provided the site for chemical immobilization of functional molecules. It should be noted that this end-functionalized PEG/PLA is a major outcome of our systematic research on heterobifunctional poly(ethylene glycols).

Recently, we have developed a facile and quantitive synthetic method for heterobifunctional poly(ethylene glycol)[21-25], which denotes PEG having different functional groups at each of both chain ends. When one of the functional end-groups in the heterobifunctional PEG selectivity initiates the polymerization of a hydrophobic monomer, a new heterobifunctional AB block copolymer can be created, keeping the other functional group at the PEG chain end available. Particularly, lactide was chosen as the hydrophobic segment because PLAs are biodegradable and non-toxic polymers, which are widely utilized as implant materials or tissue engineering scaffolds. Moreover, both PEG and PLA were approved for clinical use by the Food and Drug Administration (FDA). End-functionalized PEG/PLA forms a PEG-brushed layer on a PLA surface by simple coating, providing functional groups with high densities. Proteins and peptides can be covalently immobilized to a aldehyde-PEG coating on a PLA surface by reductive amination. For these properties of the PEG/PLA, the block copolymer of α-acethoxy-PEG/PLA can be applied as surface modifier to functionalize polylactide surfaces which have been used extensively as cell scaffolds in the field of tissue engineering. The details for this modification of a PLA surface are described in the following section.

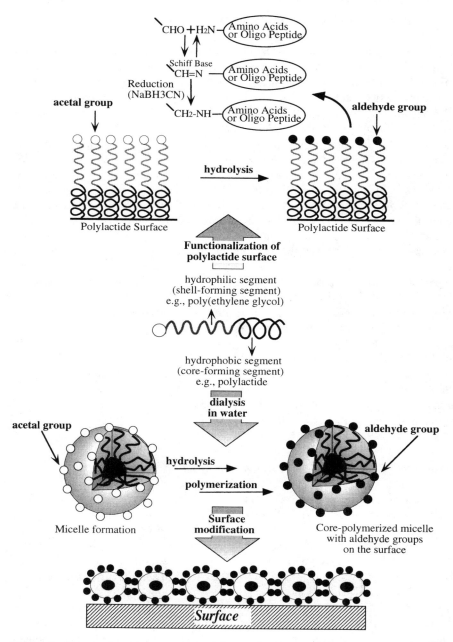

Figure 1. Schematic representation of the application to the construction of PEG layer using PEG/PLA block polymer.

Functionalization of a Polylactide Surface using Reactive PEG/PLA Block Copolymers for Tissue Engineering

A major aim of tissue engineering is to control interactions between living cells and biomaterial surfaces, allowing regeneration or artificial growth of damaged tissues[26]. Polymers for these scaffolds require good biocompatibility, suitable biodegradability, and the ability to interact specifically with appropriate cells. Polylactide, glycolide, and their copolymers have been studied as biodegradable scaffolds because they ultimately degrade to natural metabolites. These materials are reasonably biocompatible and possess appropriate biodegradability, but they lack the ability to interact biospecifically with cells. The development of novel biodegradable polymers with surfaces that are easily engineered to present high densities of ligands to adhere to cells would overcome this problem, and greatly contribute to the progress in tissue engineering[27-29]. Immobolization of the cell adhesion signal such as peptides, carbohydrates, or other bioactive moieties, is the most effective way to accomplish selective cellular attachment for tissue regeneration[30, 31]. The complexity of the in vivo surroundings, that is, fouling of polymer scaffolds with proteins can often obscure the cellular selectivity of the ligand-attached biomaterials. In an attempt to overcome this shortcoming, we have employed the end-reactive PEG as a linkage molecule for the conjugation of bioactive molecules on PLA surface to provide a selective cell adhesive surface without specific fouling with non-specific proteins.

PEG is generally well tolerated in vivo, and PEG-tetherd biomaterials are highly resistent to non-specific adsorption of cells and proteins. Therefore, PEG is well suited for the use as spacer molecule. As schematically shown in Figure 1, this paper reviews our study on novel ways of functionalization of polylactide surfaces by using heterotelechelic PEG/PLA block copolymers with an aldehyde group at the end of PEG segments[32].

Synthesis of Acetal-PEG/PLA Block Copolymers.

Reactive groups at the free end of PEG brushes are usually utilized in aqueous media for conjugation of functionality compounds such as proteins. A formyl group is very useful for this purpose due to its stability in water and its high reactivity with primary amino groups. In addition, no charge variation in protein molecules takes place by the modification because the resulting Schiff base can be easily converted to a sec-amino group by the use of an appropriate reducing agent. Since there is no change in the net-charge the original function of modified proteins might be maintained. One of the representative procedures for the preparation of α-acetal-PEG/PLA block copolymers is described[33] as follows; For an introduction of a formyl group at the PEG end, potassium 3,3-diethoxypropanolate (PDP) was used as an initiator for the polymerization of ethylene oxide (EO). The acetal group at the α-end of resultant PEG can be easily converted into an aldehyde group by acid treatment[34]. α-Acetal-PEG/PLA block copolymers have been synthesized by a one-pot anionic ring-opening polymerization of EO followed by LA initiated with PDP as an initiator at room temperature under argon (Scheme 1). One mmol (0.16 mL) of 3,3-diethoxypropanol and 1 mmol of potassium naphthalene were added to 30 mL of dry THF to form PDP. After stirring for 10 min, appropriate amounts of condensed EO were added via a cooled syringe to the formed PDP solution. The polymerization of the EO proceeded for two days at room temperature and resulted in a light brown, highly viscous solution. Potassium naphthalene (about 0.1 mmol) was added until the solution turned pale green to stabilize the living chain end, LA solution in THF (c=1.10 mol/L) was then introduced, and the polymerization proceeded for 120 min. The polymer was recovered by precipitation into a 20-fold excess of cold isopropyl alcohol (-15 ℃), stored for 2 h in the freezer and centrifuged for 30 min at 6000 rpm. The polymer was then freeze-dried with benzene. The yield of the obtained polymer was ca. 90%. The molecular weight of the PEG segment was determined by GPC at the end of the EO polymerization. The molecular weight of the PLA segment was determined using an ^{1}H NMR spectrum by examinig the ratio of methine protons in the PLA segment and methylene protons in PEG segment based on the M_n of PEG determined from the GPC results.

One of the objectives in this study is to investigate the effect of the variation in PEG chain length on surface properties. This includes assays on protein adsorption and cellular attachment to get a biochemical insight into the behavior of tethered PEG under biological conditions. For this purpose, numerous acetal-PEG/PLAs with different lengths of both PEG and PLA were synthesized. Molecular weights (MW) of PEG/PLA segments were abbreviated as follows: PEG/PLA(0.65/11.0, 1.8/7.0, 3.3/5.4, 5.0/4.6, 8.7/6.9) where the numbers in parenthesis denote the MW of the PEG segments and PLA segments in kg/mol, respectively.

Construction of a PEG Brushed-Layer with Reactive Aldehyde Group using PEG/PLA Block Copolymers

The glass substrates, which were cleaned with 1% (w/v) solution of sodium hydroxide/water and then a boiling mixture of sulfuric acid and hydrogen peroxide, were placed in 2 % (v/v) solution of 3-(Trimethoxysilyl)propyl methacrylate/ethanol for 2h. The glass substrates were dried at 160 ℃ for 24 h under vacuum. The PEG-brushed layer was constructed on this silanized glass surface by the spin coating of 4 % (w/v) solution of PLA solution/toluene, followed by the 2 % (w/v) solution of acetal-PEG/PLA/toluene[35]. To perform an on-surface conversion of the acetal-ended PEG surface into aldehyde surface, the surface was immersed in aqueous media adjusted to pH2 using hydrochloric acid for 6h.

The ζ potential of silanized glass substrates covered with various acetal-PEG/PLA block copolymers having different PEG chain length was measured in

Scheme 1

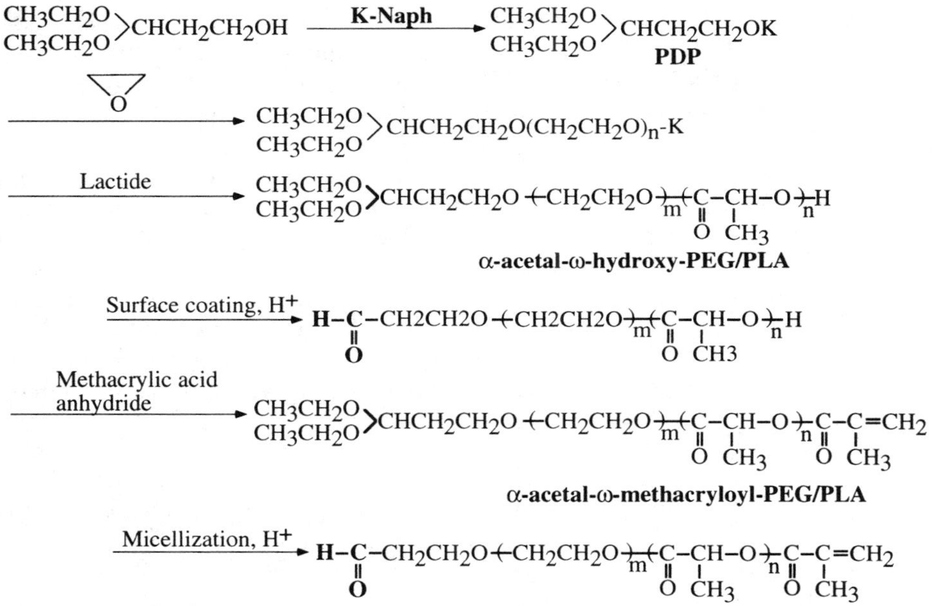

$$CH_3CH_2O \atop CH_3CH_2O \!\!> \!CHCH_2CH_2OH \quad \xrightarrow{\textbf{K-Naph}} \quad CH_3CH_2O \atop CH_3CH_2O \!\!> \!CHCH_2CH_2OK$$
PDP

$$\xrightarrow{\triangle_O} \quad CH_3CH_2O \atop CH_3CH_2O \!\!> \!CHCH_2CH_2O(CH_2CH_2O)_n\text{-}K$$

$$\xrightarrow{\text{Lactide}} \quad CH_3CH_2O \atop CH_3CH_2O \!\!> \!CHCH_2CH_2O \!\!\left(\!CH_2CH_2O\!\right)_m \!\!\left(\!\!\begin{array}{c} C-CH-O \\ \| \quad | \\ O \quad CH_3 \end{array}\!\!\right)_n \!\!H$$

α-acetal-ω-hydroxy-PEG/PLA

$$\xrightarrow{\text{Surface coating, H}^+} \quad \begin{array}{c} H-C-CH2CH2O \!\left(\!CH2CH2O\!\right)_m \!\!\left(\!\!\begin{array}{c} C-CH-O \\ \| \quad | \\ O \quad CH3 \end{array}\!\!\right)_n \!\!H \\ \| \\ \textbf{O} \end{array}$$

$$\xrightarrow[\text{anhydride}]{\text{Methacrylic acid}} \quad CH_3CH_2O \atop CH_3CH_2O \!\!> \!CHCH_2CH_2O \!\!\left(\!CH_2CH_2O\!\right)_m \!\!\left(\!\!\begin{array}{c} C-CH-O \\ \| \quad | \\ O \quad CH_3 \end{array}\!\!\right)_n \!\!\begin{array}{c} C-C=CH_2 \\ \| \quad | \\ O \quad CH_3 \end{array}$$

α-acetal-ω-methacryloyl-PEG/PLA

$$\xrightarrow{\text{Micellization, H}^+} \quad \begin{array}{c} H-C-CH_2CH_2O \!\left(\!CH_2CH_2O\!\right)_m \!\!\left(\!\!\begin{array}{c} C-CH-O \\ \| \quad | \\ O \quad CH_3 \end{array}\!\!\right)_n \!\!\begin{array}{c} C-C=CH_2 \\ \| \quad | \\ O \quad CH_3 \end{array} \\ \| \\ \textbf{O} \end{array}$$

phosphate buffer solution as a function of pH(5.8–8), the ionic strength of the medium being equal to 0.01 (Figure 2). The ζ potential of a cleaned glass surface showed clearly a negative charge(−50 to −70 mV). The silane coupling of the glass increased the ζ potential, but the still negative charge indicated incomplete silanation. When PLA homopolymer(M.W.; 20000) was spin-coated on silanized glass surface, the ζ potential was decreased again. This is due to negative charge of PLA polymer itself as reported by Dunn *et al.*[36]. Coating of PEG/PLA block copolymers onto PLA increased the ζ potential, and further a progressive increase is obvious with increasing PEG molecular weight, indicating a screening of the surface charge. This means that the formation of the PEG brush layer shifts the position of the slipping plane into the solution side. Hydrophilic PEG segments are considered to expand away from the surface due to their strong hydration power.

The wettability of the surface covered with PEG/PLA block copolymers was estimated both in air and in water by contact angle measurement (Table 1). In water-in-air measurements, coating of PEG/PLA block copolymer onto a PLA surface increased its wettability with increasing PEG molecular weight, as indicated by a decrease in static contact angle. A similar trend was observed in air-in-water measurements. The contact angle of a water droplet in air decreased remarkably in the range between PEG/PLA(0.65/1.10) and PEG/PLA(5.0/4.6). The decrease became moderate at the region with higher PEG molecular weights. Since the top few angstroms can be sensed by a contact angle measurement, the relatively high contact angles on the surfaces containing the lower molecular weight PEG is most likely to be attributed to an incomplete coverage of the uppermost surface by PEG chains. The dynamic contact angle was then measured to estimate the dynamics of the uppermost surface. The coating of PEG/PLA block copolymers reduces both the advancing and the receding angles of the substrates, although the change depends on the PEG molecular weight, which is consistent with the result of static contact angle. The maximum hysteresis was observed for the substrate with medium PEG chain length such as PEG/PLA(3.3/5.4). Hysteresis in the dynamic contact angle may be caused by the hydration of PEG segments. In the dry state, the PEG chain should assume a conformation flat to the surface experienced by the advancing contact line. Upon hydration, however, the PEG chain should extend from the surface due to the hydration of PEG chains. As a result, the receding contact line experiences a more hydrophilic surface than the advancing contact line. It is likely that this is the origin of the hysteresis observed on these surfaces.

Protein adsorption on PEG/PLA surfaces was then estimated by using bovine serum albumin (BSA) as a model protein[32]. On PLA surfaces, BSA adsorbed significantly while on PEG coated surfaces BSA adsorption is clearly decreased mainly due to the steric stabilization by PEG chains. Moreover, minimum adsorption was obtained at a medium PEG chain length, *i.e.*, PEG/PLA(3.3/5.4), note that this surface revealed the maximum hysteresis in dynamic contact angle

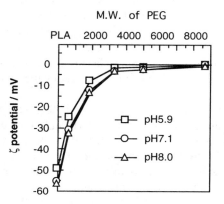

Figure 2. ζ potential variation of glass substrates as a function of MW of PEG in phosphate buffer solution.

Table.1 Static/Dynamic Contact Angle (C.A.) variations on PLA and PEG/PLA coated surfaces. The contact angle of each sample was measured on at least ten spots, and these values were averaged.

polymer code	Static C.A.		Dynamic C.A.		
	water in air	air in water	Adv.	Rec.	Hysteresis
PLA	80.2	123.4	82.1	61.9	20.2
PEG/PLA(0.65/1.1)	78.9	128.6	84.5	59.4	25.1
PEG/PLA(1.8/7.0)	75.6	131.3	73.2	46.3	26.9
PEG/PLA(3.3/5.4)	59.7	139.8	67.8	27.5	40.3
PEG/PLA(5.0/4.6)	47.2	142.3	62.2	25.3	36.9
PEG/PLA(8.7/6.9)	47.9	147.4	49.1	26.2	22.9

measurement. Protein adsorption may be related to the hysteresis observed in the dynamic contact angle, which is likely to depend on particular surface properties such as the density and mobility of tethered PEG chains on the surface.

A functionalization of the PEG chain end provides the means for attaching ligand molecules for further chemical modulation of the surface. After the construction of the PEG/PLA surface, the acetal groups at the PEG-chain-end were successfully transformed into aldehyde end groups. An aldehyde group reacts smoothly with amino groups forming a Schiff base, a chemical path which can be employed for conjugation of proteins and peptides. The conversion of the acetal end groups into aldehyde end groups was conducted directly on the surface by immersing the surface into aqueous media adjusted to pH 2 using hydrochloric acid for 6h (Figure 1). To confirm the presence as well as to examine the reactivity of aldehyde group on the surface, model reactions with 2, 2, 6, 6-tetramethyl-1-piperidinyloxy (TEMPO) derivatives as label agents were performed and ESR specta of TEMPO derivatized surfaces were successively recorded. When the acetal surface was treated with 4-amino-TEMPO, only a slight signal was observed probably due to the physical adsorption of 4-amino-TEMPO on the surface. When the aldehyde surface was treated with TEMPO having no functional (amino) group, no ESR signal was observed. Contrary to these control treatments, three typical signals were clearly observed when 4-amino-TEMPO was used as the surface modification reagent, indicating that the effective covalent-conjugation of 4-amino-TEMPO with the aldehyde group at the end of PEG on the surface took place. These results strongly suggest the high utility of these surfaces in the field of tissue engineering because the terminal aldehyde function at the end of the PEG moiety can be readily derivatized with proteins and peptides as cellular binding/activating moieties.

Surface Coating with Core-Polymerized Block Copolymer Micelles having an Aldehyde-ended PEG Shell

Tremendous research efforts focused on selective drug delivery by using polymeric devices with increased therapeutic effects and decreased undesirable side effects of drugs. Recently, colloidal carriers based on biocompatible block copolymers have been studied extensively for selective drug delivery[37]. In our previous work[38, 39], amphiphilic block copolymers consisting of PEG and PLA, both FDA-approved polymers, were studied and their formation of micellar systems in selective solvents, i.e., a good solvent for one segment and a precipitant for the other segment, was investigated aiming to use this system in the modulated drug delivery. These nanospheric particles combine the advantages provided by the hydrophobic core which can act as a reservoir for drugs[40] and the unique properties of a hydrophilic PEG shell resulting in steric stabilization effects.

Indeed, drugs can be loaded into the core by physical entrapment or covalent linkage to the side-chain functional groups of the core-forming segment directly or by means of a spacer[40, 41]. Also, many reports from our own and other groups confirm the extended plasma half-life as well as the decreased liver uptake of enzymes conjugated with PEG, without RES recognition[42]. The PEG corona prevents micelles from the foreign-body recognition. If this assembly, polymeric micelle, can be immobilized on the surface with maintaining its core-shell architecture, the advantages of micelles in solution should be also expected on the surface. However, physical coagulation forces may not be stable enough to maintain the micelle structure in the process of surface fixation. A disruption of the micelle upon attachment to the surface is reported[24] both experimentally and theoretically[43-45]. Johner and Joanny[45] used scaling arguments to show the disruption of the micelles during their adsorption process. This disruption results in the formation of loops and trains on the surface that makes the subsequent adsorption of more micelles, and consequently, the formation of a densely packed micelle layer difficult. By polymerizing or crosslinking the PLA segment, the structural stability of the micelle is expected to be facilitated. Based on this idea, a polymerizable methacryloyl group was introduced at the ω-chain end (PLA-end) of a block copolymer of α-acetal-ω-hydroxy-PEG/PLA. Such surface-functionalized and core-polymerized micelles are applicable to the coating of surfaces as in the case of linear and star polymers. Since the density of PEG is high in the shell of the micelle, surfaces grafted with the PEG-PLA micelles are eventually to have a densely tethered PEG chain, which may provide an effective non-fouling property. Further, by incorporating hydrophobic drugs into core of surface-immobilized micelles, one can create surfaces releasing drugs in a controlled manner.

The acetal moiety of the micelle prepared from a functionalized block copolymer, α-acetal-ω-methacryloyl-PEG/PLA, was converted to aldehyde groups at the end of the PEG segment, and then the methacryloyl group was polymerized after the micellization to obtain a core-stabilized micelle[46]. Thus, the core-stabilized micelle was coated onto an aminated polypropylene (PP) or glass surface as shown in Figure 1 and subjected to the surface characterization including ζ potential, contact angle, and protein adsorption measurements[47].

Preparation of reactive polymeric micelle

As described above, an acetal-PEG/PLA block copolymer having a functional group at the PEG chain end was prepared (Scheme 1). Finally, we introduced the polymerizable methacryloyl group at the ω-chain end (PLA end) for the stabilization of the polymeric micelle.

It is well known that amphiphilic block copolymers with a suitable hydrophilic/hydrophobic balance form micelle structures when exposed to a selective solvent. The block copolymer of α-acetal-ω-methacryloyl-PEG/PLA was dissolved in dimethylacetamide and the solution was dialyzed against distilled water through semi-permeable membrane tube (molecular weight cut off size 12000-14000) for 24 h. After the acetal-polymeric micelle was prepared in water, the pH of the medium was adjusted to 2 to form a polymeric micelle with aldehyde groups on its surface. The ^{1}H-NMR spectrum of PEG/PLA(5.8/4.0; in kg/mol) before and after the hydrolysis reaction are shown in Figure 3. As shown in the figure, after hydrolysis, the proton of the aldehyde end-group appears at 9.8 ppm. On the other hand, the acetal-methine proton around 4.6 ppm disappeared; retaining two vinyl protons of the methacryloyl end appearing at 5.7 and 6.2 ppm. From the ^{1}H-NMR analysis, the conversion to aldehyde was estimated to be attained up to 90 %. It should be noted that no degradation of the PLA main chain is observed as a result of the acid treatment. This apparent acid-resistant property of the PLA segment may be due to the core-shell structure of the micelle where the PLA segment is segregated from aqueous environment. Protons hardly attack the ester bond in the PLA unit of the core. For surface modification, the physical coagulation force of the hydrophobic core alone may not be sufficient to maintain the micelle structure. The core-stabilization of a micelle is considered to be a promising method to retain the micelle structure on a surface. The polymerization of the methacryloyl end group in the core of the micelle was thus performed by radical polymerization using an azo-initiator (V-65). The size of the micelle determined by DLS measurements, before and after the polymerization of the core, does not change significantly (34.8 nm in diameter for PEG/PLA(5.8/4.0)). Also, DLS measurements using a cummulant method revealed a very low polydispersity of the micelles (μ/Γ^2 = 0.090, 0.081 before and after polymerization. respectively), and no angular dependency is observed. The micelle thus obtained was stable in organic solvents and in water containing sodium dodecylsulfate (SDS) which are known as destructive environment for micelles. These results indicate the successful preparation of a core-polymerized micelle having reactive aldehyde groups on its surface. The spherical core-shell structure with a narrow polydispersity was maintained.

Coating of an aminated polypropylene surface with polymerized reactive micelles −procedure and characterization−

Before the coating of a polypropylene (PP) surface with polymerized micelles, primary amino groups were introduced onto PP by a plasma-amination reaction using a H_2/N_2 mixture as amination gas. The polymerized reactive micelle was chemically immobilized on the aminated-PP plate in HEPES buffer solution (0.1M, pH6.7) in the presence of $NaCNBH_3$ as a reducing agent. To compare the stability upon immobilization between the polymerized and non-polymerized micelle, the change in the static contact angle was measured with increasing the rinsing time of the surface. The surfaces modified by both polymerized and non-polymerized micelles showed a remarkable decrease in the contact angle value

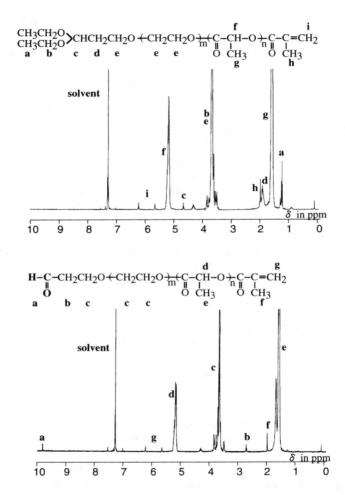

Figure 3. ^1H *NMR spectrum of* α-*acetal-*ω-*methacryloyl-PEG/PLA (a) and* α-*aldehyde-*ω-*methacryloyl-PEG/PLA (b) block copolymer.*

(θ=about 50°), indicating the surface hydrophilization by this treatment (by comparison: native amino-PP surface, θ=69°). After thorough rinsing of the surface treated with non-polymerized micelle, the contact angle remarkably increased, even higher than the native amino-PP surface, indicating the disruption of the micelle structure. The surface treated with polymerized micelles, however, did not change the contact angle value, suggesting that a micelle structure may be maintained even after the surface immobilization.

The efficacy of a surface coverage by the polymerized micelle was then examined by ζ potential measurements (Figure 4). The ζ potential on the plasma treated (aminated) PP depended strongly on the change of pH. The protonation of the amino group depends on the environment and the ζ potential of an aminated PP surface is highly positive in the acidic region. When aminated PP was coated with linear PEG and micelles, the ζ potential was significantly reduced over the pH region of 2 to 11. The grafting of a linear PEG also reduced the ζ potential, although a slight dependency of ζ potential on pH indicates the incomplete surface modification by a simple PEG treatment. On the other hand, the surface coated with core-polymerized micelles showed almost no change in the ζ potential for the whole pH region, suggesting a complete surface coverage and/or longer distance of the shear plane from the PP surface in comparison with the PEG grafted surface.

The adsorbed amount of protein was estimated at the physiological condition (protein concentration; 45 mg/ml, ionic strength; 0.15) using bovine serum albumin (BSA) as a model protein (Figure 5). BSA adsorption on a PEG modified surface (both linear PEG and micelle) was significantly reduced in comparison with native PP surface. Furthermore, protein adsorption on the micelle coated surface was significantly lower even as compared with the PEG tethered surface, presumably due to a higher density of the PEG chains. This result is in line with results of contact angle and ζ potential measurements. In fact, the AFM images confirmed a core-stabilized reactive micelle from PEG/PLA block copolymer on the surface[48].

Conclusion

We have been studying the synthesis of heterobifunctional poly(ethlene glycol). As a further extension of this method, heterobifunctional block copolymers were synthesized. A block copolymer of α-acetal-ω-hydroxy-PEG/PLA was coated on a polylactide surface to functionalize the surface, followed by the conversion of the acetal groups into aldehyde groups. The surface properties were characterized by ζ potential, static/dynamic contact angle, and ESR measurements. This polymer is expected to be useful as polymer scaffold in tissue engineering since it has a biodegradable component, L-lactide, and non-fouling and ligand-immobilizable component, PEG. Further, core-stabilized reactive micelles can be prepared from an end-derivatized block copolymer, acetal-PEG/PLA-methacrylate. The reactive micelles having polymerized cores maintained their structure even after the surface

324

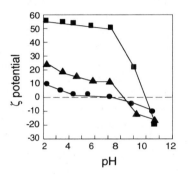

Figure 4. ζ potential variation of polypropylene (PP) substrates as a function of pH. ■;*plasma treated PP,* ▲;*linear PEG modified PP,* ●;*micelle modified PP.*

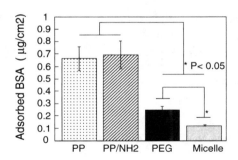

Figure 5. The adsorbed amount of BSA on various modified surfaces at room temperature after the incubation for 60 minutes.

immobilization. Such micelle-coated surfaces showed non-fouling character under physiological conditions probably due to the high density of the PEG chain on the micelle surface. The coating of PEG/PLA micelle may result in surfaces useful for medical implants with non-fouling properties.

Acknowledgements

The authors gratefully acknowledge Professors Y. Sakurai and T. Okano, Tokyo Women's Medical University, and Professor K. Kato, Science University of Tokyo. The authors would like to acknowledge Dr. K. Emoto and Mr. M. Iijima for carrying out of a part of this study. We would like to also acknowledge Dr. C. Scholz for peer and critical reading for the manuscript. This study was supported by a JSPS, The Japan Society for the Promotion of Science, "Research for the Future" Program (JSPS-RFTF96I00201).

References
1. *Polymer Surfaces and Interfaces*; Feast, W. J.; Munro, S., Eds.; John Wiley & Sons: 19 87.
2. *Polymer Surfaces and Interfaces II*; Feast, W. J.; Munro, S.; Richards, R. W., Eds.; John Wiley & Sons: 1993.
3. Osterberg, E.; Bergstrom, k.; Holmberg, K.; Schuman, T. P.; Riggs, J. A.; Burns, N. L.; Van Alstin, J. M.; Harris, J. M. *J Biomed. Mats. Res.* **1995**, *29*,741.
4. Lee, J. H.; Lee, H. B.; Andrade, L. D. *Prog. Polym. Sci.* **1995**, *20*, 1043.
5. *Poly(ethylene glycol) Chemistry, Biotechnical and Biomedical Applications;* Harris, J. M., Ed.; Plenum Press: New York, 1993.
6. Amiji, M.; Park, K. *J. Biomater. Sci., Polym. Ed.* **1993**, *4*, 217.
7. Bergstom, K.; Osterberg, E.; Holmberg, K.; Hoffman, A. S.; Schuman, T. P.; Kozlowski, A.; Harris, J. M. *J. Biomater. Sci., Polym. Ed.* **1994**, *6*,123.
8. Llanos, G. R.; Sefton, M. V. *J. Biomater. Sci., Polym. Ed.* **1993**, *4*,381.
9. Bailey, F. E.; Koleske, J. V.; *Poly(ethylene oxide);* Academic Press: New York, **1976**.
10. Kjellander, R.; Florin, E. *J. Chem. Soc. Faraday Trans I.* **1981**, *77*, 2053.
11. Merrill, E. W.; Salzman, E.W. *ASAIO* **1983**, *6*, 60.
12. Atha, D. H.; Ingham, K. C. *J. Biol. Chem.* **1981**, *256*, 12108.
13. Knoll, D.; Hermans, J. *J. Biol. Chem.* **1983**, *258*, 5710.
14. Nagaoka, S.; Mori, Y.; Takiuchi, H.; Yokota, K.; Tanzawa, H.; Nishumi, S. In *Polymers as Biomaterials*; Shalaby, S. W.; Hoffman, A. S.; Ratner, B. D.; Horbett, T. A., Eds.; Plenum Press: New York, 1984,.
15. Bergstrom, K.; Osterberg, E.; Holmberg, K.; Hoffman, A. S.; Schuman, T. P.; Kozlowski, A.; Harris, J. M. *J. Biomater. Sci., Polym. Ed.* **1994**, *6*, 123.
16. Llanos, G. R.; Sefton, M. V. *J. Biomed. Mater. Res.* **1993**, *27*,1383.

326

17. Brinkman, E.; Poot, A.; van der Does, L.; Bantjes, A. *Biomaterials* **1990**, *11*, 200.
18. Golander, C. G.; Jonsson, S.; Vladkova, T.; Stenius, P.; Eriksson, J. C. *Colloids Surf.* **1986**, *21*, 149.
19. Lee, J. H.; Kopeckova, P.; Kopecek, J.; Andrade, J. D. *Biomaterials*, **1990**, *11*, 455.
20. Andrade, J. D.; Kopecek, J.; Lee, J. H. U. S. Patent 5, 075, 400, 1991.
21. Kim, Y. J.; Nagasaki, Y.; Kataoka, K.; Kato, M.; Yokoyama, M.; Okano, T.; Sakurai, Y. *Polym. Bull.* **1994**, *33*, 1.
22. Cammas, S.; Nagasaki, Y.; Kataoka, K. *Bioconjugate Chem.* **1995**, *6*, 224.
23. Nagasaki, Y.; Kutsuna, T.; Iijima, M.; Kato, M.; Kataoka, K. *Bioconjugate Chem.* **1995**, *6*, 231.
24. Nagasaki, Y.; Iijima, M.; Kato, M.; Kataoka, K. *Bioconjugate Chem.* **1995**, *6*, 702.
25. Nagasaki, Y.; Ogawa, R.; Yamamoto, S.; Kato, M.; Kataoka, K. *Macromolecules* **1997**, *30*, 6489.
26. For reviews see:(a) Langer, R.; Vacanti, J. P. *Science*, **1993**, *260*, 920. (b) Nerem, R. M.; Sambanis, A. *Tissue Eng.* **1995**, *1*, 3.
27. Hubbell, J. A. *Bio/Technology* **1995**, *13*, 565.
28. Lysaght, M. J. *Tissue Eng.* **1995**, *1*, 221.
29. Drumheller, P. D.; Hubbell, J. A. *J. Biomed. Mater. Res.* **1995**, *29*, 207; *Anal. Biochem.* **1994**, *222*, 380.
30. Massia, S. P.; Rao, S. S.; Hubbell, J. A. *J. Biol. Chem.* **1993**, *268*, 8053.
31. Cima, L. G. *J. Cell. Biochem.* **1994**, *56*, 155.
32. Otsuka, H.; Nagasaki, Y.; Sakurai, Y.; Okano, T.; Kataoka, K. manuscript in preparaiton
33. Nagasaki, Y.; Okada, T.; Scholz, C.; Iijima, M.; Kato, M.; Kataoka, K. *Macromolecules* **1998**, *31*, 1473.
34. Greene, T. W. *Protective Group in Organic Synthesis*; John Wiley & Sons: New York, 1981.
35. Desai, N. P.; Hubbell, J. A. *Biomaterials* **1991**, *12*, 144.
36. Dunn, S. E. *J. Control. Release* **1997**, *44*, 65.
37. For reviews see: (a) Kataoka, K. *Targetable Polymeric Drugs, in Controlled Drug Delivery, The Next Generation*; Park, K., Ed.; American Chemical Society: Washington DC, 1996, 49. (b) Cammas, S.; Kataoka, K. *Site Specific Drug-Carriers: Polymeric micelles as High Potential Vehicles for Biologically Active Molecules, in Solvents and Self-Organization of Polymers* ; Webber, S. E.; Munk, P.; Tuzar, Z., Eds.; NATO ASI Series, Series E; Applied Science; Kluwer Academic Publishers: Dordrecht, 1996; Vol. 327.
38. La, S. B.; Nagasaki, Y.; Kataoka, K. *Poly(ethylene glycol)-Based Micelles for Drug Delivery, in Poly(ethylene glycol) Chemistry*; Harris, J. M.; Zalipsky, S. Eds.; Plenum Press, 1997, 99.

39. Kataoka, K.; Kwon, G. S.; Yokoyama, M.; Okano, T.; Sakurai, Y. *J. Controlled Release* **1993**, *24*, 119.
40. Kataoka, K. *J. Macromol. Sci. Pure Apl. Chem.* **1994**, *A31*, 1759.
41. Abuchowski, A.; McCoy, J. R.; Palczuk, N. C.; van Es, T.; Davis, F. F. *J. Biol. Chem.* **1977**, *252*, 3582.
42. Kamisaki, Y.; Wada, H.; Yagura, H.; Matsushima, A.; Inada, Y. *J. Pharm. Exp. Ther.* **1981**, *216*, 410.
43. Farinha, J. P. S.; d'Oliveira, J. M. R.; Martinoho, J. M.; Xu, R.; Winnik, M. A. *Langmuir* **1998**, *14*, 2291.
44. Bijsterbosch, H. D.; Cohen Stuart, M. A.; Fleer, G. J. *Macromolecules* **1998**, *31*, 9281.
45. Jonner, A.; Joanny, J. F. *Macromolecules* **1990**, *26*, 5299.
46. Iijima, M.; Nagasaki, Y.; Okada, T.; Kato, M.; Kataoka, K. *Macromolecules* **1999**, *32*, 1140.
47. Iijima, M.; Nagasaki, Y.; Kato, M.; Kataoka, K. Manuscript in Preparation.
48. Emoto, K.; Nagasaki, Y.; Kataoka, K. Manuscript in Preparation.

Chapter 21

Poly(β-hydroxyalkanoates) as Potential Biomedical Materials: An Overview

Carmen Scholz

Department of Chemistry, University of Alabama in Huntsville, 301 Sparkman Drive, MSB 333, Huntsville, AL 35899

Poly(β-hydroxyalkanoates), (PHAs), are produced by a wide variety of microorganism. Due to their natural origin these biopolyesters are biodegradable and biocompatible. The inherent biocompatibility of the material made it an ideal candidate for biomedical applications. PHAs, in particular poly(β-hydroxybutyrate), (PHB), and poly(β-hydroxybutyrate-co-valerate), (PHBV), have been studied and tested extensively as implant, implant coating and drug delivery systems. Since there is no depolymerase enzyme expressed in mammals, degradation is based solely on hydrolysis and therefore a very slow process. Thus, this material is particularly interesting for long-term applications, that is, bone and nerve repair and bone treatment. PHB-based implants produced favorable bone-tissue adaptation response. Thermoplasiticity of the material is advantageous, since implants can be molded into desired shapes. Drug delivery systems based on PHBV were successfully tested in the treatment of bone related diseases.

PHAs are less suited for implants that are in immediate and extensive blood contact, since PHA causes, like other hydrophobic surfaces, protein adsorption that leads eventually to thrombi formation. The formation of natural-synthetic block copolymers consisting of PHB and poly(ethylene glycol) will contribute to an increased blood compatibility and furnish the biopolyester with a stealth character.

Introduction

Poly(β-hydroxyalkanoates), (PHAs), were first discovered by Lemoigne[1] in 1925 while he was examining *Bacillus megaterium*. From those observations Lemoigne concluded that these cellular inclusion bodies play a role in the survival mechanism of the organism. They form when one or more nutrients is limited, but a carbon source is readily available, and they are consumed when nutrient availability worsens and an external carbon source is no longer available. The nature of these inclusion bodies was determined to be poly(β-hydroxybutyrate), (PHB), a polyester. Instead of storing small molecules that would alter the osmotic state of the cell and, would leak out, these bacteria polymerize the soluble intermediates into insoluble polymers, keeping them as an internal carbon and energy storage. Due to enzymatic synthesis, PHAs are isotactic, with the chiral carbon atoms in the R-configuration. A wide variety of microorganisms produces PHAs, by choosing the microbial strain and fermentation conditions judiciously the primary structure of the PHA can be influenced, resulting in the formation of functionalized biopolyesters[2]. The material is biodegradable, it was produced by a polymerase enzyme, which is counteracted by a depolymerase enzyme. Depolymerase enzymes are expressed in bacteria and fungi.

Commercialization of PHAs

Since their renaissance in the 1980's, PHAs have been considered for a variety of applications. Early on, anticipation was high that PHAs could eventually replace packaging material. PHAs would be an ideal material for packaging, it is a thermoplastic, and the properties can be tailored to range from a brittle material to an elastomer. After their function as package material is exhausted, the material could be discarded, and would eventually degrade in a composting or landfill facility – an environmentally benign alternative to existing packaging materials. However, PHAs never reached the level of actually competing with conventional packaging materials due to their high costs. Even though continous fermentation techniques might bring the price down to about $ 3.50/kg, it is still much higher than that of traditional polyolefins. Many avenues were pursued to try to make biopolyesters less expensive. Approaches ranged from over-expressing polymerase genes in *Escherichia coli*[3] to genetically engineering strains to turn them into super-producers[4]. *E. coli* does not produce PHAs naturally and therefore has no pathways for intracellular PHA degradation. Thus, an increase in yield was expected since the biological path for natural polymer consumption is not existent. To this day there continues to exist a debate on the cost-effectiveness of commercial PHA production. Whereas some scholars strongly advocate PHAs as polymers of the future because of their production from renewable resources and inherent biodegradability, others argue that microbial fermentation is a very energy-consuming process with a low overall yield relative to the amount of starting material. In addition, aerobic fermentation processes generate the green-house gas carbon dioxide[5].

As is so often the case, the truth might actually lay in the middle. PHAs have their raison d'être, even though they probably will never replace the market leaders in packaging supplies. PHAs need to be recognized for their unique properties. PHAs are isotactic polymers with all chiral carbon atoms in R-configuration. PHB can also be synthesized chemically from butyrolactone and, synthetic PHB with a high degree of isotacticity can be obtained by choosing a catalyst judiciously[6]. A fault-free, truly isotactic polymer, however, is currently only available through whole cell catalysis, requiring the use of microorganisms. The question arises, whether isotactic PHB is needed. Is isotacticity a prerequisite for biodegradability and biocompatibility?

PHAs for biomedical applications

Biocompatibility is a unique attribute inherent to PHAs. Biocompatibility has been shown for short chain PHAs, PHB and PHBV[7] and for long-chain PHA, such as poly(β-hydroxyoctanoate)[8]. PHAs are envisioned to play a major role in the biomedical field in the future. Several studies have shown that PHAs are suitable for biomedical applications since no adverse tissue reaction was observed. Depolymerase enzymes are not expressed in mammals and *in vivo* PHA degradation relies solely on hydrolysis. Hence, a very slow degradation of the implant material is observed, ranging from several months to a year or more. Thus, implants based on PHAs are excellent candidates where long-time implantation is sought, e.g. in bone replacement. Thermoplasticity of PHAs is another very advantageous feature with respect to bone implants, since implants can be easily molded into the desired shape. This is particularly important for reconstructive surgery. Implantation studies using PHB films and molded objects did not cause an inflammatory reaction[9].

In addition, blending of different PHAs exhibiting a variety of mechanical properties, presents itself as an option to design materials of distinct mechanical properties which can closely resemble the strength and elasticity of the body part that needs to be replaced. This is an area of research, which will gain increasing consideration in the future. Mimicking exactly the mechanical properties of bodily tissue that has been lost due to disease or accident and that is replaced by an implant is as important a consideration as biocompatibility and eventual biodegradability. Artificial cartilage, for example, achieves full biological function and fibrocartilage ingrowth only when the modulus of the implant matches that of natural cartilage[10].

PHAs have been tested intensively for biomedical uses. They were investigated as drug delivery systems, implants, sutures, staples, screws, clips, fixation rods, cardiovascular stents and others. Most of the studies involving PHAs as biomedical devices reported good biocompatibility and little or no immunological responses. The polymer itself is not toxic and the hydrolysis product, R-β-hydroxybutyric acid, itself is a mammalian metabolite that occurs in low concentrations in humans. van der Giessen[11] et al., however, studied PHBV among other polymers as stent material to be implanted into the coronary arteries and observed severe thrombosis. This is a typical response of the blood to a foreign, hydrophobic surface on which unfolding of blood proteins occurs, followed by platelet adhesion and subsequent thrombi formation. Thus, PHBV does not appear to be a good material for immediate and extended blood contact. The strength of the material in biological applications appears to be in the realm of bone-implant. Moreover, PHB and PHBV seem to be advantageous in a variety of bone-related treatments.

Doyle[12] et al. studied PHB composites reinforced with hydroxyapatite and found that bone is rapidly formed close to the composite material, and it subsequently becomes highly organized as new bone tissue. They found no evidence of a structural break down of the PHB over an implantation time of 12 months. Holland[13] et al. suggested that an accelerated degradation of PHBV occurs at implantation times beyond one year using *in vitro* studies. This degradation phase is preceded by a surface modification phase that leads to a slight increase in crystallinity due to the erosion of the amorphous domains and a subsequent progressive increase of the polymer porosity. Water can then eventually penetrate into the now porous polymer matrix, initiating hydrolysis and subsequent degradation. Materials based on PHB produced a consistently favorable bone tissue adaptation response with no evidence of undesirable chronic inflammation. The advantageous influence of PHB and PHBV biopolymers is believed to be based on their piezoelectric properties, which lead to osteoconductivity. Bone inflammation occurs often in connection with implant procedures due to *Staphylococcus aureus* infections, a common skin bacterium. Yagmurlu[14] et al. reported a positive effect on treating osteomyelitis with antibiotics administered by using PHBV as a drug delivery system. In a similar manner, PHBV microspheres were tested as drug delivery systems for the treatment of periodontal diseases. An antibiotic drug was loaded into PHBV microcapsules and the release patterns were studied *in vitro*, indicating a complete drug release prior to disintegration of the microcapsules[15].

Drug carrier systems in the form of microcapsules were formed from blends of PHB and PHBV with poly(ε-caprolacton) (PCL). It was shown that surface morphologies could be rendered by the amount of PCL in the blend[16]. PHB has been suggested as a drug delivery system as early as 1989. PHB drug delivery vehicles carrying particulates were suggested for intramuscular administration and colloidal carriers were suggested for drug targeting in intravenous administration[17].

Recently, PHB has been tested for nerve repair, using PHB films as support to wrap around the transected nerve. The PHB films were used as bioabsorbable sheets, that were supposed to degrade eventually after the healing of the nerve was complete, and the structural support would not be needed anymore. Nerve ends did heal within this PHB-wrapping, restoring the nerve function[18].

As indicated above, PHB and PHBV block copolymers seem to be very beneficial for bone-related treatments. As stated earlier, (to van der Giessen[11] et al.) blood compatibility of PHBV is problematic. Severe adverse effects were reported when PHBV was used as stent material in the coronary system. This observation prompted the question of the usefulness of PHBV based drug delivery systems for intravenous applications. Interaction between the hydrophobic drug delivery system (PHB, PHBV) and proteins can be expected. This interaction can cause the activation of the reticuloendothelial system and an accelerated clearing of the delivery system and the drug from the blood stream. Applying a stealth character to the delivery system could prevent this. By forming PHB-PEG natural synthetic block copolymers blood compatibility could be conferred to a material that is already superior in terms of non-toxicity and biocompatibility. Shi[19] et al. showed the feasibility for the formation of AB block copolymers consisting of the natural PHB-block and a synthetic PEG-block. Whole cell catalysis can be employed for the synthesis of natural-synthetic hybrid block copolymers. By adding PEG to the fermentation broth, block copolymerization is initiated. PEG is highly biocompatible, it penetrates the microbial cell wall and interacts with the active site of the polymerase enzyme. An ester bond is formed between the carboxylate chain end of the growing PHA chain and the PEG segment (see figure 1).

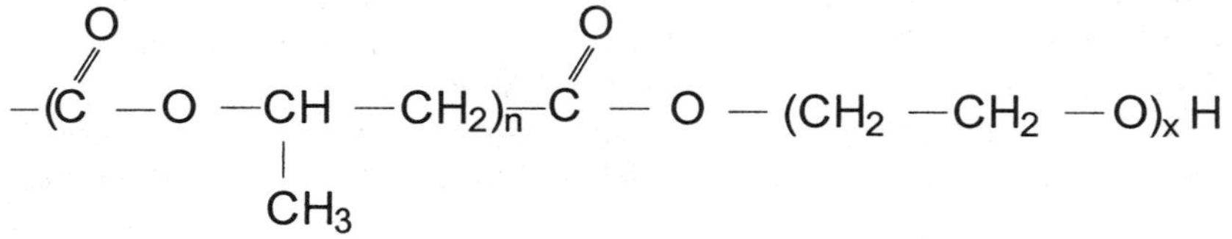

$$-(\overset{O}{\underset{}{\overset{\|}{C}}}-O-\underset{\underset{CH_3}{|}}{CH}-CH_2)_n-\overset{O}{\overset{\|}{C}}-O-(CH_2-CH_2-O)_x H$$

Figure 1: natural-synthetic PHB-b-PEG hybrid block copolymer

This reaction, which basically exhibits an interference with the natural microbial process of carbon storage preparation, leads to a new class of polymers with intriguing possibilities for their application. The end group structure of the natural polymers can be modified according to predetermined needs. Esterification of the PHA-terminus results in a reduction of the molecular weight of the PHB-block, by up to 74 %[20]. In summary, a natural-synthetic telechelic block copolymer was prepared by whole cell catalysis.

The PEG segment depicted in figure 1 can be replaced by a multiarm PEG[21] (see figure 2). Multiarm PEG, for example Pentaerythritol derivatives, generate not only amphiphilicity, but also introduce multifunctionality at one chain end of the block copolymer. The resulting block copolymers exhibit a high hydrophilic density since a large number of hydroxyl groups is confined in a small volume. The

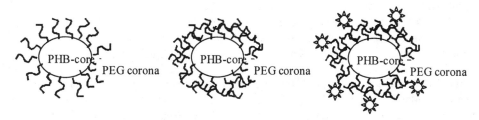

Figure 2: Polymeric micelle consisting of PHB core and a, linear PEG segment;
b, multiarm PEG segment and c, functionalized PEG segment

synthesis of this new kind of natural-synthetic block copolymers opens a new research field which will focus on employing the functionality of the hydrophilic segment for unique modifications of the molecule.

PHB-b-PEG block copolymers are envisioned, among other applications, for drug delivery systems. Exposing an amphiphilic AB block copolymer to a selective solvent leads to the formation of polymeric micelles which can act as vehicles for therapeutic drugs. Generating the polymeric micelle in an aqueous environment will result in the formation of a micelle with a hydrophobic core, consisting of the PHB segment, which is surrounded by a hydrophilic PEG-corona. The PHB segment is suited to accommodate a hydrophobic drug. The PEG segment provides the stealth character that permits a long half-life in the circulation. Adjustment of the molecular weight of the PHB block might be necessary, in order to maintain an appropriate ratio of the A- and B-block, to guarantee optimum micelle formation[22]. Due to it's hydrophilicity and large excluded volume, PEG prevents protein interactions and detection by the reticuloendothelial system. Blood compatibility will be enhanced as the inner, hydrophobic core is covered with PEG-chains. This can be achieved by using high molecular weight PEG. However, this poses a problem with microbial polymerization since the molar density of hydroxyl groups decreases with increasing chain length. This might be circumvented by using a highly branched, multiarm PEG. The latter provides the advantage of introducing a high concentration of hydroxyl groups within a small volume. These hydroxyl groups can undergo further functionalization, focusing on drug targeting (figure 2). The chemical reaction that introduces a homing device can be conducted after micellization occurs, or a functionalized PEG can be added to the fermentation for subsequent copolymerization.

The biomedical field is in need of tissue adhesives and tissue sealants, soft tissue fillers to replace collagen, and tissue adhesion preventives for the treatment of severe burns to mention just a few of the current challenges. PHAs and modified PHAs will be considered in the future as candidates for an increasing variety of biomedical challenges.

Research efforts are now devoted to supply a wider range of materials with physical and functional properties that can be adjusted to meet specific needs of the biomedical market.

References

1   Lemoigne, M. *CR Acad. Sci.* **1925**, 180, 1539; Lemoigne, M. *Ann. Inst. Pasteur* **1925**, 39, 144; Lemoigne, M. *Bull. Soc. Chim. Biol.* **1926**, 8, 770

2   Lenz, R.W.; Fuller, R.C. Scholz, C.; Touraud, F. in: *Biodegradable Plastics and Polymers* (eds.: Y. Doi, K. Fukuda), Elsevier, New York, **1994**, 109

3   Schubert, P. Krüger, N.; Steinbüchel, A. *J. Bacteriol.* **1988**, 170, 5837

4   Park, J.S.; Huh, T.L.; Lee, Y.H. *Enzyme Microb. Technol.* **1997**, 21, 85

5   Gerngross, T.U. *Nature Biotechnol.* **1999**, 17, 541-544

6   Jedlinski, Z.; Kurcok, P.; Lenz, R.W. *Macromolecules* **1998**, 31, 6718

334

7   Gogolewski, S.; Jovanovich, M.; Perren, S.M., Dillon, J.G.; Hughes, M.K. J. *Biomed. Mat. Res.* **1993**, 27 (9) 1135
8   Lenz, R.W. personal communication, University of Massachusetts at Amherst 1996
9   Saito, T.; Tomota, K.; Juni, K.; Ooba, K. *Biomaterials* **1991**, 12 309
10 de Groot, J.H.; Zijlstra, F.M.; Kuipers, H.W.; Pennings, A.J.; Klompmaker, J.; Veth, R.P.; Jansen, H.W. *Biomaterials* **1997**, 18, 613
11 van der Giessen, W.J.; Lincoff, A.M.; Schwartz, R.S.; van Beusekom, H.M.; Serruys, P.W.; Holmes, D.R. Jr.; Ellis, S.G.; Topol, E.J. *Circulation* **1996**, 94 1690
12 Doyle, C.; Tanner, E.T.; Bonfield, W. *Biomaterials* **1991**, 12, 841
13 Holland, S.J.; Yasin, M.; Tighe, B.J. *Biomaterials* **1990**, 11, 206
14 Yagmurlu, M.F.; Korkusuz, F.; Gursel, I.; Korkusuz P.; Ors, U.; Hasirci, V. *J. Biomed. Mater. Res.* **1999**, 46, 494
15 Sendil D.; Gursel, I.; Wise, D.L. Hasirci, V. *J. Controlled Release* **1999**, 59, 207
16 Embleton, J.K.; Tighe, B.J. *J. Microencapsul.* **1993**, 10, 341
17 Koosha, F.; Muller, R.H., Davis, S.S. *Crit. Rev. Ther. Drug Carrier Syst.* **1989**, 6, 117
18 Hazari, A.; Johansson-Ruden G.; Junemo-Bostrom, K.; Ljungberg, C.; Terenghi, G.; Green C.; Wiberg, M. *J. Hand Surg.* **1999**, 24, 291
19 Shi, F. Gross, R.A.; Rutherford, D.R. *Macromolecules* **1996**, 29, 10
20 Shi, F.; Ashby, R.; Gross, R.A. *Macromolecules* **1996**, 29, 7753
21 Shi, F., Scholz, C., Deng, F., Gross, R.A. *Polymer Preprints* **1998**, 39 102
22 Scholz C., Iijima, M., Nagasaki, Y., Kataoka, K. *Polym.Adv.Technol.* **1998**, 9, 1

Author Index

335

Subject Index

A

Acid end-group
  autocatalysis, 232, 235–236
  concentration effect on hydrolysis,
    231–232
  See also Degradation kinetics of poly(-
    hydroxy) acids
Actynomycetes, biodegradation of aliphatic-aromatic copolyesters, 6
Adipic anhydride, copolymer with DL-lactide, 216
Alanine, copolymer with DL-lactic acid,
  214
Alcaligenes eutrophus, intracellular depolymerase associated with poly(3-hydro-
  xybutyrate) (PHB) inclusion bodies, 58
Alcaligenes faecalis, enzymatic degradation of ultra-high-molecular-weight
  P(3HB) films, 73–74, 75f
Alcaligenes latus
  comparison of fermentation perfor-
    mance and annual operating costs for
    poly(3-hydroxybutyrate) (P(3HB))
    production, 83, 86t
  economic evaluation for process em-
    ploying recombinant *Escherichia coli*
    harboring *A. latus* poly(hydroxyalka-
    noate) (PHA) biosynthesis genes, 83
  fed-batch cultures of recombinant *E.
    coli* strains harboring *A. latus* PHA
    biosynthesis genes, 81, 83
  superior ability of recombinant *E. coli*
    strains harboring *A. latus* PHA bio-
    synthesis genes, 81
  See also Recombinant *Escherichia coli*
Aldehyde group
  construction of poly(ethylene glycol)
    (PEG) brushed-layer with reactive, us-
    ing PEG/poly(lactide) block copoly-
    mers, 317, 319–321
  surface coating with core-polymerized
    block copolymer micelles with alde-
    hyde-ended PEG shell, 321–325

Aliphatic-aromatic copolyesters
  applications, 6
  general formula, 4f
Aliphatic polyesters
  general formula, 4f
  lactone ring-opening polymerization, 3,
    5f
Alkylmetal alkoxides, initiation mechanism, 168–169
Amino acid-containing polyphosphazene.
  See Bioresorbable matrices
Anaerobic biodegradation, synthetic and
  bacterial polyesters, 270, 274t, 275t
Anaerobic degradation
  gas analysis, 265
  landfill reactor, 265–266
  See also Biodegradability
Animal fats. *See* Medium-chain-length
  poly(hydroxyalkanoates) (mcl-PHA)
Aromatic polyesters, general formula, 4f
Aspergillus fumigatus
  ability to hydrolyze poly(3-hydroxybu-
    tyrate) (PHB), 258–259
  microbial growth studies, 266
Autocatalysis
  derivation of isothermal half-order auto-
    catalysis equation, 248–250
  derivation of isothermal quadratic auto-
    catalysis, 246–248
  hydrolysis of poly(lactic acid) (PLA),
    232, 235–236
  hydrolysis of poly(lactic acid) using half
    order power of [COOH], 241f
  quadratic isothermal, 239–240
  rate constants for hydrolysis of PLA us-
    ing half order power of [COOH] at
    50°C and 60°C, 242t
  rate equation, 232

B

Bacillus cereus
  electron micrograph of thin section of
    disrupted *B. cereus* cell showing mem-